**Hybrid Intelligence for Image Analysis
and Understanding**

Hybrid Intelligence for Image Analysis and Understanding

Edited by

Siddhartha Bhattacharyya
RCC Institute of Information Technology
India

Indrajit Pan
RCC Institute of Information Technology
India

Anirban Mukherjee
RCC Institute of Information Technology
India

Paramartha Dutta
Visva-Bharati University
India

Registered Office(s)
John Wiley & Sons, Inc., 111 River Street, Hoboken, NJ 07030, USA
John Wiley & Sons Ltd, The Atrium, Southern Gate, Chichester, West Sussex, PO19 8SQ, UK

Editorial Office
The Atrium, Southern Gate, Chichester, West Sussex, PO19 8SQ, UK

For details of our global editorial offices, customer services, and more information about Wiley products visit us at www.wiley.com.

Wiley also publishes its books in a variety of electronic formats and by print-on-demand. Some content that appears in standard print versions of this book may not be available in other formats.

Library of Congress Cataloging-in-Publication Data:

Names: Bhattacharyya, Siddhartha, 1975- editor. | Pan, Indrajit, 1983-
    editor. | Mukherjee, Anirban, 1972- editor. | Dutta, Paramartha, editor.
Title: Hybrid intelligence for image analysis and understanding / edited by
    Siddhartha Bhattacharyya, Indrajit Pan, Anirban Mukherjee, Paramartha
    Dutta.
Description: Hoboken, NJ : John Wiley & Sons, 2017. | Includes index. |
    Identifiers: LCCN 2017011673 (print) | LCCN 2017027868 (ebook) | ISBN
    9781119242932 (pdf) | ISBN 9781119242956 (epub) | ISBN 9781119242925
    (cloth)
Subjects: LCSH: Image analysis. | Computational intelligence.
Classification: LCC TA1637 (ebook) | LCC TA1637 .H93 2017 (print) | DDC
    621.36/7028563–dc23
LC record available at https://lccn.loc.gov/2017011673

Cover design by Wiley
Cover images: (Background) © Yakobchuk/Gettyimages; (From left to right)
© zmeel/Gettyimages; © tkemot/Shutterstock; © Semnic/Shutterstock;
© Callista Images/Gettyimages; © Karl Ammann/Gettyimages

Set in 10/12pt Warnock by SPi Global, Chennai, India
Printed and bound in Malaysia by Vivar Printing Sdn Bhd
10 9 8 7 6 5 4 3 2 1

Dedication

Dedicated to my parents, the late Ajit Kumar Bhattacharyya and the late Hashi Bhattacharyya; my beloved wife, Rashni; my elder sisters, Tamali, Sheuli, and Barnali; my cousin sisters, Sutapa, Mousumi, and Soma; and all my students, who have made this journey enjoyable.

Dr. Siddhartha Bhattacharyya

Dedicated to all my students.

Dr. Indrajit Pan

Dedicated to my respected teachers.

Dr. Anirban Mukherjee

Dedicated to my parents, the late Arun Kanti Dutta and Mrs. Bandana Dutta.

Dr. Paramartha Dutta

Contents

Editor Biographies   *xvii*
List of Contributors   *xxi*
Foreword   *xxvii*
Preface   *xxxi*
About the Companion website   *xxxv*

1   **Multilevel Image Segmentation Using Modified Genetic Algorithm (MfGA)-based Fuzzy C-Means**   *1*
         Sourav De, Sunanda Das, Siddhartha Bhattacharyya, and Paramartha Dutta
1.1   Introduction   *1*
1.2   Fuzzy C-Means Algorithm   *5*
1.3   Modified Genetic Algorithms   *6*
1.4   Quality Evaluation Metrics for Image Segmentation   *8*
1.4.1   Correlation Coefficient (ρ)   *8*
1.4.2   Empirical Measure $Q(I)$ *8*
1.5   MfGA-Based FCM Algorithm   *9*
1.6   Experimental Results and Discussion   *11*
1.7   Conclusion   *22*
         References   *22*

2   **Character Recognition Using Entropy-Based Fuzzy C-Means Clustering**   *25*
         B. Kondalarao, S. Sahoo, and D.K. Pratihar
2.1   Introduction   *25*
2.2   Tools and Techniques Used   *27*
2.2.1   Fuzzy Clustering Algorithms   *27*
2.2.1.1 Fuzzy C-means Algorithm   *28*
2.2.1.2 Entropy-based Fuzzy Clustering   *29*
2.2.1.3 Entropy-based Fuzzy C-Means Algorithm   *29*
2.2.2   Sammon's Nonlinear Mapping   *30*
2.3   Methodology   *31*
2.3.1   Data Collection   *31*
2.3.2   Preprocessing   *31*
2.3.3   Feature Extraction   *32*

2.3.4   Classification and Recognition   *34*
2.4   Results and Discussion   *34*
2.5   Conclusion and Future Scope of Work   *38*
         References   *39*
         Appendix   *41*

3   A Two-Stage Approach to Handwritten Indic Script Identification   47
         Pawan Kumar Singh, Supratim Das, Ram Sarkar, and Mita Nasipuri
3.1   Introduction   *47*
3.2   Review of Related Work   *48*
3.3   Properties of Scripts Used in the Present Work   *51*
3.4   Proposed Work   *52*
3.4.1   Discrete Wavelet Transform   *53*
3.4.1.1 Haar Wavelet Transform   *55*
3.4.2   Radon Transform (RT)   *57*
3.5   Experimental Results and Discussion   *63*
3.5.1   Evaluation of the Present Technique   *65*
3.5.1.1 Statistical Significance Tests   *66*
3.5.2   Statistical Performance Analysis of SVM Classifier   *68*
3.5.3   Comparison with Other Related Works   *71*
3.5.4   Error Analysis   *73*
3.6   Conclusion   *74*
         Acknowledgments   *75*
         References   *75*

**4   Feature Extraction and Segmentation Techniques in a Static Hand
         Gesture Recognition System   79**
         Subhamoy Chatterjee, Piyush Bhandari, and Mahesh Kumar Kolekar
4.1   Introduction   *79*
4.2   Segmentation Techniques   *81*
4.2.1   Otsu Method for Gesture Segmentation   *81*
4.2.2   Color Space–Based Models for Hand Gesture Segmentation   *82*
4.2.2.1 RGB Color Space–Based Segmentation   *82*
4.2.2.2 HSI Color Space–Based Segmentation   *83*
4.2.2.3 YCbCr Color Space–Based Segmentation   *83*
4.2.2.4 YIQ Color Space–Based Segmentation   *83*
4.2.3   Robust Skin Color Region Detection Using K-Means Clustering and
         Mahalanobish Distance   *84*
4.2.3.1 Rotation Normalization   *85*
4.2.3.2 Illumination Normalization   *85*
4.2.3.3 Morphological Filtering   *85*
4.3   Feature Extraction Techniques   *86*
4.3.1   Theory of Moment Features   *86*
4.3.2   Contour-Based Features   *88*
4.4   State of the Art of Static Hand Gesture Recognition Techniques   *89*
4.4.1   Zoning Methods   *90*
4.4.2 F-Ratio-Based Weighted Feature Extraction   *90*

4.4.3   Feature Fusion Techniques   *91*
4.5   Results and Discussion   *92*
4.5.1   Segmentation Result   *93*
4.5.2   Feature Extraction Result   *94*
4.6   Conclusion   *97*
4.6.1   Future Work   *99*
           Acknowledgment   *99*
           References   *99*

5   **SVM Combination for an Enhanced Prediction of Writers' Soft**
           Biometrics   *103*
           Nesrine Bouadjenek, Hassiba Nemmour, and Youcef Chibani
5.1   Introduction   *103*
5.2   Soft Biometrics and Handwriting Over Time   *104*
5.3   Soft Biometrics Prediction System   *106*
5.3.1   Feature Extraction   *107*
5.3.1.1   Local Binary Patterns   *107*
5.3.1.2   Histogram of Oriented Gradients   *108*
5.3.1.3   Gradient Local Binary Patterns   *108*
5.3.2   Classification   *109*
5.3.3   Fuzzy Integrals–Based Combination Classifier   *111*
5.3.3.1 g_λ Fuzzy Measure   *111*
5.3.3.2   Sugeno's Fuzzy Integral   *113*
5.3.3.3   Fuzzy Min-Max   *113*
5.4   Experimental Evaluation   *113*
5.4.1   Data Sets   *113*
5.4.1.1   IAM Data Set   *113*
5.4.1.2   KHATT Data Set   *114*
5.4.2   Experimental Setting   *114*
5.4.3   Gender Prediction Results   *117*
5.4.4   Handedness Prediction Results   *117*
5.4.5   Age Prediction Results   *118*
5.5   Discussion and Performance Comparison   *118*
5.6   Conclusion   *120*
           References   *121*

6   **Brain-Inspired Machine Intelligence for Image Analysis: Convolutional**
           Neural Networks   *127*
           Siddharth Srivastava and Brejesh Lall
6.1   Introduction   *127*
6.2   Convolutional Neural Networks   *129*
6.2.1   Building Blocks   *130*
6.2.1.1   Perceptron   *134*
6.2.2   Learning   *135*
6.2.2.1   Gradient Descent   *136*
6.2.2.2   Back-Propagation   *136*
6.2.3   Convolution   *139*

6.2.4   Convolutional Neural Networks: The Architecture   *141*
6.2.4.1   Convolution Layer   *142*
6.2.4.2   Pooling Layer   *145*
6.2.4.3   Dense or Fully Connected Layer   *146*
6.2.5   Considerations in Implementation of CNNs   *146*
6.2.6   CNN in Action   *147*
6.2.7   Tools for Convolutional Neural Networks   *148*
6.2.8   CNN Coding Examples   *148*
6.2.8.1   MatConvNet   *148*
6.2.8.2   Visualizing a CNN   *149*
6.2.8.3   Image Category Classification Using Deep Learning   *153*
6.3   Toward Understanding the Brain, CNNs, and Images   *157*
6.3.1   Applications   *157*
6.3.2   Case Studies   *158*
6.4   Conclusion   *159*
          References   *159*

7   Human Behavioral Analysis Using Evolutionary Algorithms and Deep Learning   *165*
          Earnest Paul Ijjina and Chalavadi Krishna Mohan
7.1   Introduction   *165*
7.2   Human Action Recognition Using Evolutionary Algorithms and Deep Learning   *167*
7.2.1   Evolutionary Algorithms for Search Optimization   *168*
7.2.2   Action Bank Representation for Action Recognition   *168*
7.2.3   Deep Convolutional Neural Network for Human Action Recognition   *169*
7.2.4   CNN Classifier Optimized Using Evolutionary Algorithms   *170*
7.3   Experimental Study   *170*
7.3.1   Evaluation on the UCF50 Data Set   *170*
7.3.2   Evaluation on the KTH Video Data Set   *172*
7.3.3   Analysis and Discussion   *176*
7.3.4   Experimental Setup and Parameter Optimization   *177*
7.3.5   Computational Complexity   *182*
7.4   Conclusions and Future Work   *183*
          References   *183*

8   Feature-Based Robust Description and Monocular Detection: An Application to Vehicle Tracking   *187*
          Ramazan Yíldíz and Tankut Acarman
8.1   Introduction   *187*
8.2   Extraction of Local Features by SIFT and SURF   *188*
8.3   Global Features: Real-Time Detection and Vehicle Tracking   *190*
8.4   Vehicle Detection and Validation   *194*
8.4.1   X-Analysis   *194*
8.4.2   Horizontal Prominent Line Frequency Analysis   *195*
8.4.3   Detection History   *196*

8.5   Experimental Study   *197*
8.5.1   Local Features Assessment   *197*
8.5.2   Global Features Assessment   *197*
8.5.3   Local versus Global Features Assessment   *201*
8.6   Conclusions   *201*
         References   *202*

9   **A GIS Anchored Technique for Social Utility Hotspot Detection**   *205*
         Anirban Chakraborty, J.K. Mandal, Arnab Patra, and Jayatra Majumdar
9.1   Introduction   *205*
9.2   The Technique   *207*
9.3   Case Study   *209*
9.4   Implementation and Results   *221*
9.5   Analysis and Comparisons   *224*
9.6   Conclusions   *229*
         Acknowledgments   *229*
         References   *230*

10   **Hyperspectral Data Processing: Spectral Unmixing, Classification, and Target Identification**   *233*
         Vaibhav Lodhi, Debashish Chakravarty, and Pabitra Mitra
10.1   Introduction   *233*
10.2   Background and Hyperspectral Imaging System   *234*
10.3   Overview of Hyperspectral Image Processing   *236*
10.3.1   Image Acquisition   *237*
10.3.2   Calibration   *237*
10.3.3   Spatial and Spectral preprocessing   *238*
10.3.4   Dimension Reduction   *239*
10.3.4.1 Transformation-Based Approaches   *239*
10.3.4.2 Selection-Based Approaches   *239*
10.3.5   postprocessing   *240*
10.4   Spectral Unmixing   *240*
10.4.1   Unmixing Processing Chain   *240*
10.4.2   Mixing Model   *241*
10.4.2.1 Linear Mixing Model (LMM)   *242*
10.4.2.2 Nonlinear Mixing Model   *242*
10.4.3   Geometrical-Based Approaches to Linear Spectral Unmixing   *243*
10.4.3.1 Pure Pixel-Based Techniques   *243*
10.4.3.2 Minimum Volume-Based Techniques   *244*
10.4.4   Statistics-Based Approaches   *244*
10.4.5   Sparse Regression-Based Approach   *245*
10.4.5.1 Moore–Penrose Pseudoinverse (MPP)   *245*
10.4.5.2 Orthogonal Matching Pursuit (OMP)   *246*
10.4.5.3 Iterative Spectral Mixture Analysis (ISMA)   *246*
10.4.6   Hybrid Techniques   *246*
10.5   Classification   *247*
10.5.1   Feature Mining   *247*

10.5.1.1 Feature Selection (FS) *248*
10.5.1.2 Feature Extraction *248*
10.5.2   Supervised Classification *248*
10.5.2.1 Minimum Distance Classifier *249*
10.5.2.2 Maximum Likelihood Classifier (MLC) *250*
10.5.2.3 Support Vector Machines (SVMs) *250*
10.5.3   Hybrid Techniques *250*
10.6   Target Detection *251*
10.6.1   Anomaly Detection *251*
10.6.1.1 RX Anomaly Detection *252*
10.6.1.2 Subspace-Based Anomaly Detection *253*
10.6.2   Signature-Based Target Detection *253*
10.6.2.1 Euclidean distance *254*
10.6.2.2 Spectral Angle Mapper (SAM) *254*
10.6.2.3 Spectral Matched Vilter (SMF) *254*
10.6.2.4 Matched Subspace Detector (MSD) *255*
10.6.3   Hybrid Techniques *255*
10.7   Conclusions *256*
         References *256*

11   **A Hybrid Approach for Band Selection of Hyperspectral Images** *263*
            Aditi Roy Chowdhury, Joydev Hazra, and Paramartha Dutta
11.1   Introduction *263*
11.2   Relevant Concept Revisit *266*
11.2.1   Feature Extraction *266*
11.2.2   Feature Selection Using 2D PCA *266*
11.2.3   Immune Clonal System *267*
11.2.4   Fuzzy KNN *268*
11.3   Proposed Algorithm *271*
11.4   Experiment and Result *271*
11.4.1   Description of the Data Set *272*
11.4.2   Experimental Details *274*
11.4.3   Analysis of Results *275*
11.5   Conclusion *278*
         References *279*

12   **Uncertainty-Based Clustering Algorithms for Medical Image
            Analysis** *283*
            Deepthi P. Hudedagaddi and B.K. Tripathy
12.1   Introduction *283*
12.2   Uncertainty-Based Clustering Algorithms *283*
12.2.1   Fuzzy C-Means *284*
12.2.2   Rough Fuzzy C-Means *285*
12.2.3   Intuitionistic Fuzzy C-Means *285*
12.2.4   Rough Intuitionistic Fuzzy C-Means *286*
12.3   Image Processing *286*

12.4   Medical Image Analysis with Uncertainty-Based Clustering
          Algorithms   *287*
12.4.1   FCM with Spatial Information for Image Segmentation   *287*
12.4.2   Fast and Robust FCM Incorporating Local Information for Image
          Segmentation   *290*
12.4.3   Image Segmentation Using Spatial IFCM   *291*
12.4.3.1 Applications of Spatial FCM and Spatial IFCM on Leukemia Images   *292*
12.5   Conclusions   *293*
          References   *293*

**13   An Optimized Breast Cancer Diagnosis System Using a Cuckoo Search
          Algorithm and Support Vector Machine Classifier**   *297*
          Manoharan Prabukumar, Loganathan Agilandeeswari, and Arun Kumar Sangaiah
13.1   Introduction   *297*
13.2   Technical Background   *301*
13.2.1   Morphological Segmentation   *301*
13.2.2   Cuckoo Search Optimization Algorithm   *302*
13.2.3   Support Vector Machines   *303*
13.3   Proposed Breast Cancer Diagnosis System   *303*
13.3.1   Preprocessing of Breast Cancer Image   *303*
13.3.2   Feature Extraction   *304*
13.3.2.1 Geometric Features   *304*
13.3.2.2 Texture Features   *305*
13.3.2.3 Statistical Features   *306*
13.3.3   Features Selection   *306*
13.3.4   Features Classification   *307*
13.4   Results and Discussions   *307*
13.5   Conclusion   *310*
13.6   Future Work   *310*
          References   *310*

14   Analysis of Hand Vein Images Using Hybrid Techniques   *315*
          R. Sudhakar, S. Bharathi, and V. Gurunathan
14.1   Introduction   *315*
14.2   Analysis of Vein Images in the Spatial Domain   *318*
14.2.1   Preprocessing   *318*
14.2.2   Feature Extraction   *319*
14.2.3   Feature-Level Fusion   *320*
14.2.4   Score Level Fusion   *320*
14.2.5   Results and Discussion   *322*
14.2.5.1 Evaluation Metrics   *323*
14.3   Analysis of Vein Images in the Frequency Domain   *326*
14.3.1   Preprocessing   *326*
14.3.2   Feature Extraction   *326*
14.3.3   Feature-Level Fusion   *330*
14.3.4   Support Vector Machine Classifier   *331*
14.3.5   Results and Discussion   *331*

14.4   Comparative Analysis of Spatial and Frequency Domain Systems   *332*
14.5   Conclusion   *335*
         References   *335*

15   Identification of Abnormal Masses in Digital Mammogram Using Statistical Decision Making   *339*
         Indra Kanta Maitra and Samir Kumar Bandyopadhyay
15.1   Introduction   *339*
15.1.1   Breast Cancer   *339*
15.1.2   Computer-Aided Detection/Diagnosis (CAD)   *340*
15.1.3   Segmentation   *340*
15.2   Previous Works   *341*
15.3   Proposed Method   *343*
15.3.1   Preparation   *343*
15.3.2   Preprocessing   *345*
15.3.2.1 Image Enhancement and Edge Detection   *346*
15.3.2.2 Isolation and Suppression of Pectoral Muscle   *348*
15.3.2.3 Breast Contour Detection   *351*
15.3.2.4 Anatomical Segmentation   *353*
15.3.3   Identification of Abnormal Region(s)   *354*
15.3.3.1 Coloring of Regions   *354*
15.3.3.2 Statistical Decision Making   *355*
15.4   Experimental Result   *358*
15.4.1   Case Study with Normal Mammogram   *358*
15.4.2   Case Study with Abnormalities Embedded in Fatty Tissues   *358*
15.4.3   Case Study with Abnormalities Embedded in Fatty-Fibro-Glandular Tissues   *359*
15.4.4   Case Study with Abnormalities Embedded in Dense-Fibro-Glandular Tissues   *359*
15.5   Result Evaluation   *360*
15.5.1   Statistical Analysis   *361*
15.5.2   ROC Analysis   *361*
15.5.3   Accuracy Estimation   *365*
15.6   Comparative Analysis   *366*
15.7   Conclusion   *366*
         Acknowledgments   *366*
         References   *367*

16   Automatic Detection of Coronary Artery Stenosis Using Bayesian Classification and Gaussian Filters Based on Differential Evolution   *369*
         Ivan Cruz-Aceves, Fernando Cervantes-Sanchez, and Arturo Hernandez-Aguirre
16.1   Introduction   *369*
16.2   Background   *370*
16.2.1   Gaussian Matched Filters   *371*
16.2.2   Differential Evolution   *371*
16.2.2.1 Example: Global Optimization of the Ackley Function   *373*

16.2.3   Bayesian Classification   *375*
16.2.3.1   Example: Classification Problem   *375*
16.3   Proposed Method   *377*
16.3.1   Optimal Parameter Selection of GMF Using Differential Evolution   *377*
16.3.2   Thresholding of the Gaussian Filter Response   *378*
16.3.3   Stenosis Detection Using Second-Order Derivatives   *378*
16.3.4   Stenosis Detection Using Bayesian Classification   *379*
16.4   Computational Experiments   *381*
16.4.1   Results of Vessel Detection   *382*
16.4.2   Results of Vessel Segmentation   *382*
16.4.3   Evaluation of Detection of Coronary Artery Stenosis   *384*
16.5   Concluding Remarks   *386*
            Acknowledgment   *388*
            References   *388*

**17   Evaluating the Efficacy of Multi-resolution Texture Features for
            Prediction of Breast Density Using Mammographic Images**   *391*
            Kriti, Harleen Kaur, and Jitendra Virmani
17.1   Introduction   *391*
17.1.1   Comparison of Related Methods with the Proposed Method   *397*
17.2   Materials and Methods   *398*
17.2.1   Description of Database   *398*
17.2.2   ROI Extraction Protocol   *398*
17.2.3   Workflow for CAD System Design   *398*
17.2.3.1   Feature Extraction   *400*
17.2.3.2   Classification   *407*
17.3   Results   *410*
17.3.1   Results Based on Classification Performance of the Classifiers (Classification
            Accuracy and Sensitivity) for Each Class   *411*
17.3.1.1   Experiment I: To Determine the Performance of Different FDVs Using SVM
            Classifier   *411*
17.3.1.2   Experiment II: To Determine the Performance of Different FDVs Using SSVM
            Classifier   *412*
17.3.2   Results Based on Computational Efficiency of Classifiers for Predicting 161
            Instances of Testing Dataset   *412*
17.4   Conclusion and Future Scope   *413*
            References   *415*

            Index   *423*

Editor Biographies

Dr Siddhartha Bhattacharyya earned his bachelor's in Physics, bachelor's in Optics and Optoelectronics, and master's in Optics and Optoelectronics from University of Calcutta, India, in 1995, 1998, and 2000, respectively. He completed a PhD in computer science and engineering from Jadavpur University, India, in 2008. He is the recipient of the University Gold Medal from the University of Calcutta for his master's in 2012. He is also the recipient of the coveted **ADARSH VIDYA SARASWATI RASHTRIYA PURASKAR** for excellence in education and research in 2016. He is the recipient of the **Distinguished HoD Award** and **Distinguished Professor Award** conferred by Computer Society of India, Mumbai Chapter, India in 2017. He is also the recipient of the coveted **Bhartiya Shiksha Ratan Award** conferred by Economic Growth Foundation, New Delhi in 2017.

He is currently the Principal of RCC Institute of Information Technology, Kolkata, India. In addition, he is serving as the Dean of Research and Development of the institute from November 2013. Prior to this, he was the Professor and Head of Information Technology of RCC Institute of Information Technology, Kolkata, India, from 2014 to 2017. Before this, he was an Associate Professor of Information Technology in the same institute, from 2011 to 2014. Before that, he served as an Assistant Professor in Computer Science and Information Technology of University Institute of Technology, The University of Burdwan, India, from 2005 to 2011. He was a Lecturer in Information Technology of Kalyani Government Engineering College, India, during 2001–2005. He is a coauthor of four books and the coeditor of eight books, and has more than 175 research publications in international journals and conference proceedings to his credit. He has got a patent on intelligent colorimeter technology. He was the convener of the AICTE-IEEE National Conference on Computing and Communication Systems (CoCoSys-09) in 2009. He was the member of the Young Researchers' Committee of the WSC 2008 Online World Conference on Soft Computing in Industrial Applications. He has been the member of the organizing and technical program committees of several national and international conferences. He served as the Editor-in-Chief of *International Journal of Ambient Computing and Intelligence* (IJACI) published by IGI Global (Hershey, PA, USA) from July 17, 2014, to November 6, 2014. He was the General Chair of the IEEE International Conference on Computational Intelligence and Communication Networks (ICCICN 2014) organized by the Department of Information Technology, RCC Institute of Information Technology, Kolkata, in association with Machine Intelligence Research Labs, Gwalior, and IEEE Young Professionals, Kolkata Section; it was held at Kolkata, India, in 2014. He is the Associate Editor of *International Journal of Pattern Recognition Research*. He is the member of the editorial board of *International Journal of Engineering, Science and Technology* and *ACCENTS Transactions on Information Security* (ATIS). He is also the member of the editorial advisory board of HETC *Journal of Computer Engineering and Applications*. He has been the Associate Editor of the *International Journal of BioInfo Soft Computing* since 2013. He is the Lead Guest Editor of the Special Issue on *Hybrid Intelligent Techniques for Image Analysis and Understanding of Applied Soft Computing* (Elsevier, Amsterdam). He was the General Chair of the 2015 IEEE International Conference on Research in Computational Intelligence and Communication Networks (ICRCICN 2015) organized by the Department of Information Technology, RCC Institute of Information Technology, Kolkata, in association with IEEE Young Professionals, Kolkata Section and held at Kolkata, India, in 2015. He is the Lead Guest Editor of the Special Issue on Computational Intelligence and Communications in *International Journal of Computers and Applications* (IJCA) (Taylor & Francis, London) in 2016. He has been the Issue Editor of *International Journal of Pattern Recognition Research* since January 2016. He was the General Chair of the 2016 International Conference on Wireless Communications, Network Security and Signal Processing (WCNSSP2016) held during June 26–27, 2016, at Chiang Mai, Thailand. He is the member of the editorial board of *Applied Soft Computing* (Elsevier, Amsterdam).

He has visited several leading universities in several countries like China, Thailand, and Japan for delivering invited lectures. His research interests include soft computing, pattern recognition, multimedia data processing, hybrid intelligence, and quantum computing. Dr Bhattacharyya is a Fellow of Institute of Electronics and Telecommunication Engineers (IETE), India. He is a senior member of Institute of Electrical and Electronics Engineers (IEEE), USA; Association for Computing Machinery (ACM), USA; and International Engineering and Technology Institute (IETI), Hong Kong. He is a member of International Rough Set Society, International Association for Engineers (IAENG), Hong Kong; Computer Science Teachers Association (CSTA), USA; International Association of Academicians, Scholars, Scientists and Engineers (IAASSE), USA; Institution of Engineering and Technology (IET), UK; and Institute of Doctors Engineers and Scientists (IDES), India. He is a life member of Computer Society of India, Optical Society of India, Indian Society for Technical Education, and Center for Education Growth and Research, India.

Dr Indrajit Pan has done his BE in Computer Science and Engineering with Honors from The University of Burdwan in 2005; M.Tech. in Information Technology from Bengal Engineering and Science University, Shibpur, in 2009; and PhD (Engg.) from Indian Institute of Engineering Science and Technology, Shibpur, in 2015. He is the recipient of BESU, University Medal for securing first rank in M.Tech. (IT). He has a couple of national- and international-level research publications and book chapters to his credit. He has attended several international conferences, national-level faculty development programs, workshops, and symposiums.

In this Institute, his primary responsibility is teaching and project guidance at UG (B.Tech.) and PG (M.Tech. and MCA) levels as Assistant Professor of Information Technology (erstwhile Lecturer since joining in February 2006). Apart from this, he has carried out additional responsibility of Single Point of Contact (SPoC) for Infosys Campus Connect Initiative in 2009–2011, and Coordinator of Institute-level UG Project Committee in 2008–2010. At present, his additional responsibility includes Nodal Officer of Institutional Reforms for TEQIP–II Project since 2011 and Member Secretary of

Academic Council since 2013. Apart from these, he has actively served as an organizing member of several Faculty Development Programs and International Conferences (ICCICN 2014) for RCCIIT. He has also acted as the Session Chair in an International Conference (ICACCI 2013) and Member of Technical Committee for FICTA 2014. Before joining RCCIIT, Indrajit served Siliguri Institute of Technology, Darjeeling, as a Lecturer in CSE/IT from 2005 to 2006.

Indrajit is a Member of Institute of Electrical and Electronics Engineers (IEEE), USA; and Association for Computing Machinery (ACM), USA.

Dr Anirban Mukherjee did his bachelor's in Civil Engineering in 1994 from Jadavpur University, Kolkata. While in service, he achieved a professional Diploma in Operations Management (PGDOM) in 1998 and completed his PhD on *Automatic Diagram Drawing based on Natural Language Text Understanding* from Indian Institute of Engineering, Science and Technology (IIEST), Shibpur, in 2014. Serving RCC Institute of Information Technology (RCCIIT), Kolkata, since its inception (in 1999), he is currently an Associate Professor and Head of the Department of Engineering Science & Management at RCCIIT. Before joining RCCIIT, he served as an Engineer in the Scientific & Technical Application Group in erstwhile RCC, Calcutta, for six years. His research interest includes computer graphics, computational intelligence, optimization, and assistive technology. He has coauthored two UG engineering textbooks: one on *Computer Graphics and Multimedia* and another on *Engineering Mechanics.* He has also coauthored more than 18 books on Computer Graphics/Multimedia for distance learning courses He holds BBA/MBA/BCA/MCA/B.Sc (Comp.Sc.)/M.Sc (IT) of different universities of India. He has a few international journal articles, book chapters, and conference papers to his credit. He is in the editorial board of *International Journal of Ambient Computing and Intelligence* (IJACI).

Dr Paramartha Dutta, born 1966, did his bachelor's and master's in Statistics from the Indian Statistical Institute, Calcutta, in the years 1988 and 1990, respectively. He afterward completed his master's degree of technology in Computer Science from the same Institute in 1993, and PhD in engineering from the Bengal Engineering and Science University, Shibpur, in 2005, respectively. He has served in the capacity of research personnel in various projects funded by Government of India, which include DRDO, CSIR, Indian Statistical Institute, Calcutta, and others. Dr Dutta is now a Professor in the Department of Computer and System Sciences of the Visva Bharati University, West Bengal, India. Prior to this, he served Kalyani Government Engineering College and College of Engineering in West Bengal as a full-time faculty member. Dr Dutta remained associated as Visiting/Guest Faculty of several universities and institutes, such as West Bengal University of Technology, Kalyani University, and Tripura University. He has coauthored eight books and has also five edited books to his credit. He has published about 185 papers in various journals and conference proceedings, both international and national; as well as several book chapters in edited volumes of reputed international publishing houses like Elsevier, Springer-Verlag, CRC Press, and John Wiley, to name a few.

Dr Dutta has guided three scholars who already had been awarded their PhD. Presently, he is supervising six scholars for their PhD program. Dr Dutta has served as editor of special volumes of several international journals published by publishers of international repute such as Springer. Dr Dutta, as investigator, could implement successfully projects funded by AICTE, DST of the Government of India.

Prof. Dutta has served/serves in the capacity of external member of boards of studies of relevant departments of various universities encompassing West Bengal University of Technology, Kalyani University, Tripura University, Assam University, and Silchar, to name a few. He had the opportunity to serve as the expert of several interview boards conducted by West Bengal Public Service Commission, Assam University, Silchar, National Institute of Technology, Arunachal Pradesh, Sambalpur University, and so on.

Dr Dutta is a Life Fellow of the Optical Society of India (OSI); Computer Society of India (CSI); Indian Science Congress Association (ISCA); Indian Society for Technical Education (ISTE); Indian Unit of Pattern Recognition and Artificial Intelligence (IUPRAI) – the Indian affiliate of the International Association for Pattern Recognition (IAPR); and senior member of Associated Computing Machinery (ACM); IEEE Computer Society, USA; and IACSIT.

List of Contributors

Tankut Acarman
Computer Engineering Department
Galatasaray University
Istanbul
Turkey

Loganathan Agilandeeswari
School of Information
Technology & Engineering
VIT University
Vellore
Tamil Nadu
India

Samir Kumar Bandyopadhyay
Department of Computer
Science and Engineering
University of Calcutta
Salt Lake Campus
Kolkata
West Bengal
India

Piyush Bhandari
Indian Institute of Technology Patna
Bihta, Patna
Bihar, India

S. Bharathi
Department of Electronics and
Communication Engineering
Dr. Mahalingam College of Engineering
and Technology, Coimbatore
Tamil Nadu, India

Sangita Bhattacharjee
Department of Computer
Science and Engineering
University of Calcutta
Salt Lake Campus
Kolkata
West Bengal
India

Siddhartha Bhattacharyya
Department of Information Technology
RCC Institute of Information Technology
Kolkata
West Bengal
India

Nesrine Bouadjenek
Laboratoire d'Ingénierie des Systèmes
Intelligents et Communicants (LISIC)
Faculty of Electronics and
Computer Sciences
University of Sciences and Technology
Houari Boumediene (USTHB), Algiers
Algeria

Fernando Cervantes-Sanchez
Centro de Investigación
en Matemáticas (CIMAT)
A.C., Jalisco S/N
Col. Valenciana
Guanajuato
México

Anirban Chakraborty
Department of Computer Science
Barrackpore Rastraguru
Surendranath College
Barrackpore
Kolkata
West Bengal
India

Debashish Chakravarty
Indian Institute of Technology
Kharagpur
West Bengal
India

Subhamoy Chatterjee
Indian Institute of Technology Patna
Bihta, Patna
Bihar, India

Youcef Chibani
Laboratoire d'Ingénierie des Systèmes
Intelligents et Communicants (LISIC)
Faculty of Electronics and
Computer Sciences
University of Sciences and Technology
Houari Boumediene (USTHB), Algiers
Algeria

Aditi Roy Chowdhury
Department of Computer
Science and Technology
Women's Polytechnic
Jodhpur Park
Kolkata
West Bengal
India

Ivan Cruz-Aceves
CONACYT – Centro de Investigación
en Matemáticas (CIMAT)
A.C., Jalisco S/N
Col. Valenciana
Guanajuato
México

Sunanda Das
Department of Computer
Science & Engineering
University Institute of Technology
The University of Burdwan
Burdwan
West Bengal
India

Supratim Das
Department of Computer
Science and Engineering
Jadavpur University
Kolkata
West Bengal
India

Sourav De
Department of Computer
Science & Engineering
Cooch Behar Government
Engineering College
Cooch Behar
West Bengal
India

Paramartha Dutta
Department of Computer Science
and System Sciences
Visva-Bharati University
Santiniketan
West Bengal
India

V. Gurunathan
Department of Electronics and
Communication Engineering
Dr. Mahalingam College of
Engineering and Technology
Tamil Nadu
India

Joydev Hazra
Department of Information Technology
Heritage Institute of Technology
Kolkata
West Bengal
India

Arturo Hernandez-Aguirre
Centro de Investigación
en Matemáticas (CIMAT)
A.C., Jalisco S/N
Col. Valenciana
Guanajuato
México

Deepthi P. Hudedagaddi
School of Computer Science and
Engineering (SCOPE)
VIT University
Vellore
Tamil Nadu
India

Earnest Paul Ijjina
Visual Learning and
Intelligence Group (VIGIL)
Department of Computer
Science and Engineering
Indian Institute of Technology
Hyderabad (IITH)
Hyderabad
Telangana
India

Harleen Kaur
Electrical and Instrumentation
Engineering Department
Thapar University
Patiala
Punjab
India

Mahesh Kumar Kolekar
Indian Institute of Technology Patna
Bihta, Patna
Bihar, India

B. Kondalarao
Department of Mechanical Engineering
Indian Institute of Technology
Kharagpur
West Bengal
India

Kriti
Electrical and Instrumentation
Engineering Department
Thapar University
Patiala
Punjab
India

Brejesh Lall
Department of Electrical Engineering
Indian Institute of Technology Delhi
India

Vaibhav Lodhi
Indian Institute of Technology
Kharagpur
West Bengal
India

Indra Kanta Maitra
Department of Information Technology
B.P. Poddar Institute of Management and
Technology
Kolkata
West Bengal
India

Jayatra Majumdar
Department of Computer Science
Barrackpore Rastraguru Surendranath
College
Barrackpore
Kolkata
West Bengal
India

J.K. Mandal
Department of Computer
Science & Engineering
University of Kalyani
Kalyani
Nadia
West Bengal
India

Pabitra Mitra
Indian Institute of Technology
Kharagpur
West Bengal
India

Chalavadi Krishna Mohan
Visual Learning and
Intelligence Group (VIGIL)
Department of Computer
Science and Engineering
Indian Institute of Technology
Hyderabad (IITH)
Hyderabad
Telangana
India

Mita Nasipuri
Department of Computer Science and
Engineering
Jadavpur University
Kolkata
West Bengal
India

Hassiba Nemmour
Laboratoire d'Ingénierie des Systèmes
Intelligents et Communicants (LISIC)
Faculty of Electronics and
Computer Sciences
University of Sciences and Technology
Houari Boumediene (USTHB) Algiers
Algeria

Arnab Patra
Department of Computer Science
Barrackpore Rastraguru
Surendranath College
Barrackpore
Kolkata
West Bengal
India

Manoharan Prabukumar
School of Information
Technology & Engineering
VIT University
Vellore
Tamil Nadu
India

D.K. Pratihar
Department of Mechanical Engineering
Indian Institute of Technology
Kharagpur
West Bengal
India

S. Sahoo
Department of Mechanical Engineering
Indian Institute of Technology
Kharagpur
West Bengal
India

Arun Kumar Sangaiah
School of Computing
Science & Engineering
VIT University
Vellore
Tamil Nadu
India

Ram Sarkar
Department of Computer
Science and Engineering
Jadavpur University
Kolkata
West Bengal
India

Pawan Kumar Singh
Department of Computer Science and
Engineering
Jadavpur University
Kolkata
West Bengal
India

Siddharth Srivastava
Department of Electrical Engineering
Indian Institute of Technology Delhi
India

R. Sudhakar
Department of Electronics and
Communication Engineering
Dr. Mahalingam College of
Engineering and Technology
Coimbatore
Tamil Nadu
India

B.K. Tripathy
School of Computer Science and
Engineering (SCOPE)
VIT University
Vellore
Tamil Nadu
India

Jitendra Virmani
CSIR-CSIO
Sector-30C
Chandigarh
India

Ramazan Yíldíz
Computer Engineering Department
Galatasaray University
Istanbul
Turkey

Foreword

Image analysis and understanding have been daunting tasks in computer vision given the high level of uncertainty involved therein. At the same time, a proper analysis of images plays a key role in many real-life applications. Examples of applications include image processing, image mining, image inpainting, video surveillance, and intelligent transportation systems, to name a few. Albeit there exists a plethora of classical techniques for addressing the problem of image analysis, which include filtering, hierarchical morphologic algorithms, 2D histograms, mean shift clustering, and graph-based segmentation, most of these techniques often fall short owing to their incapability in handling inherent real-life uncertainties. In past decades, researchers have been able to address different types of uncertainties prevalent in real-world images, thanks to the evolving state of the art of intelligent tools and techniques such as convolutional neural networks (CNNs) and deep learning. In this direction, computational intelligence techniques deserve special mention owing to their flexibility, application-free usability, and adaptability. Of late, hybridization of different computational intelligence techniques has come up with promising avenues in that these are more robust and offer more efficient solutions in real time.

This book aims to introduce the readers with the basics of image analysis and understanding, with recourse to image thresholding, image segmentation, and image and multimedia data analysis. The book also focuses on the foundations of hybrid intelligence as it applies to image analysis and understanding. As a sequel to this, different state-of-the-art hybrid intelligent techniques for addressing the problem of image analysis will be illustrated to enlighten the readers of upcoming research trends.

As an example of the recent trends in image analysis and understanding, albeit aging mitigates the glamor in human beings, wrinkles in face images can often be used for estimation of age progression in human beings. This can be further utilized for tracing unknown or missing persons. Images exhibit varied uncertainty and ambiguity of information, and hence understanding an image scene is far from being a general procedure. The situation becomes even graver when the images become corrupt with noise artifacts.

In this book, the editors have attempted to deliver some of the recent trends in hybrid intelligence as it applies to image analysis and understanding. The book contains 17 well-versed chapters illustrating diversified areas of application of image analysis using hybrid intelligence. These include multilevel image segmentation, character recognition, image analysis, video image processing, hyperspectral image analysis, and medical image analysis.

The first chapter deals with multilevel image segmentation. The authors propose a modified genetic algorithm (MfGA) to generate the optimized class levels of the multilevel images, and those class levels are employed as the initial input in the fuzzy c-means (FCM) algorithm. A performance comparison is depicted between the MfGA-based FCM algorithm, the conventional genetic algorithm (GA)-based FCM algorithm, and the well-known FCM algorithm with the help of three real-life multilevel grayscale images. The comparison revealed the superiority of the proposed method over the other two image segmentation algorithms.

Chapters 2 to 5 address the issue of character recognition and soft biometrics. Chapter 2 shows pros and cons of an entropy-based FCM clustering technique to classify huge training data for English character recognition. In chapter 3, the authors propose a two-stage word-level script identification technique for eight handwritten popular scripts, namely, Bangla, Devanagari, Gurumukhi, Oriya, Malayalam, Telugu, Urdu, and Roman. Firstly, discrete wavelet transform (DWT) is applied on the input word images to extract the most representative information, whereas in the second stage, radon transform (RT) is applied to the output of the first stage to compute a set of 48 statistical features from each word image. Chapter 4 presents a skin color region segmentation method based on K-means clustering and Mahalanobis distance for static hand gesture recognition. The final chapter in this group deals with soft-biometrics prediction. There, three prediction systems are developed using a support vector machine (SVM) classifier associated to various gradient and textural features. Since different features yield different aspects of characterization, the authors investigate classifier combination in order to improve the prediction accuracy. As a matter of fact, the fuzzy integral is used to produce a robust soft-biometrics prediction.

Chapters 6 and 7 focus on image analysis applications. More specifically, in chapter 6, the authors draw an analogy and comparison between the working principle of CNNs and the human brain. In chapter 7, the authors propose a framework for human action recognition that is trained using evolutionary algorithms and deep learning. A CNN classifier designed to recognize human actions from action bank features is initialized by evolutionary algorithms and trained using back-propagation algorithms.

Chapter 8 is targeted to video image processing applications. The authors propose a technique using Haar-like simple features to describe object models in chapter 8. This technique is applied with Adaboost classifier for object detection on the video records. The tracking method is described and illustrated by fusing global and local features.

Chapter 9 deals with a GIS-based application. The proposed GIS-anchored system extends a helping hand toward common and innocent people and exposes the trails for releasing themselves from the clutches of criminals. The chief intent of the proposed work is to not only check out the hot-spot areas (the crime-prone areas), but also give a glance to the flourishing of criminal activities in future. The process of determination of hot-spots is carried out by associating rank (an integer value) to each ward/block on the digitized map. The process of hooking up rank to a specific region is carried out on the basis of criminal activity at that particular region.

Chapters 10 and 11 deal with hyperspectral image analysis. Chapter 10 covers the hyperspectral data analysis and processing algorithms organized into three topics: spectral unmixing, classification, and target identification. In chapter 11, the authors deal with the band selection problem. They use the fuzzy k-nearest neighbors (KNN) technique to calculate the affinity of the band combination based on features extracted by 2D principal component analysis (PCA).

The remaining chapters deal with medical image analysis using different hybrid intelligent techniques. Chapter 12 aims to introduce several uncertainty-based hybrid clustering algorithms and their applications in medical image analysis. In chapter 13, the authors propose a diagnosis system for early detection of breast cancer tissues from digital mammographic breast images using a Cuckoo search optimization algorithm and SVM classifiers. In general, the complete diagnosis process involves various stages, such as preprocessing of images, segmentation of the breast cancer region from its surroundings, extracting tissues of interest and then determining the associated features that may be vital, and, finally, classifying the tissue into either benign or malignant. Chapter 14 proposes a new approach for biometric recognition, using dorsal, palm, finger, and wrist veins of the hand. The analysis of these vein modalities is done in both the spatial domain and frequency domain. In the spatial domain, a modified 2D Gabor filter is used for feature extraction, and these features are fused at both the feature level and score level for further analysis. Similarly in the frequency domain, a contourlet transform is used for feature extraction, a multiresolution singular value decomposition technique is utilized for fusing these features, and further classification is done with the help of an SVM classifier. An automated segmentation technique has been proposed in chapter 15 for digital mammograms to detect abnormal mass/masses [i.e., tumor(s)]. The accuracy of an automatic segmentation algorithm requires standardization (i.e., preparation of images and preprocessing of medical images that are mandatory, distinct, and sequential). The detection method is based on a modified seeded region growing algorithm (SRGA) followed by a step-by-step statistical elimination method. Finally a decision-making system is proposed to isolate mass/masses in mammograms. Chapter 16 presents a novel method for the automatic detection of coronary stenosis in X-ray angiograms. In the first stage, Gaussian matched filters (GMFs) are applied over the input angiograms for the detection of coronary arteries. The GMF method is tuned in a training step applying differential evolution (DE) for the optimization process, which is evaluated using the area under the receiver operating characteristic (ROC) curve. In the second stage, an iterative thresholding method is applied to extract vessel-like structures from the background of the Gaussian filter response. The authors of chapter 17 introduce an efficient CAD system, based on multiresolution texture descriptors using 2D wavelet transform, which has been implemented using a smooth SVM (SSVM) classifier. The standard Mammographic Image Analysis Society (MIAS) dataset has been used for classifying the breast tissue into one of three classes, namely, fatty (F), fatty-glandular (FG), and dense-glandular (DG). The performance of the SSVM-based CAD system has been compared with SVM-based CAD system design.

From the varied nature of the case studies treated within the book, I am fully confident and can state that the editors have done a splendid job in bringing forward the facets of hybrid intelligence to the scientific community. This book will, hence, not only serve as a good reference for senior researchers, but also stand in good stead for novice researchers and practitioners who have started working in this field. Moreover, the book will also serve as a reference book for some parts of the curriculum for postgraduate and undergraduate disciplines in computer science and engineering, electronics and communication engineering, and information technology.

I invite the readers to enjoy the book and take advantage of its benefits. One could join the team of computational intelligence designers and bring new insights into this developing and challenging enterprise.

Italy, November 2016 *Cesare Alippi*

Preface

Image processing happens to be a wide subject encompassing various problems therein. The very understanding of different image-processing applications is not an easy task considering the wide variety of contexts they represent. Naturally, the better the understanding, the better the analysis thereof is expected to be. The uncertainty and/or imprecision associated inherently in this field make the problem even more challenging. The last two decades have witnessed several soft computing techniques for addressing such issues. Each such soft computing method has strengths as well as shortcomings. As a result, no individual soft computing technique offers a solution to such problems in a comprehensive sense that is uniformly applicable. The concept of hybridization pervades from this situation, where more than one soft computing technique are subjected to coupling with the hope that the weakness of one may be overcome by the other. Present-time reporting is more oriented toward that. In fact, the present volume is also aimed at inviting the contributions toward adequate justification for hybrid methods in different aspects of image processing.

The authors of chapter 1 try to demonstrate the significance of using modified genetic algorithms in obtaining optimized class levels, which when subsequently applied as initial input in the fuzzy c-means (FCM) algorithm overcome the limitation of possible premature convergence in FCM-based clustering.

The authors justify through extensive experimentation in chapter 2 as to how FCM-based clustering may be effectively hybridized with entropy-based clustering for English character recognition. The main uniqueness of the contribution of chapter 3 is the development of a two-stage system capable of recognizing texts containing multilingual handwritten scripts comprising eight Indian languages. The authors substantiate the effectiveness of their method by evaluating an extensive database of 16,000 handwritten words with a reported achievement of 97.69%.

In chapter 4, the authors propose a robust technique based on K-means clustering using a Mahalanobis distance metric to achieve static hand gesture recognition. In their proposed method, they consider zoning-based shape feature extraction to overcome the problem of misclassification associated to other techniques for this purpose.

The use of a support vector machine (SVM) for classification and effective soft-biometric prediction on the basis of three gradient and texture features, and how these features are combined to attain higher prediction accuracy, are the cardinal contributions of chapter 5. The authors validate their results by extensive experimentation.

In chapter 6, the authors attempt to draw an effective comparison of the functioning of a convolutional neural network (CNN) with that of the human brain by highlighting the fundamental explanation for the functioning of CNNs.

In chapter 7, the authors demonstrate how hybridization of evolutionary optimized deep learning may be effective in human behavior analysis.

The authors in chapter 8 explore how effectively various feature extraction techniques may be used for vehicle tracking purposes with requisite validation by experimental results.

Chapter 9 reports a GIS-anchored system that offers a helping hand toward common and innocent people by identifying crime hotspot regions and thereby alerting them to criminals' possible whereabouts.

In chapter 10, authors offer hyper-spectral data analysis and processing algorithms consisting of spectral unmixing, classification, and target identification. The effectiveness is validated by adequate test results.

In chapter 11, the authors demonstrate the effectiveness of artificial immune systems (AISs) in identifying important bands in hyperspectral images. The affinity of band combination, derived based on 2D principal component analysis (2DPCA), is computed by fuzzy k-nearest neighbor (KNN), reporting convincing results.

In chapter 12, the authors aim to study several hybrid clustering algorithms capable of tackling uncertainties and their applications in medical image analysis.

In chapter 13, the authors hybridize Otsu thresholding and morphological segmentation algorithms to achieve accurate segmentation of breast cancer images. Subsequently, relevant features are optimized using Cuckoo search, which in turn equips SVMs to classify breast images into benign or malignant. Their finding has been substantiated by the accuracy reported.

Chapter 14 proposes a new approach for biometric recognition of a person, using dorsal, palm, finger, and wrist veins of the hand. The analysis of these vein modalities is done in both the spatial domain and frequency domain with encouraging results reported.

In chapter 15, the authors are able to achieve a convincing performance in terms of metrics like accuracy, sensitivity, and specificity of their proposed method, based on computer-aided automated segmentation of mammograms comprising a seeded region growing algorithm followed by statistical elimination and eventually decision making.

In chapter 16, authors provide a novel method for the automatic detection of coronary stenosis in X-ray angiograms, comprising three stages: (1) differential evolution-based optimization of Gaussian matching filters (GMFs), (2) application of iterative thresholding of the Gaussian filter response, followed by (3) naive Bayes classification for determining vessel stenosis. Their technique is justified by encouraging experimental results.

In the work reported in chapter 17, the authors substantiate the effectiveness of an efficient CAD system based on multiresolution texture descriptors (derived from ten different compact support wavelet filters) using 2D wavelet transform using a smooth SVM (SSVM) classifier. The proposed technique is validated with adequate experimentation reporting very encouraging results in the context of classification of mammogram images.

The chapters appearing in this volume show a wide spectrum of applications pertaining to image processing through which the use of hybrid techniques is vindicated. It is needless to mention that the reporting of various chapters in the present scope indicates very limited applicability of hybridization in the field of various image-processing problems. There are, of course, many more to come in days ahead. From that point of view, the editors of the present volume want to avail this opportunity to express their gratitude to the contributors of different chapters, without whose participation the present treatise would not have arrived in the present form. The editors feel encouraged to expect many more contributions from the present authors in this field in their future endeavours also. This is just a limited effort rendered by the editors to accommodate what the relevant research community is thinking today. In fact, from that perspective, it is just the beginning of such an endeavour and definitely not the end. Last but not the least, the editors also thank Wiley for all the cooperation they rendered as a professional publishing house toward making the present effort a success.

India, November 2016

Siddhartha Bhattacharyya
Indrajit Pan
Anirban Mukherjee
Paramartha Dutta

About the Companion website

Don't forget to visit the companion website for this book:

www.wiley.com/go/bhattacharyya/hybridintelligence

There you will find valuable material designed to enhance your learning, including:

- Videos

Scan this QR code to visit the companion website.

1

Multilevel Image Segmentation Using Modified Genetic Algorithm (MfGA)-based Fuzzy C-Means

Sourav De[1], Sunanda Das[2], Siddhartha Bhattacharyya[3], and Paramartha Dutta[4]

[1] Department of Computer Science and Engineering, Cooch Behar Government Engineering College, Cooch Behar, West Bengal, India
[2] Department of Computer Science and Engineering, University Institute of Technology, The University of Burdwan, Burdwan, West Bengal, India
[3] Department of Information Technology, RCC Institute of Information Technology, Kolkata, West Bengal, India
[4] Department of Computer Science and System Sciences, Visva-Bharati University, Santiniketan, West Bengal, India

1.1 Introduction

Image segmentation plays a key role in image analysis and pattern recognition and also has other application areas, like machine vision, biometric measurements, medical imaging, and so on for the purposes of detecting, recognzing or tracking an object. The objective of image segmentation is to segregate an image into a set of disjoint regions on the basis of uniform and homogeneous attributes such as intensity, color, texture, and so on. The attributes in the different regions are heterogeneous to each other, but the attributes in the same region are homogeneous to each other. On the basis of the inherent features of an image and some *a priori* knowledge and/or presumptions about the image, the pixels of that image can be classified successfully. Let an image be represented by I, and that image can be partitioned into n number of subimages (i.e., $I_1, I_2, ..I_n$), such that:

- $\bigcup_{i=1}^{K} I_i = I$;
- I_i is a connected segment;
- $I_i \neq \emptyset$ for $i = 1, \cdots, K$; and
- $I_i \cap I_j = \emptyset$ for $i = 1, \cdots, K, j = 1, \cdots, K,$ and $i \neq j$.

Different image segmentation techniques have been developed on the basis of the discontinuity and similarity of the intensity levels of an image. Multilevel grayscale and color image segmentation became a perennial research area due to the diversity of the grayscale and color-intensity gamuts. In different research articles, the image segmentation problem is handled by different types of classical and nonclassical image-processing algorithms.

Among different types of classical image segmentation techniques [1–3], edge detection and region growing, thresholding, normalized cut, and others are well-known image segmentation techniques to segment the multilevel grayscale images. The image

Hybrid Intelligence for Image Analysis and Understanding, First Edition.
Edited by Siddhartha Bhattacharyya, Indrajit Pan, Anirban Mukherjee, and Paramartha Dutta.
© 2017 John Wiley & Sons Ltd. Published 2017 by John Wiley & Sons Ltd.
Companion Website: www.wiley.com/go/bhattacharyya/hybridintelligence

segmentation by the edge detection techniques are done by finding out the boundaries of the objects in that image. But an incorrect segmentation may happen with edge detection algorithms as these processes are not assistive for segmenting any complicated images or the blur images. Region-growing techniques are not efficiently applied for multilevel image segmentation, as the different regions of an image are not fully separated. The image segmentation using thresholding techniques are fully dependent on the histogram of that image. The images that have the distinctive background and objects can be segmented efficiently by the thresholding techniques. This process may not work if the distribution of the pixels in the image is very complex in nature.

To solve the clustering problems, the k-means [4] and the fuzzy c-means [5] are two very renowned clustering algorithms. The common feature of these two algorithms is that both start with a fixed number of predefined cluster centroids. The meaning of k in the k-means algorithm is the same as that of c in the FCM algorithm, and both k and c signify the number of clusters. At the initial stage of the k-means algorithm, the k number of cluster centroids are generated randomly. The clusters are formed with the patterns on the basis of the minimum distance between the pattern and the cluster centroids. The cluster centriods update their positions iteratively by minimizing the sum of the squared error in between the corresponding members within the clusters. This clustering algorithm is considered as hard clustering as it is assumed that each pattern is a member of a single cluster. In FCM, the belongingness of a pattern within a cluster is defined by the degree of membership value of that pattern. The well-known least-squares objective function is used to minimize, and the degree of membership of the pattern will be updated to optimize the clustering problem. That is why FCM is considered a soft clustering algorithm.

Quality improvement of the k-means algorithm has been reported in different research articles. Luo *et al.* [6] proposed a spatially constrained k-means approach for image segmentation. Initially, the k-means algorithm is applied in feature space, and after that, the spatial constraints are considered in the image plane. Khan and Ahmad [7] tried to improve the k-means algorithm by modifying the initialization of the cluster centers. The high-resolution images are segmented by the k-means algorithm after optimization by the Pillar algorithm [8]. The k-means algorithm along with the improved watershed segmentation algorithm are employed to segment the medical images [9]. The noisy images can be segmented efficiently by the modified FCM algorithm, which is proposed by Ahmed *et al.* [10, 11]. In this process, the objective function for the standard FCM algorithm has been modified to allow the labels in the immediate neighborhood of a pixel to influence its labeling [10]. The spatial information of the image is incorporated in the traditional FCM algorithm to segment the noisy grayscale images [12]. In this method, the spatial function is considered as the sum of all the membership functions within the neighborhood of the pixel under consideration. Noisy medical images can be segmented by FCM [13]. In this approach, the input images are converted into multiscale images by smoothing them in different scales. The scaling operation, from high scale to low scale, is performed by FCM. The image with the highest scale is processed with the FCM, and after that, the membership of the current scale is determined by the cluster centers of the previous scale. Noordam *et al.* [14] presented a geometrically guided FCM (GG-FCM) algorithm in which geometrical information is added with the FCM algorithm during clustering. In this process, the clustering is guided by the condition of each pixel, which is decided by the local neighborhood of that pixel. Different types of modified objective functions are applied to segment the corrupted images to derive a lower error rate [10, 15]. A modified FCM algorithm, considering spatial neighborhood information, is presented by Chen *et al.* [16]. In this method, a simple and effective 2D distance metric function is projected by considering the relevance of the current pixel to its neighbor pixels, and an objective function is also formed to update the cluster centers simultaneously in two dimensions of the pixel value and its neighboring value.

However, the k-means algorithm and fuzzy c-means algorithm have some common disadvantages. Both the algorithms need a predefined number of clusters, and the centroid initialization is also a problem. It may not be feasible to know the exact number of clusters beforehand in a large data set. It has been observed in different research articles that different initial numbers of clusters with different initializations had been applied to determine the exact number of clusters in a large data set [17]. Moreover, it may not be effective if the algorithm starts with a limited number of center of clusters. This type of problem is known as the *number of clusters dependency problem* [18]. Ultimately, it may happen that the solutions may be stuck in local minima instead of obtaining a global optimal solution.

In this real-world scenario, most of the problems can be considered as optimization problems. The differentiable functions are solved by the traditional heuristic algorithms, but the real-world optimization problems are nondifferentiable. It is rare that the nondifferentiable optimization functions are properly solved by a heuristic algorithm. In many research articles, it has been found that nondifferentiable optimization functions are solved by different types of metaheuristic approaches, and these algorithms are now becoming more popular in the research arena. Applying stochastic principles, evolutionary algorithms are now becoming an alternative way to solve optimization problems as well as clustering problems. These types of algorithms work on the basis of probabilistic rules to search for a near-optimal solution from a global search space. By inspiring the principle of natural genetics, genetic algorithms (GAs), differential evolution, particle swarm optimization, and so on are some of the examples of evolutionary algorithms.

GAs [19], as randomized search and optimization techniques, are applied efficiently to solve different types of image segmentation problems. Inspired by the principles of evolution and natural genetics, GAs apply three operators (viz., selection, crossover, and mutation over a number of generations) to derive potentially better solutions. Without having any *a priori* knowledge about the probable solutions to the problem or difficulties to formulate the problem, GAs can be the solution to solve those types of problems. Due to the generality characteristic, GAs demand a segmentation quality measurement to evaluate the segmentation technique. Another important characteristic of GAs is that they can derive the optimal or near-optimal solutions for their balanced global and local search capability.

The GA in combination with wavelet transformation is applied to segment multilevel images [20]. At first, the wavelet transform is applied to reduce the size of the original histogram, and after that, the number of thresholds and the threshold values are resolved with the help of a GA on the basis of the lower resolution version of the histogram. Fu *et al.* [21] presented an image segmentation method using a multilevel threshold of gray-level and gradient-magnitude entropy based on GAs. A hybrid GA is proposed by Hauschild *et al.* [22] for image segmentation on the basis of the q-state Potts spin glass model to a grayscale image. In this approach, a set of weights for a q-state spin glass is formed from the test image, and after that, the candidate segmented images are generated using the GA until a suitable candidate solution is detected. The hierarchical local search is applied to speed up the convergence to an adequate solution [22]. The GA in combination with the multilayer self-organizing neural network (MLSONN) architecture are employed to segment the multilevel grayscale images [11]. In [11], it is claimed that the MLSONN architecture, with the help of multilevel sigmoidal (MUSIG) activation function, is not capable of incorporating the image heterogeneity property in the multilevel grayscale image segmentation process. To induce the image content heterogeneity in the segmentation process, the authors employed a GA to generate the optimized class levels for designing the optimized MUSIG (OptiMUSIG) activation function [11]. Now, the OptiMUSIG activation function based MLSONN architecture is capable of segmenting the multilevel grayscale images. The variable threshold–based OptiMUSIG activation function is also efficient for grayscale image segmentation [23]. To decrease the effect of isolated points on the k-means algorithm, a GA is applied to enhance the k-means clustering algorithm [24].

The performance and the shortcomings of FCM algorithms can be improved by different types of evolutionary algorithms. Biju *et al.* [25] proposed a genetic algorithm–based fuzzy c-means (GAFCM) technique to segment spots of complementary DNA (cDNA) microarray images for finding gene expression. An improved version of the GA is presented by Wei *et al.* [26]. In this approach, the genetic operators are modified to enhance the global searching capability of GAs. To improve the convergence speed, the improvised GA applies FCM optimization after each generation of genetic operation [26]. The GA inspired with the FCM algorithm is capable of segmenting the grayscale images [27]. In this approach, the population of GAs is generated with the help of FCM. Jansi and Subashini [28] proposed a GA integrated with FCM for medical image segmentation. The resultant best chromosome, derived by the GA, is applied as the input in the FCM. The drawback of the FCM algorithm (i.e., convergence to the local optima solution,) is overcome in this approach.

The main concern of this chapter is to overcome the shortcomings of the FCM algorithm as this algorithm is generally stuck to a local minima point. To eliminate this drawback, an evolutionary algorithm is the better choice to deal with these types of problems. A GA has the capability to find the global optimal or near–global optimal solutions in a large search space. That is why the authors considered the GA to handle this problem. In this chapter, a modified genetic algorithm (MfGA)-based FCM algorithm is proposed. For that reason, some modifications are made in the traditional GA. A weighted mean approach is introduced for the chromosome generation in the population initialization stage. Like the traditional GA, the crossover probability is not fixed throughout the generations in the MfGA. As crossover probability plays a vital role in GAs, the value of the crossover probability decreases as the number of generations increases. The resultant class levels, derived by the MfGA, are applied as the input of the FCM algorithm. The proposed MfGA-based FCM algorithm is compared with the traditional FCM algorithm [5] and the GA-based FCM algorithm [28]. The above-mentioned algorithms are employed on three benchmark images to determine the performance of those algorithms. Two standard efficiency measures, the correlation coefficient (ρ) [11] and the empirical measure Q [29], are applied as the evaluation functions to measure the quality of the segmented images. The experimental results show that the proposed

MfGA-based FCM algorithm outperforms the other two algorithms to segment the multilevel grayscale images.

This chapter is organized in the following ways. Section 1.2 discusses the traditional FCM algorithm. Two quality evaluation metrics for image segmentation, the correlation coefficient (ρ) and the empirical measure (Q), are narrated in Section 1.4. Before that, in Section 1.3, the proposed MfGA is illustrated. After that, the MfGA-based FCM is discussed in Section 1.5. The experimental results and the comparison with the FCM algorithm [5] and the GA-based FCM [28] algorithm are included in Section 1.6. Finally, Section 1.7 concludes the chapter.

1.2 Fuzzy C-Means Algorithm

The FCM, introduced by Bezdek [5], is the most widely applied fuzzy clustering method. Basically, this algorithm is an extension of the hard c-means clustering method and is based on the concept of fuzzy c-partition [30]. Being an unsupervised clustering algorithm, FCM is associated with the fields of clustering, feature analysis, and classifier design. The wide application areas of FCM are in astronomy, geology, chemistry, image processing, medical diagnosis, target recognition, and others.

The main objectives of the FCM algorithm are to update the cluster centers and to calculate the membership degree. The clusters are formed on the basis of the distance calculation between data points, and the cluster centers are formed for each cluster. The membership degree is applied to show the belongingness of each data point with each cluster, and the cluster centers are also updated with this information. This means that every data point in the data set is associated with every cluster, and among them, a data point may be considered in a cluster when it has high membership value with that cluster and low membership value with the rest of the clusters in that data set.

Let $X = x_1, x_2, \ldots, x_N, x_k \in R^n$ be a set of unlabeled patterns, where N is the number of patterns and each pattern has n number of features. This algorithm tries to minimize the value of an objective function that calculates the quality of the partitioning that divides the data set into C clusters. The distance between a data point x_k to a cluster center T_i is calculated by the well-known Euclidean distance, and it is represented as [5]:

$$D_{ik}^2 = \sum_{j=1}^{n} (x_{kj} - T_{ij})^2, \quad 1 \leq k \leq N, 1 \leq i \leq C \tag{1.1}$$

where the squared Euclidian distance in n-dimensional space is denoted as D_{ik}^2. The membership degree is calculated as [5]:

$$U_{ik} = \frac{1}{\sum_{j=1}^{C} \left(\frac{D_{ik}}{D_{jk}}\right)^{\frac{2}{(m-1)}}}, \quad 1 \leq k \leq N, 1 \leq i \leq C \tag{1.2}$$

where U_{ik} represents the degree of membership of x_k in the ith cluster. The degree of fuzziness is controlled by the parameter $m > 1$. It signifies that every data point has a degree of membership in every cluster.

The centroids are updated according to the following equation [5]:

$$T_{ij} = \frac{\sum\limits_{k=1}^{N} (U_{ik})^m x_k}{\sum\limits_{k=1}^{N} (U_{ik})^m}, \quad 1 \leq i \leq C \tag{1.3}$$

Ultimately, the membership degree of each point is measured using equation (1.2) with the help of new centroid values.

1.3 Modified Genetic Algorithms

Conventional GAs [19] are well-known, efficient, adaptive heuristic search and global optimization algorithms. The basic idea of GAs took from the evolutionary ideas of natural selection and genetics. A GA provides a near-optimal solution in a large, complex, and multimodal problem space. It contains a fixed population size over search space. In GAs, the performance of the entire population can be upgraded instead of improving the performance of the individual members in the population-based optimization procedure. The evolution starts by generating the population through random creation of individuals. These individual solutions are known as *chromosomes*. The quality or goodness of each chromosome or individual is assessed by its *fitness* value. The potentially better solutions are evolved after applying three well-known genetically inspired operators like *selection, crossover*, and *mutation*.

In this chapter, a MfGA is proposed to improve the performance of the conventional GA. In the chromosome representation of the conventional GA, each chromosome contains N number of centroids to cluster a data set into N number of clusters. The initial values of the chromosomes are selected randomly. The clusters are formed between the cluster centroids and the individual data points on the basis of some criteria. In most of the cases, the distance between the data points is the selection criterion. A data point is included in a cluster when the distance between that data point and the particular cluster centroid is minimum rather than the same with other cluster centroids. It may happen that a cluster may contain very few data points compared to other clusters in that data set. The relevance of that small cluster may decrease in that data set. At the same time, the spatial information of the data points in the data set is also not considered. To overcome that, $N+1$ number of cluster centroids are selected in the initial stage to generate the N number of cluster centroids. Steps of MfGA are discussed here:

- *Population initialization*: In MfGA, the population size is fixed, and the lengths of individual chromosomes are fixed. At the initial stage, $N+1$ number of cluster centroids are generated randomly within the range of minimum and maximum values of the data set, and the cluster centroids are denoted as $R_1, R_2, R_3,..,R_i, ..., R_{n+1}$. These cluster centroids are temporary, and they are used to generate the actual cluster centroids. The weighted mean between the temporary cluster centroids of R_i and R_{i+1} is applied to generate the actual cluster centroids L_i, and it is represented as:

$$L_i = \frac{\sum_{I=R_i}^{R_{i+1}} f_I \times I}{\sum_{I=R_i}^{R_{i+1}} f_I} \tag{1.4}$$

where I represents the value of the data point and f_I denotes the frequency of the Ith data point. For example, the cluster centroid L_1 is generated after taking the weighted mean between the temporary cluster centroids R_1 and R_2. The cluster centroid L_i, having the most frequency value in between the cluster centroids R_i and R_{i+1}, will be selected by this process, and this will make a good chromosome for future generation. Ultimately, the spatial information is taken into consideration at the population initialization time.

- *Selection*: After population initialization, the effectiveness of the individual chromosomes is determined by a proper fitness function so that better chromosomes are taken for further usage. This is done in the perception that the better chromosomes may develop by transmitting the superior genetic information to new generations, and they will persist and generate offspring. In the selection step, the individuals are selected for mating on the basis of the fitness value of each individual. This fitness value is used to associate a probability of selection with each individual chromosome. Some of the well-known selection methods are roulette wheel selection, stochastic universal selection, Boltzmann selection, rank selection, and so on [19].

- *Crossover*: The randomly selected parent chromosomes exchange their information by changing parts of their genetic information. Two chromosomes, having the same probability of crossover rate, are selected to generate offspring for the next generation. The crossover probability plays a vital role in this stage, and it is used to show a ratio of how many parents will be picked for mating. In conventional GA, the crossover probability is same throughout the process. It may happen that a good chromosome may be mated with a bad chromosome, and that good chromosome is not stored in the next stage. To overcome this drawback, it is suggested in this process that the crossover probability rate will decrease as the number of iterations will increase. Here, the crossover probability is inversely proportional to the number of iterations so that the better chromosomes will remain unchanged and will go to the next generation of population. Mathematically, it is represented as:

$$C_p = C_{max} - \left(\frac{C_{max} - C_{min}}{IT_{max} - IT_{cur}} \right) \tag{1.5}$$

where C_p is the current crossover probability, and C_{max} and C_{min} are the maximum and minimum crossover probability, respectively. The maximum number of iterations and the current number of iterations are represented as IT_{max} and IT_{cur}, respectively.

- *Mutation*: The sole objective of the mutation stage is to introduce the genetic diversity into the population. Being a divergence operation, the frequency of the mutation operation is much less, and so the members of the population are affected much less. Mutation probability is taken as a very small value for this reason. Generally, a value in the randomly selected position in the chromosome is flipped.

1.4 Quality Evaluation Metrics for Image Segmentation

It is absolutely necessary to measure the quality of the final segmented images after segmenting the images by different types of segmentation algorithms. Usually, different statistical mathematical functions are employed to evaluate the results of the segmentation algorithms. Here, two unsupervised approaches are provided to measure the goodness of the segmented images in Sections 1.4.1 and 1.4.2.

1.4.1 Correlation Coefficient (ρ)

The degree of the similarity between the original and segmented images can be measured by using the standard measure of correlation coefficient (ρ) [11], and it is represented as [11]:

$$\rho = \frac{\frac{1}{n^2}\sum\limits_{i=1}^{n}\sum\limits_{j=1}^{n}(R_{ij} - \overline{R})(G_{ij} - \overline{G})}{\sqrt{\frac{1}{n^2}\sum\limits_{i=1}^{n}\sum\limits_{j=1}^{n}(R_{ij} - \overline{R})^2}\sqrt{\frac{1}{n^2}\sum\limits_{i=1}^{n}\sum\limits_{j=1}^{n}(G_{ij} - \overline{G})^2}} \tag{1.6}$$

where $R_{ij}, 1 \leq i, j \leq n$ and $G_{ij}, 1 \leq i, j \leq n$ are the original and the segmented images, respectively, each of dimensions $n \times n$. The respective mean intensity values of R_{ij} and G_{ij} are denoted as $\overline{R}$ and $\overline{G}$, respectively. The higher value of ρ signifies the better quality of image segmentation [11].

1.4.2 Empirical Measure Q(*I*)

A empirical measure, $Q(I)$, is used to evaluate the goodness of the segmented images, and it is suggested by Borsotti *et al.* [29, 31]. It is denoted as [29]:

$$Q(I) = \frac{1}{1000.S_I}\sqrt{N}\sum\limits_{g=1}^{N}\left[\frac{e_g^2}{1 + \log S_g} + \left(\frac{N(S_g)}{S_g}\right)^2\right] \tag{1.7}$$

where the area of the gth region of an image (I) is denoted as S_g, and the number of regions having an area S_g is signified by $N(S_g)$. S_I is the area of an image (I) to be segmented. The squared color error of region g, e_g^2, is noted as [29, 31]:

$$e_g^2 = \sum\limits_{v\in(r,g,b)}\sum\limits_{pl\in RI_g}(C_v(pl) - \hat{C}_v(RI_g))^2 \tag{1.8}$$

The number of pixels in region g is presented as RI_g. The average value of feature v (red, green, or blue) of a pixel pl in region g is referred to as [29, 31]:

$$\hat{C}_v(RI_g) = \frac{\sum\limits_{pl\in RI_g} C_v(pl)}{S_g} \tag{1.9}$$

where $C_v(pl)$ signifies the value of component v for pixel pl.

A smaller value of Q implicates better quality of segmentation [29]. In this chapter, the quality of the segmented images is evaluated by these measures, which are applied as different objective functions.

1.5 MfGA-Based FCM Algorithm

FCM has three major drawbacks: there must be *a priori* information of the number of classes to be segmented, this algorithm can only be applied to hyperspherical-shaped clusters, and it often converges to local minima [6]. In this chapter, we have tried to overcome the drawback about the convergence to the local minima. In many cases, it has been observed that a FCM algorithm will easily converge to the local minimum point and the clustering will be affected if the initial centroid values are not initialized correctly. The spatial relative information is not considered in the process of the FCM algorithm. The selection of the initial cluster centers and/or the initial membership value plays a vital role in the performance of the FCM algorithm. The quick convergence and drastic reduction of processing time can be possible in FCM if the selection of the initial cluster center is very close to the actual final cluster center.

The drawback of FCM can be overcome with the help of the proposed MfGA algorithm. The optimized class levels, obtained by the proposed MfGA algorithm, are applied as the initial class levels in the FCM algorithm. Another advantage of this proposed method is that the image content heterogeneity is also considered. The pixels, having most occurrence in the image, have the higher probability for being selected as the class levels in the initial stage. The flowchart in Figure 1.1 shows the steps of the MfGA-based FCM algorithm.

The pseudo-code of the GA-based FCM is as follows:

1. *Pop* ← Generate *P* number of feasible solutions randomly with *N* number of class levels.
2. Calculate fitness (*P*).
3. For i=1 to *itr*, do the following:
   A. Apply selection on *Pop*.
   B. Apply crossover operation with crossover probability C_p.
   C. Apply mutation operation with mutation probability μ_m.
   D. End for.
4. Return the best chromosome with the class levels.
5. Initialize cluster centers of the FCM algorithm with the best solution derived from GA.
6. For i=0 to itr2, do the following:
   A. Update the membership matrix (U_{ik}) using equation (1.2).
   B. Update the centroids [equation (1.3)].
7. Return the ultimate cluster centroids/class levels.

The pseudo-code of the MfGA-based FCM is as follows:

1. Generate *M* number of chromosomes randomly with $N + 1$ number of temporary cluster centroids individually.
2. Using weighted mean equation (1.4), generate *M* number of chromosomes with *N* number of actual cluster centroids individually.
3. Calculate the fitness computation of population using a relevant fitness function.
4. Repeat (for a predefined number of iterations or until a certain condition is satisfied).
   A. Select parents from population, on the basis of fitness values.

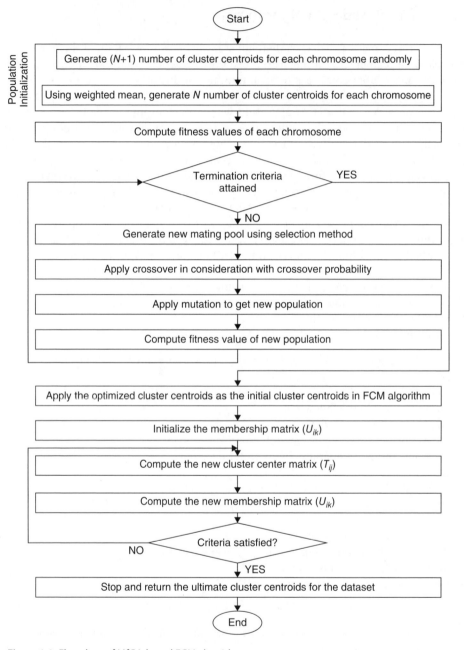

Figure 1.1 Flowchart of MfGA-based FCM algorithm.

    B. Execute crossover and mutation to create a new population [crossover probability is applied using equation (1.5)].

    C. Compute fitness of new population.

5. Return the best chromosome with the class levels.

6. Initialize cluster centers of the FCM algorithm with the best solution derived from the GA.
7. For i=0 to itr2, do the following:
   A. Update the membership matrix (U_{ik}) using equation (1.2).
   B. Update the centroids [equation (1.3)].
8. Return the ultimate cluster centroids/class levels.

1.6 Experimental Results and Discussion

Multilevel grayscale image segmentation with the MfGA-based FCM algorithm was demonstrated with three real-life grayscale images (viz., Lena, baboon, and peppers) of dimensions 256×256. Experimental results in quantitative and qualitative terms are reported in this section. The multilevel images are segmented in $K = \{6, 8, 10\}$ classes, but the results are reported for $K = \{6, 8\}$ classes. Results are also presented for the segmentation of the multilevel grayscale images with the GA-based FCM algorithm [5] and also with the FCM algorithm [28]. To measure the efficiency of the different algorithms qualitatively, two evaluation functions (ρ, Q), presented in equations (1.6) and (1.7), respectively, have been used in this chapter.

In the initial stage, the class levels are generated by the proposed MfGA algorithm, and after that the obtained class levels are supplied as the input in the FCM algorithm. For that, the pixel intensity levels of the multilevel grayscale images and the number of classes (K) to be segmented are supplied as inputs to this process. The randomly generated real numbers within the range of the minimum and maximum intensity values of the input image are applied to create the chromosomes for this process. These components of the chromosomes are treated as the class levels or class boundaries to segment the input image. For example, to segment an image into eight segments, the chromosomes with nine class levels are generated at the starting point. The process of generating the chromosomes and creating the chromosome pool in the GA-based FCM method is the same as the proposed method. A population size of 50 has been used for the proposed and GA-based FCM method. At the initial stage, the class levels are also generated randomly in the FCM method.

The GA operators (viz., selection, crossover, and mutation) are used in the proposed and GA-based FCM [28] approaches. The roulette wheel approach is employed for the selection phase for both the GA-based approaches. Afterward, the crossover and mutation operators are applied to generate the new population. Though the crossover probability value is fixed in the GA-based FCM approach, the crossover probability value in the MfGA-based FCM method is dependent on the number of iterations, and that works according to equation (1.5). The crossover probability value is 0.9 for all the stages in the GA-based FCM algorithm. The maximum crossover probability value (C_{max}) and minimum crossover probability value (C_{min}) are applied as 0.9 and 0.5, respectively, in the proposed MfGA-based FCM method. In both GA-based approaches, the single-point crossover operation is used in the crossover stage, and 0.01 is considered as the mutation probability. The new population is formed after the mutation operation.

The class levels generated by the proposed algorithm on the basis of the two fitness functions $(\rho$ and $Q)$ for different numbers of classes are tabulated in Tables 1.1 and 1.2 for Lena images. In this type of tables, the number of segments (# segments), the name

Table 1.1 Class boundaries and evaluated segmentation quality measures ρ by different algorithms for different classes of the Lena image

# Segments	Algorithm	#	Class levels	Fitness value
6	FCM	1	7, 61, 107, 143, 179, 229	0.9133
		2	6, 64, 109, 144, 177, 229	0.9213
		3	9, 64, 107, 143, 179, 229	0.9194
		4	**7, 64, 110, 145, 180, 227**	**0.9227**
	GA-based FCM	1	47, 76, 105, 135, 161, 203	0.9823
		2	**50, 93, 124, 148, 173, 206**	**0.9824**
		3	47, 78, 105, 134, 161, 203	0.9823
		4	47, 77, 105, 135, 161, 203	0.9823
	Proposed	1	**47, 76, 105, 135, 161, 203**	**0.9828**
		2	50, 92, 123, 148, 173, 206	0.9824
		3	50, 94, 125, 149, 173, 204	0.9824
		4	**47, 76, 105, 135, 161, 203**	**0.9828**
8	FCM	1	**4, 37, 71, 105, 131, 155, 188, 231**	**0.9559**
		2	5, 37, 70, 105, 131, 155, 188, 231	0.9273
		3	4, 37, 71, 105, 131, 154, 188, 230	0.9251
		4	6, 37, 71, 105, 131, 155, 188, 231	0.9266
	GA-based FCM	1	42, 58, 84, 104, 130, 152, 178, 207	0.9893
		2	**42, 58, 83, 105, 131, 152, 176, 207**	**0.9894**
		3	46, 72, 99, 122, 140, 156, 179, 208	0.9893
		4	**42, 58, 83, 105, 130, 152, 176, 207**	**0.9894**
	Proposed	1	46, 71, 98, 122, 140, 157, 179, 208	0.9897
		2	**45, 72, 99, 124, 141, 158, 179, 208**	**0.9898**
		3	46, 72, 99, 123, 140, 156, 177, 207	0.9896
		4	**46, 72, 98, 122, 142, 157, 179, 208**	**0.9898**

of the algorithm (Algorithm), the serial number (#), class levels, and the fitness value (*fit val*) are accounted in the columns. The class levels evaluated by the FCM algorithm and the GA-based FCM algorithm are also reported in these tables. Each type of experiments is performed 50 times, but only four good results are tabulated in these tables in respect to the number of segments and in respect to the individual algorithms. The best results obtained by any process for each number of segments are highlighted by **boldface** type.

The mean and standard deviations are evaluated for different algorithm-based fitness values using different types of fitness functions, and these results for Lena images are reported in Table 1.3. The time taken by different algorithms is also preserved, and the mean of the time taken by the different algorithms is also mentioned in this table. The good results are marked in **bold**.

For segmenting the Lena image, the proposed algorithm outperforms the other two algorithms. In Table 1.1, the ρ values, derived by the MfGA-based FCM algorithm, are better than the other two algorithms for a different number of segments. The same observation can be obtained if anyone goes through the reported results in Table 1.2. The Q-based fitness values, obtained by the proposed algorithm, are much better than the results derived by the other two processes. In Table 1.3, the reported mean and standard

Table 1.2 Class boundaries and evaluated segmentation quality measures Q by different algorithms for different classes of the Lena image

# Segments	Algorithm	#	Class levels	Fitness value
6	FCM	1	7, 62, 107, 144, 179, 228	24169.019
		2	8, 64, 107, 143, 177, 229	19229.788
		3	7, 64, 106, 140, 179, 231	24169.018
		4	**5, 66, 110, 142, 180, 229**	**15017.659**
	GA-based FCM	1	**47, 77, 105, 135, 161, 203**	**10777.503**
		2	50, 93, 124, 148, 173, 206	16017.944
		3	47, 74, 105, 135, 161, 203	12194.313
		4	47, 76, 108, 134, 161, 203	10909.195
	Proposed	1	47, 78, 105, 135, 161, 203	7117.421
		2	**47, 77, 105, 134, 161, 204**	**7020.245**
		3	50, 93, 124, 148, 173, 206	13146.389
		4	50, 94, 126, 158, 179, 204	10577.836
8	FCM	1	4, 38, 71, 105, 131, 155, 188, 229	64095.622
		2	5, 37, 69, 105, 132, 155, 188, 231	83743.507
		3	6, 37, 71, 105, 129, 155, 187, 230	108324.494
		4	**4, 37, 71, 105, 131, 155, 184, 231**	**48229.038**
	GA-based FCM	1	46, 72, 99, 122, 140, 157, 179, 208	49382.388
		2	46, 73, 100, 127, 149, 169, 193, 212	90707.0511
		3	46, 72, 99, 122, 140, 157, 179, 208	47419.664
		4	**46, 73, 99, 122, 140, 158, 180, 208**	**46906.552**
	Proposed	1	42, 58, 83, 105, 130, 152, 176, 207	33531.256
		2	42, 59, 84, 107, 129, 152, 175, 207	32630.511
		3	**42, 58, 83, 105, 131, 152, 176, 207**	**32496.323**
		4	46, 72, 99, 122, 140, 157, 179, 208	47429.683

Table 1.3 Different algorithm-based means and standard deviations using different types of fitness functions and mean of time taken by different algorithms for the Lena image

Fitness function	# Segments	Algorithm	Mean ± standard deviation	Mean time
ρ	6	FCM	0.9311 ± 0.02033	31 sec
		GA-based FCM	0.9623 ± 0.00003	23 sec
		Proposed	**0.9824 ± 0.00002**	**20 sec**
	8	FCM	0.9506 ± 0.01814	41 sec
		GA-based FCM	0.9817 ± 0.00021	39 sec
		Proposed	**0.9898 ± 2.34E-16**	**37 sec**
Q	6	FCM	13119.427 ± 7046.408	39 sec
		GA-based FCM	12018.495 ± 2149.968	25 sec
		Proposed	**8966.715 ± 2635.109**	**22 sec**
	8	FCM	47943.902 ± 29441.925	49 sec
		GA-based FCM	47417.776 ± 16557.246	47 sec
		Proposed	**43233.689 ± 7179.615**	**43 sec**

Table 1.4 Class boundaries and evaluated segmentation quality measures ρ by different algorithms for different classes of the peppers image

# Segments	Algorithm	#	Class levels	Fitness value
6	FCM	**1**	**28, 76, 108, 140, 178, 201**	**0.9291**
		2	26, 75, 103, 144, 169, 191	0.9312
		3	23, 69, 98, 148, 171, 201	0.7860
		4	25, 71, 107, 139, 170, 204	0.8916
	GA-based FCM	**1**	**29, 76, 108, 144, 173, 201**	**0.9829**
		2	28, 75, 108, 144, 170, 201	0.9828
		3	29, 76, 108, 145, 173, 203	0.9828
		4	30, 76, 110, 144, 173, 201	0.9827
	Proposed	1	29, 78, 108, 144, 173, 203	0.9830
		2	28, 76, 110, 144, 172, 203	0.9830
		3	29, 78, 110, 146, 173, 202	0.9830
		4	**28, 75, 108, 144, 172, 201**	**0.9831**
8	FCM	1	26, 69, 94, 111, 136, 160, 181, 203	0.9268
		2	17, 46, 79, 106, 128, 156, 178, 203	0.9317
		3	**40, 79, 104, 121, 143, 160, 176, 199**	**0.9585**
		4	17, 46, 79, 106, 127, 155, 178, 202	0.9309
	GA-based FCM	1	27, 71, 98, 115, 146, 170, 193, 213	0.9887
		2	**17, 45, 79, 106, 128, 156, 178, 203**	**0.9890**
		3	27, 71, 98, 114, 146, 170, 193, 213	0.9887
		4	18, 47, 81, 108, 140, 191, 167, 211	0.9889
	Proposed	1	18, 47, 81, 108, 140, 167, 191, 211	0.9890
		2	27, 71, 98, 114, 146, 170, 193, 213	0.9890
		3	17, 46, 79, 106, 128, 156, 179, 203	0.9891
		4	**17, 46, 78, 106, 129, 156, 179, 203**	**0.9892**

deviation of both fitness values (ρ and Q) are best for the proposed algorithm compared to the same reported by the other two algorithms. It is also observed that the proposed method has taken less time than the other two approaches.

The same experiment is repeated for the peppers image. The ρ-based and Q-based class boundaries and the fitness values are reported in Table 1.4 and Table 1.5, respectively. The better results are highlighted in **bold**. From these tables, it is detected that the fitness values derived by the proposed algorithm are better than the same results obtained by the other two algorithms. This observation will remain unchanged for the peppers image whether it is segmented in six segments or eight segments.

The peppers image is segmented by the proposed approach 50 times for each fitness function. The mean and standard deviation of those different function-based fitness values are reported in Table 1.6. The mean of the time taken by the proposed algorithm is also tabulated in this table. The same results are also saved for the other two algorithms. The mean and standard deviation of the fitness values and the mean time taken by the other two algorithms are also reported in Table 1.6. The MfGA-based FCM algorithm is much better than the other two algorithms if anyone considers the accounted results in

Table 1.5 Class boundaries and evaluated segmentation quality measures Q by different algorithms for different classes of the peppers image

# Segments	Algorithm	#	Class levels	Fitness value
6	FCM	1	28, 76, 108, 145, 173, 201	42696.252
		2	29, 76, 110, 144, 175, 202	37371.362
		3	27, 75, 105, 143, 170, 201	31326.604
		4	**26, 71, 108, 144, 173, 201**	**9813.211**
	GA-based FCM	1	28, 78, 106, 144, 169, 203	17130.923
		2	**29, 76, 108, 140, 173, 201**	**9130.923**
		3	26, 75, 109, 144, 174, 202	15065.869
		4	27, 76, 108, 144, 171, 203	14531.759
	Proposed	1	29, 76, 108, 144, 173, 201	5693.583
		2	29, 78, 110, 156, 174, 194	4907.482
		3	29, 76, 107, 144, 172, 201	5693.583
		4	**29, 76, 108, 145, 173, 201**	**4892.024**
8	FCM	1	18, 47, 81, 108, 140, 167, 191, 211	181883.131
		2	**27, 71, 98, 115, 146, 170, 193, 213**	**102247.881**
		3	27, 71, 98, 115, 146, 172, 198, 211	198840.268
		4	17, 46, 79, 106, 128, 156, 179, 203	283016.901
	GA-based FCM	1	27, 71, 98, 115, 146, 170, 193, 213	53784.935
		2	27, 76, 97, 118, 140, 167, 191, 211	44311.485
		3	28, 71, 98, 115, 146, 170, 193, 213	53908.007
		4	**27, 71, 98, 115, 147, 175, 193, 212**	**27230.946**
	Proposed	1	18, 47, 81, 108, 125, 142, 170, 199	20251.096
		2	17, 46, 79, 106, 128, 156, 178, 202	21406.658
		3	27, 71, 99, 115, 147, 171, 199, 213	24926.826
		4	**17, 46, 79, 106, 128, 156, 178, 202**	**19632.432**

Table 1.6 Different algorithm-based mean and standard deviation using different types of fitness functions and mean of time taken by different algorithms for the peppers image

Fitness function	# Segments	Algorithm	Mean $\pm$ standard deviation	Mean time
ρ	6	FCM	0.8890 $\pm$ 0.06096	36 sec
		GA-based FCM	0.9727 $\pm$ 2.686E-06	28 sec
		Proposed	**0.9831 $\pm$ 3.789E-07**	**24 sec**
	8	FCM	0.9472 $\pm$ 0.01815	58 sec
		GA-based FCM	0.9824 $\pm$ 0.00021	56 sec
		Proposed	**0.9890 $\pm$ 0.00016**	**54 sec**
Q	6	FCM	14739.441 $\pm$ 15846.203	30 sec
		GA-based FCM	14681.419 $\pm$ 2967.542	26 sec
		Proposed	**5534.818 $\pm$ 334.728**	**24 sec**
	8	FCM	111303.905 $\pm$ 91628.397	59 sec
		GA-based FCM	40479.148 $\pm$ 15736.887	56 sec
		Proposed	**32567.705 $\pm$ 15566.113**	**55 sec**

Table 1.7 Class boundaries and evaluated segmentation quality measures ρ by different algorithms for different classes of the baboon image

# Segments	Algorithm	#	Class levels	Fitness value
6	FCM	1	**37, 70, 95, 121, 140, 173**	**0.9204**
		2	37, 69, 95, 118, 144, 175	0.9154
		3	37, 69, 89, 116, 142, 174	0.9099
		4	35, 68, 98, 118, 145, 170	0.9188
	GA-based FCM	1	**37, 69, 96, 119, 146, 173**	**0.9785**
		2	36, 69, 95, 118, 144, 175	0.9783
		3	37, 69, 95, 120, 144, 173	0.9784
		4	37, 68, 94, 118, 145, 175	0.9783
	Proposed	1	38, 70, 96, 120, 154, 173	0.9784
		2	43, 79, 110, 133, 158, 174	0.9785
		3	**37, 69, 98, 120, 145, 173**	**0.9786**
		4	38, 68, 96, 118, 144, 174	0.9785
8	FCM	1	**36, 66, 82, 101, 125, 141, 159, 180**	**0.9514**
		2	31, 60, 86, 96, 118, 130, 159, 179	0.9183
		3	29, 58, 81, 100, 116, 128, 161, 181	0.9204
		4	30, 59, 79, 99, 119, 138, 159, 178	0.9498
	GA-based FCM	1	32, 61, 81, 101, 118, 137, 160, 178	0.9869
		2	**32, 59, 81, 100, 119, 137, 159, 177**	**0.9870**
		3	33, 59, 82, 99, 118, 135, 158, 178	0.9869
		4	31, 60, 81, 101, 120, 137, 158, 177	0.9861
	Proposed	1	33, 59, 81, 99, 121, 137, 159, 178	0.9869
		2	32, 59, 81, 101, 117, 138, 159, 178	0.9870
		3	**33, 59, 81, 100, 118, 137, 158, 178**	**0.9871**
		4	32, 59, 81, 101, 118, 137, 159, 178	0.9870

Table 1.6. The fitness values and the execution time of the proposed algorithm are best compared to the same for the other two algorithms.

The class boundaries and the ρ-based fitness values for segmenting the baboon image are reported in Table 1.7. In Table 1.8, the class levels for segmenting the same image and the Q-based fitness values are presented. The best results derived by the individual algorithm are highlighted in **bold**.

The proposed approach is applied 50 times for segmenting the baboon image on the basis of two fitness functions separately. The mean and standard deviation of those different function-based fitness values are tabulated in Table 1.9. The mean of the time taken by the proposed algorithm is also reported in this table. The same thing is also done for the other two algorithms. The mean and standard deviation of the fitness values and the mean time taken by the other two algorithms are also accounted in Table 1.9. If anyone goes through the reported results in Table 1.9, they can observe that the other two algorithm-based results are far behind the results derived by the MfGA-based FCM algorithm.

The segmented grayscale output images obtained for the $K = \{6, 8\}$ classes, with the proposed approach vis-à-vis with the FCM [5] and the GA-based FCM [28] algorithms,

Table 1.8 Class boundaries and evaluated segmentation quality measures Q by different algorithms for different classes of the baboon image

# Segments	Algorithm	#	Class levels	Fitness value
6	FCM	1	38, 71, 95, 116, 140, 173	16413.570
		2	**37, 69, 95, 114, 134, 165**	**13422.376**
		3	36, 69, 92, 114, 141, 170	18223.109
		4	37, 68, 95, 112, 143, 176	16808.168
	GA-based FCM	1	37, 69, 98, 120, 144, 173	7933.554
		2	36, 69, 95, 119, 146, 173	15368.872
		3	**38, 71, 98, 118, 144, 174**	**5532.344**
		4	37, 73, 95, 118, 145, 173	15368.872
	Proposed	**1**	**38, 69, 95, 120, 146, 173**	**4376.556**
		2	38, 70, 94, 118, 144, 173	4571.008
		3	43, 79, 110, 133, 158, 174	5053.403
		4	37, 68, 94, 118, 144, 173	4985.565
8	FCM	1	32, 49, 80, 99, 118, 131, 159, 178	95888.688
		2	33, 62, 81, 101, 118, 137, 152, 178	73697.780
		3	32, 60, 81, 100, 113, 137, 159, 179	102621.172
		4	**29, 59, 81, 97, 118, 137, 159, 178**	**70320.873**
	GA-based FCM	**1**	**32, 59, 82, 101, 118, 137, 159, 178**	**27848.183**
		2	33, 60, 82, 102, 119, 138, 160, 179	31323.896
		3	34, 60, 82, 101, 118, 138, 159, 178	32329.115
		4	33, 59, 81, 101, 118, 137, 161, 178	28488.563
	Proposed	1	33, 61, 84, 104, 121, 147, 164, 175	21661.820
		2	32, 59, 80, 99, 117, 135, 156, 177	24137.965
		3	**34, 61, 84, 104, 121, 143, 160, 172**	**21198.127**
		4	33, 60, 81, 101, 118, 137, 158, 178	26805.747

Table 1.9 Different algorithm-based mean and standard deviation using different types of fitness functions and mean of time taken by different algorithms for the baboon image

Fitness function	# Segments	Algorithm	Mean ± standard deviation	Mean time
ρ	6	FCM	0.9211 ± 0.01114	25 sec
		GA-based FCM	0.9680± 8.198E-07	22 sec
		Proposed	**0.9788 ± 7.971E-07**	**14 sec**
	8	FCM	0.9489 ± 0.01751	59 sec
		GA-based FCM	0.9762 ± 2.135E-05	56 sec
		Proposed	**0.9870 ± 2.519E-05**	**54 sec**
Q	6	FCM	9777.519 ± 5695.687	42 sec
		GA-based FCM	9671.766 ± 4961.474	40 sec
		Proposed	**4894.895 ± 628.569**	**32 sec**
	8	FCM	52408.164 ± 30991.141	53 sec
		GA-based FCM	28137.835 ± 2594.486	51 sec
		Proposed	**24343.203 ± 8091.355**	**43 sec**

are demonstrated afterward. The segmented grayscale test images with $K = 8$ segments are presented in this chapter. The segmented multilevel grayscale test images derived by the FCM algorithm and the GA-based FCM algorithm are depicted in the first and second rows of each figure. In the third row of each figure, the segmented multilevel grayscale test images by the proposed MfGA-based FCM algorithm are shown. For easy recognition in each figure, the GA-based FCM and the MfGA-based FCM algorithms are noted as GA FCM and MfGA FCM, respectively. The segmented images by the FCM algorithm are presented in (a–d), the segmented images by the GA-based FCM algorithm are shown in (e–h), and the MfGA-based FCM-based multilevel segmented images are pictured in (i–l) of each figure, respectively. In Figure 1.2, the class levels (K=8) of Table 1.1, obtained by the proposed approach and other two algorithms, are employed to get the segmented output image of the Lena image. The fitness function, ρ, is applied in this case. Figure 1.3 is generated with the class levels obtained by each algorithm. These images are segmented into eight segments, and Q is applied as the evaluation function in this case.

It is observed from Figures 1.2 and 1.3 that the proposed approach gives better segmented outputs than the same derived by the other two approaches.

The multilevel segmented outputs of the peppers image are shown in Figures 1.4 and 1.5. The four class levels (K=8) of Table 1.4, obtained by the proposed algorithm as

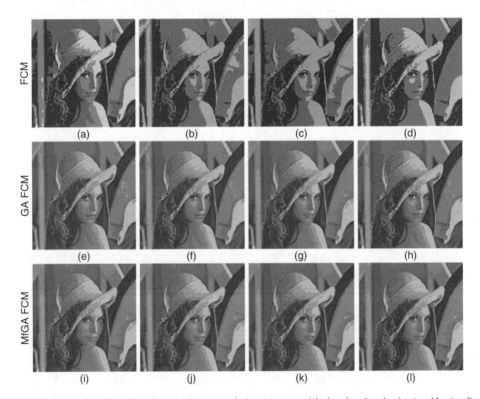

Figure 1.2 8-class segmented 256×256 grayscale Lena image with the class levels obtained by (a–d) FCM, (e–h) GA-based FCM, and (i–l) MfGA-based FCM algorithms of four results of Table 1.1, with ρ as the quality measure.

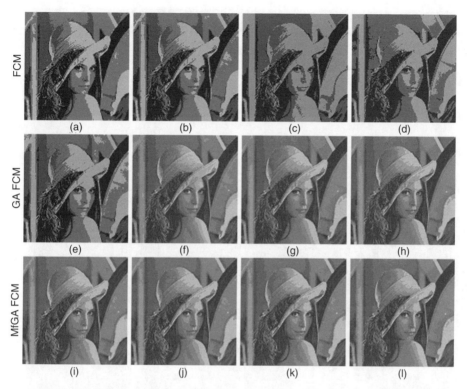

Figure 1.3 8-class segmented 256×256 grayscale Lena image with the class levels obtained by (a–d) FCM, (e–h) GA-based FCM, and (i–l) MfGA-based FCM algorithms of four results of Table 1.2, with Q as the quality measure.

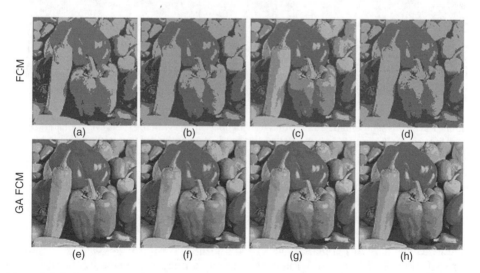

Figure 1.4 8-class segmented 256×256 grayscale peppers image with the class levels obtained by (a–d) FCM, (e–h) GA-based FCM, and (i–l) MfGA-based FCM algorithms of four results of Table 1.4, with ρ as the quality measure.

MfGA FCM

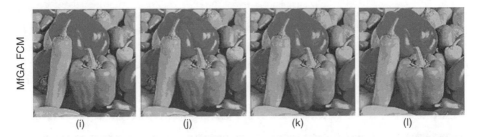

(i) (j) (k) (l)

Figure 1.4 (*Continued*)

FCM

GA FCM

MfGA FCM

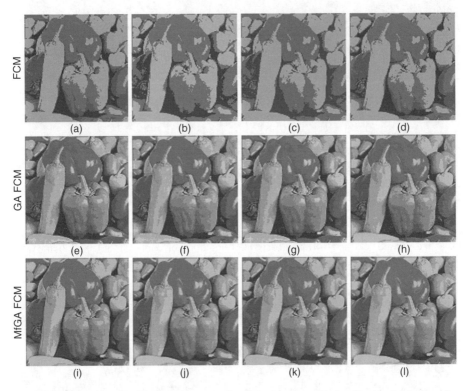

(a) (b) (c) (d)

(e) (f) (g) (h)

(i) (j) (k) (l)

Figure 1.5 8-class segmented 256×256 grayscale peppers image with the class levels obtained by (a–d) FCM, (e–h) GA-based FCM, and (i–l) MfGA-based FCM algorithms of four results of Table 1.5, with Q as the quality measure.

well as other two approaches, are employed to derive the segmented output images of the peppers image in Figure 1.4. In this case, ρ is employed as the evaluation function. In Figure 1.5, the multilevel segmented outputs of the peppers image are yielded using the Q fitness function based on four results from Table 1.5 with $K=8$ class levels. The multilevel segmented peppers images by the proposed MfGA-based FCM algorithm are segmented in a better way than the segmented images deduced by the other two approaches, and this is clear from Figures 1.4 and 1.5.

The ρ is applied as the fitness function to generate the segmented baboon image, which is shown in Figure 1.6 using $K=8$ class levels of Table 1.7. The class levels ($K=8$) of

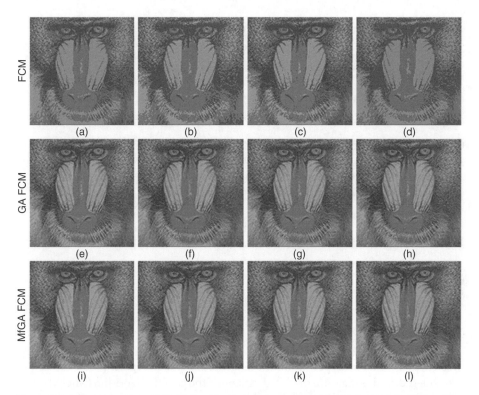

Figure 1.6 8-class segmented 256×256 grayscale baboon image with the class levels obtained by (a–d) FCM, (e–h) GA-based FCM, and (i–l) MfGA-based FCM algorithms of four results of Table 1.7, with ρ as the quality measure.

Table 1.8 are employed to generate the segmented baboon image that is depicted in Figure 1.7. In this case, the empirical measure Q is applied as the quality measure.

From Figures 1.6 and 1.7, it can be said that the segmented multilevel baboon images are better segmented by the proposed algorithm than by the FCM and GA-based FCM algorithms.

At the end, it can be concluded that the proposed MfGA algorithm overwhelms the FCM [5] and GA-based FCM [28] algorithms quantitatively and qualitatively.

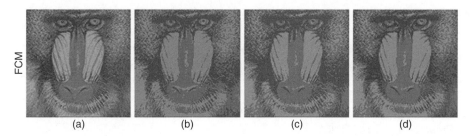

Figure 1.7 8-class segmented 256×256 grayscale baboon image with the class levels obtained by (a–d) FCM, (e–h) GA-based FCM, and (i–l) MfGA-based FCM algorithms of four results of Table 1.8, with Q as the quality measure.

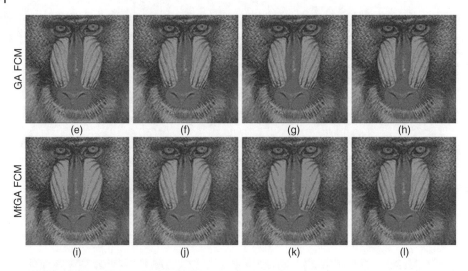

Figure 1.7 *(Continued)*

1.7 Conclusion

In this chapter, different types of multilevel grayscale image segmentation techniques are considered. In this regard, a modified version of the genetic algorithm (MfGA)-based FCM algorithm is proposed to segment the multilevel grayscale images. The FCM and the GA-based FCM algorithms are also recounted briefly, and they are also used to segment the same multilevel grayscale images. The drawback of the original FCM algorithm is also pointed out in this chapter quite efficiently. The way to get rid of the drawback of the FCM algorithm is also discussed elaborately by proposing the proposed algorithm. The solutions derived by the MfGA-based FCM algorithm are globally optimized solutions. To derive the optimized class levels in this procedure, different image segmentation quality measures are used. The performance of the proposed MfGA-based FCM algorithm for real-life multilevel grayscale image segmentation is superior in most of the cases as compared to the other two segmentation algorithms.

References

1 Zaitoun, N.M. and Aqel, M.J. (2015) Survey on image segmentation techniques, in *Proceedings of International Conference on Communications, management, and Information technology (ICCMIT'2015)*, vol. 65, Procedia Computer Science, vol. 65, pp. 797–806.

2 Khan, W. (2013) Image segmentation techniques: a survey. *Journal of Image and Graphics*, **1** (4), 166–170.

3 Chauhan, A.S., Silakari, S., and Dixit, M. (2014) Image segmentation methods: a survey approach, in *Proceedings of 2014 Fourth International Conference on Communication Systems and Network Technologies (CSNT)*, pp. 929–933.

4 MacQueen, J. (1967) Some methods for classification and analysis of multivariate observations, in *Fifth Berkeley Symposium on Mathematics, Statistics and Probability*, pp. 281–297.

5 Bezdek, J. (1981) *Pattern Recognition with Fuzzy Objective Function Algorithms*, Plenum Press, New York.

6 Luo, M., Ma, Y.F., and Zhang, H.J. (2003) A spatial constrained K-means approach to image segmentation, in *Proceedings of the 2003 Joint Conference of the Fourth International Conference on Information, Communications and Signal Processing, 2003 and Fourth Pacific Rim Conference on Multimedia*, vol. 2, pp. 738–742.

7 Khan, S.S. and Ahamed, A. (2004) Cluster center initialization algorithm for K-means clustering. *Pattern Recognition Letters*, **25** (11), 1293–1302.

8 Barakbah, A.R. and Kiyoki, Y. (2009) A new approach for image segmentation using Pillar-kmeans algorithm. *World Academy of Science, Engineering and Technology*, **59**, 23–28.

9 Ng, H.P., Ong, S.H., Foong, K.W.C., Goh, P.S., and Nowinski, W.L. (2006) Medical image segmentation using k-means clustering and improved watershed algorithm, in *2006 IEEE Southwest Symposium on Image Analysis and Interpretation*, pp. 61–65.

10 Ahmed, M.N., Yamany, S.M., Mohamed, N., Farag, A.A., and Moriarty, T. (2002) A modified fuzzy c-means algorithm for bias field estimation and segmentation of MRI data. *IEEE Transactions on Medical Imaging*, **21** (3), 193–199.

11 De, S., Bhattacharyya, S., and Dutta, P. (2010) Efficient grey-level image segmentation using an optimised MUSIG (OptiMUSIG) activation function. *International Journal of Parallel, Emergent and Distributed Systems*, **26** (1), 1–39.

12 Tripathy, B.K., Basu, A., and Govel, S. (2014) Image segmentation using spatial intuitionistic fuzzy C means clustering, in *2014 IEEE International Conference on Computational Intelligence and Computing Research (ICCIC)*, pp. 1–5.

13 Balafar, M.A., Ramli, A.R., Saripan, M.I., Mahmud, R., Mashohor, S., and Balafar, M. (2008) New multi-scale medical image segmentation based on fuzzy c-mean (FCM), in *2008 IEEE Conference on Innovative Technologies in Intelligent Systems and Industrial Applications*, pp. 66–70.

14 Noordam, J.C., van den Broek, W.H.A.M., and Buydens, L.M.C. (2000) Geometrically guided fuzzy c-means clustering for multivariate image segmentation, in *Proceedings of 15th International Conference on Pattern Recognition*, pp. 462–465.

15 Pham, D.L. and Prince, J.L. (1999) Adaptive fuzzy segmentation of magnetic resonance images. *IEEE Transastions on Medical Imaging*, **18** (9), 737–752.

16 Chen, L., Cui, B., Han, Y., Guan, Y., and Luo, Y. (2015) Two-dimensional fuzzy clustering algorithm (2DFCM) for metallographic image segmentation based on spatial information, in *2nd International Conference on Information Science and Control Engineering*, pp. 519–521.

17 Hruschka, E.R., Campello, R.J.G.B., Freitas, A.A., and Leon, P. (2009) A survey of evolutionary algorithms for clustering. *IEEE Transactions on System Man Cybernatics, C: Applcations and Reviews*, **39**, 133–155.

18 Ripon, K.S.N., Tsang, C.H., and Kwong, S. (2006) Multi-objective data clustering using variable-length real jumping genes genetic algorithm and local search method, in *IEEE International Joint Conference on Neural Networks*, pp. 3609–3616.

19 Goldberg, D.E. (1989) *Genetic Algorithms: Search, Optimization and Machine Learning*, Addison-Wesley, New York.

20 Hammouche, K., Diaf, M., and Siarry, P. (2008) A multilevel automatic thresholding method based on a genetic algorithm for a fast image segmentation. *Computer Vision and Image Understanding*, **109** (2), 163–175.

21 Fu, Z., He, J.F., Cui, R., Xiang, Y., Yi, S.L., Cao, S.J., Bao, Y.Y., Du, K.K., Zhang, H., and Ren, J.X. (2015) Image segmentation with multilevel threshold of gray-level & gradient-magnitude entropy based on genetic algorithm, in *International Conference on Artificial Intelligence and Industrial Engineering*, pp. 539–542.

22 Hauschild, M., Bhatia, S., and Pelikan, M. (2012) Image segmentation using a genetic algorithm and hierarchical local search, in *Proceedings of the 14th Annual Conference on Genetic and Evolutionary Computation*, pp. 633–640.

23 De, S., Bhattacharyya, S., and Dutta, P. (2009) Multilevel image segmentation using OptiMUSIG activation function with fixed and variable thresholding: a comparative study, in *Applications of Soft Computing: From Theory to Praxis, Advances in Intelligent and Soft Computing* (J. Mehnen, M. Koppen, A. Saad, and A. Tiwari), Springer-Verlag, Berlin, pp. 53–62.

24 Min, W. and Siqing, Y. (2010) Improved k-means clustering based on genetic algorithm, in *International Conference on Computer Application and System Modeling (ICCASM)*, pp. 636–639.

25 Biju, V.G. and Mythili, P. (2012) A genetic algorithm based fuzzy c-mean clustering model for segmenting microarray images. *International Journal of Computer Applications*, **52** (11), 42–48.

26 Wei, C., Tingjin, L., Jizheng, W., and Yanqing, Z. (2010) An improved genetic FCM clustering algorithm, in *2nd International Conference on Future Computer and Communication*, vol. 1, pp. 45–48.

27 Halder, A., Pramanik, S., and Kar, A. (2011) Dynamic image segmentation using fuzzy c-means based genetic algorithm. *International Journal of Computer Applications*, **28** (6), 15–20.

28 Jansi, S. and Subashini, P. (2014) Modified FCM using genetic algorithm for segmentation of MRI brain images, in *IEEE International Conference on Computational Intelligence and Computing Research*, pp. 1–5.

29 Borsotti, M., Campadelli, P., and Schettini, R. (1998) Quantitative evaluation of color image segmentation results. *Pattern Recognition Letters*, **19** (8), 741–747.

30 Ruspini, E. (1969) A new approach to clustering. *Information and Control*, **15**, 22–32.

31 Zhang, H., Fritts, J., and Goldman, S. (2004) An entropy-based objective evaluation method for image segmentation, in *Proceedings of SPIE Storage and Retrieval Methods and Applications for Multimedia*, pp. 38–49.

2

Character Recognition Using Entropy-Based Fuzzy C-Means Clustering

B. Kondalarao, S. Sahoo, and D.K. Pratihar

Department of Mechanical Engineering, Indian Institute of Technology, Kharagpur, West Bengal, India

2.1 Introduction

Recognition is a unique ability of the human brain. People and animals effortlessly identify the objects around them even when the number of possibilities is vast. A lot of research had been carried out for recognition of different characters like Bengali, English, Hindi, Tamil, Chinese, and so on [1]. The challenging area of research in recognition is handwritten characters, but most of the studies have been on multi-font and fixed-font reading. The automatic recognition of handwritten characters is a difficult task, because of the fact that it has noise problem and large variety of handwriting styles. This noise problem could arise due to the quality of paper or pressure applied while writing.

Clustering technique for character recognition had been used for a long time. Necognitron network [2] was specifically designed to recognize handwritten characters. Two other unsupervised methods have become popular in recent literature of character recognition; those are the adaptive reasoning theory classifier [3] and the nearest-neighbor clustering algorithm [4]. To overcome the problem that is caused due to the noise present in the image, in the above-mentioned methods, a Fourier coefficient as a feature and adaptive fuzzy leader clustering as classifier were used [5]. However, because of more changes in the word arrangement with respect to spacing, size, and orientation of the component, character recognition of complete words, using Fourier coefficient as a feature, is not successful. Adaptive clustering had been used [6] to improve the performance of the recognition system further for overlapping characters.

Extensively used classifiers for character recognition are neural network, statistical classifiers, template matching, support vector machine, and so on. Template matching is the simplest way of character recognition, where a template is previously stored and the similarity between the stored data and the template gives the degree of recognition. Michael Ryan *et al.* [7] used a template matching-technique for character recognition on ID cards with proper preprocessing technique and segmentation. Many other researchers also worked on template matching [8, 9], but the technique was more sensitive to noise and had a poor recognition efficiency. Many statistical methods like nearest

Hybrid Intelligence for Image Analysis and Understanding, First Edition.
Edited by Siddhartha Bhattacharyya, Indrajit Pan, Anirban Mukherjee, and Paramartha Dutta.
© 2017 John Wiley & Sons Ltd. Published 2017 by John Wiley & Sons Ltd.
Companion Website: www.wiley.com/go/bhattacharyya/hybridintelligence

neighbor [10], Bayes' classifier [11], and hidden Markov model [12] were also used for character recognition. Neural network is an efficient soft computing technique that is extensively used for recognition by training a massively parallel network. The problem with neural network is its computational complexity and it requires a large data set for training. The problem with individual classification techniques is that none of these methods is independently sufficient for a good classification; that's why hybrid techniques came into the picture. Hybrid soft computing is a combination of two or more soft computing tools, where advantages of both the tools are utilized to eliminate their limitations for betterment of the classification. A combined neuro-fuzzy technique showed better results with respect to the neural network itself [13].

Fuzzy clustering techniques had also been used by many researchers for character recognition. A lot of fuzzy clustering algorithms had been proposed that dealt with the classification of a large data set. The logic behind differentiating the large data set into clusters is finding the cluster centers of a data set formed by the samples of similar characteristic. Still, there are different approaches where the clusters are formed using the membership function approach [14]. Zadeh *et al.* [15] used a different fuzzy clustering technique called FANNY [16] for recognizing Twitter hashtags. The renowned clustering techniques are fuzzy ISODATA [17], Fuzzy *c*-means (FCM) [18], fuzzy *k*-nearest neighbors algorithm [19], and entropy-based clustering [20]. Among the fuzzy clustering tools described above, FCM is the most popular clustering method used for different applications. Sometimes, the traditional FCM clustering technique is unable to form distinct clusters. Several modified versions of FCM had been developed to encounter different drawbacks; a few of them are Conditional FCM [21], quadratic function-based FCM [22], and Modied FCM [23]. These techniques are data dependent; each method is proposed for solving a specific problem. To deal with difficulties like uneven distribution of centers, local minima problem, and dependency on the random values of membership (μ) taken initially, entropy-based clustering technique was proposed. The entropy-based fuzzy clustering technique can automatically detect the number of clusters depending upon the user-defined threshold value of similarity (β). But the compactness of the clusters formed by entropy-based clustering is not good enough, and the number of clusters formed by this technique is highly sensitive to β value. Pratihar *et al.* [24] proposed a combined technique called entropy-based FCM algorithm for obtaining distinct and compact clusters. The working principle of this algorithm will be discussed in the next section of this chapter.

Preprocessing and feature extraction are important parts of classification, used for character recognition. The features of an image can be divided into three broad categories: global features, statistical features, and geometrical and topological features [25]. Global features include Fourier transform, wavelet transform, Gabor transform, moments, and so on. Global features do not change with transforms or rotation of the original image. Major statistical features like zoning, crossing, and projections are used for reducing the dimensionality of the image data to improve the speed of recognition in case of online processing. There are various geometrical features available that give some local or global characteristic of an image. A combination of two or more features might help the recognition process. Preprocessing helps the classification process by improving the quality of an image or by reducing the amount of noise incorporated in the image. Preprocessing followed by feature extraction is always preferable for character recognition. The popular preprocessing techniques, such as smoothening, thinning or broadening, and size normalization, are generally used. As the test image can be taken from any imaging system, distortion of the image is obvious for such cases. Smoothening generally deals with any random change in the image. Thinning or broadening helps to bring all images under observation into a standard shape that helps the classification process. Size normalization is a method of converting all the input images to a standard matrix (pixel value) format, such as 64×64, to reduce the variation of size of different images collected from different sources. Some popular features like zoning [26], projection profile [27], chain codes [28], and longest run [29] are frequently used by the researchers for character recognition. In zoning, the image is divided into some finite number of sections or zones, and the average image intensity value of individual zones is calculated and used as the training data for classification. For a binary image, the projection profile is the sum of either dark or white (depends upon the user) pixels along the rows or columns. The chain code or boundary descriptor is an important feature for character recognition that gives the information about the shape of the character. Four-directional chain codes and eight-directional chain codes are the two types of chain codes available to define complex boundaries of an image. Most of the studies had been done on Devnagari [25], Bengali [29], Malayalam [30] Tamil [31], Chinese [32], and Arabic [33] letter recognition. Due to its extensive variety, these letters are comparatively easy to classify with respect to English letters. For example, English letters like 'A' and 'V', 'E' and 'B', and 'E' and 'F' give almost the same feature values, so the classification algorithm often confuses to clearly classify the cluster centers. Thus, selection of good features is an important task for proper character recognition. For template matching, preprocessing and segmentation are important, and feature extraction is not required in this case. But template matching does not provide good recognition results as mentioned earlier, and it also has several other disadvantages.

This study made an approach to introduce the entropy-based fuzzy C-means (EFCM) technique for the recognition of printed English alphabets. For training, five different types of letters (A to Z) of different fonts are created in Microsoft Word, and the extracted features are fed to the classifier for training. Fonts used here are the popular letters used in newspapers, ID cards, and the internet. Three different fuzzy classifiers are used to draw a comparison and for validating the results of EFCM to classify and recognize English alphabets correctly. The study is specifically aimed at finding the success rate of the EFCM technique for classification of English letters and its applications like a fully automated system for ID card detection that can be used in government offices, airports, and libraries. With some improvements, this methodology can be implemented for helping visually impaired people to read newspapers, books, and so on.

2.2 Tools and Techniques Used

2.2.1 Fuzzy Clustering Algorithms

Clustering is one of the methods of data mining that is used to extract useful information from a set of data. Clusters are generally of two types in nature. The clusters that have well-defined and fixed boundaries are said to be crisp clusters, whereas those with vague boundaries are said to be fuzzy clusters [34]. In this study, recognition of English alphabets is done using some of the algorithms of fuzzy clustering, namely the FCM algorithm, entropy-based fuzzy clustering (EFC) algorithm, and EFCM algorithm.

2.2.1.1 Fuzzy C-means Algorithm

FCM is one of the methods of fuzzy clustering algorithms. Two or more clusters may contain the same data, so a membership value is used to define the belongingness of a data point to the clusters. This means that each data point of a data set belongs to different clusters with different membership values. It is an iterative algorithm, in which a dissimilarity measure is going to be minimized by updating centers of the clusters and data point membership values with the clusters. The dissimilarity values are measured in terms of Euclidean distance between data points and predefined cluster centers. The data points residing in the same cluster have high similarity value.

The FCM algorithm has the following steps, considering N data points represented by $(x_1, x_2, x_3 \ldots x_N)$, each of which has L dimensions.

Step 1: Let C be the number of clusters to be created, where $2 \leq C \leq N$.

Step 2: Assume a right value of cluster fuzziness ($g \geq 1$).

Step 3: Assign random values to the $N \times C$ sized membership matrix $[\mu]$, such that

$$\mu \in [0.0\ 1.0] \text{ and } \sum_{j=1}^{C} \mu_{ij} = 1.0 \text{ for each i.}$$

Step 4: Compute the kth dimension of the jth cluster center CC_{jk} using the following expression:

$$CC_{jk} = \frac{\sum_{i=1}^{N} \mu_{ij}^{g} x_{ik}}{\sum_{i=1}^{N} \mu_{ij}^{g}}$$

Step 5: The Euclidean distance is computed as follows between the ith data point and j^{th} cluster center:

$$d_{ij} = \|CC_j - x_i\|$$

Step 6: Update membership matrix $[\mu]$ using the following expression according to d_{ij}. If $d_{ij} \geq 0$, then

$$\mu = \frac{1}{\sum_{m=1}^{C} \left(\frac{d_{ij}}{d_{im}} \right)^{\frac{2}{g-1}}}$$

Step 7: Repeat Step 4 through Step 6 until the changes in $[\mu]$ becomes less than some prespecified values.

Step 8: Form the clusters using the similarities in membership values of data points.

Step 9: If the cluster contains more than $\gamma\%$ of the total data points, then only the cluster is said to be valid.

2.2.1.2 Entropy-based Fuzzy Clustering

In this algorithm, the entropy value of each data point is determined based on a similarity measure. This similarity measure value depends on the Euclidean distance between two data points. Generally, a data point with low entropy value is selected as the center of a cluster. The cluster thus formed is considered as valid, if and only if the number of data points contained in a cluster is found to be more than $\gamma\%$ of total data points.

EFC consists of the following steps, considering an input data set matrix [T], which consists of N data points having L dimensions.

Step 1: Calculate the Euclidean distance between points i and j as follows:

$$d_{ij} = \sqrt{\sum_{k=1}^{L} (x_{ik} - x_{jk})^2}$$

Step 2: Determine the similarity S_{ij} between two data points i and j, as given below:

$$S_{ij} = e^{-\alpha d_{ij}},$$

where α is a constant and similarity S_{ij} varies in the range of (0.0, 1.0).

Step 3: Compute the entropy value of a data point x_i with reference to all other points as follows:

$$E_i = -\sum_{\substack{j\in x}}^{j\neq1} (S_{ij}log_2 S_{ij} + (1 - S_{ij})log_2(1 - S_{ij}))$$

Step 4: The data point with the minimum entropy value is selected as the cluster center.

Step 5: The data points having similarity with the selected cluster center greater than a prespecified threshold value (β) will be put in the cluster, and the same will be removed from [T].

Step 6: If [T] is empty, terminate the program; otherwise, go to Steps 4 and 5.

2.2.1.3 Entropy-based Fuzzy C-Means Algorithm

Both the above-discussed algorithms have their inherent limitations to generate the clusters, which have good distinctness as well as compactness. In this algorithm, both the above-discussed algorithms are combined to get both the distinct as well as compact clusters simultaneously.

The EFCM algorithm consists of the following steps. Consider N data points with L dimensions each.

Step 1: Compute the Euclidean distance between points i and j, as follows:

$$d_{ij} = \sqrt{\sum_{k=1}^{L} (x_{ik} - x_{jk})^2}$$

Step 2: Compute the similarity value S_{ij} between the data points i and j, as follows:

$$S_{ij} = e^{-\alpha d_{ij}},$$

where α is a constant that represents the relation between distance and similarity of two data points.

Step 3: Compute entropy of the ith data point with respect to all other data points, as given here:

$$E_i = -\sum_{\substack{j\in x}}^{j\neq 1}(S_{ij}log_2 S_{ij} + (1 - S_{ij})log_2(1 - S_{ij}))$$

Step 4: The data point that has the minimum entropy value is selected as the cluster center.

Step 5: The data points having similarity with the selected cluster center greater than a prespecified threshold value (β) will be put in the cluster, and the same will be removed from [T]. The cluster is declared as invalid if the number of data points present in that cluster is found to be less than a prespecified fraction of total data points. If the data set becomes empty, then go to Step 6; otherwise, go to Step 3.

Step 6: Compute the Euclidean distance between i^{th} and j^{th} cluster centers using the following expression:

$$d_{ij} = \|CC_j - x_i\|$$

Step 7: Update the fuzzy membership matrix as follows:

$$\mu = \frac{1}{\sum\limits_{m=1}^{C}\left(\dfrac{d_{ij}}{d_{im}}\right)^{\frac{2}{g-1}}}$$

Step 8: Update the coordinates of the cluster centers using the information of the fuzzy membership matrix as follows:

$$CC_{jk} = \frac{\sum\limits_{i=1}^{N}\mu_{ij}^g x_{ik}}{\sum\limits_{i=1}^{N}\mu_{ij}^g}$$

Step 9: Put the data points into the clusters obtained in Step 5, depending on the similarity of their membership values.

2.2.2 Sammon's Nonlinear Mapping

This is used to map the data from a higher dimension to lower dimension for the purpose of visualization. Let us consider N vectors having L-dimensions each, which are denoted by X_i where i=1,2, ..., N. Our aim is to map these N vectors from L-dimensional space to 2D space. Let us assume that the 2D data is denoted by Y_i, where i = 1,2, ..., N. The mapping process is as follows:

Step 1: Generate N vectors in the 2D plane at random.

Step 2: The following condition is to be satisfied for perfect mapping:

$$d_{ij}^* = d_{ij},$$

where d_{ij}^* is the Euclidean distance between two vectors X_i and X_j in higher dimensional space; and d_{ij} is the distance between the corresponding two mapped points Y_i and Y_j in a 2D plane.

Step 3: The mapping error is calculated using the following expression:

$$E(m) = \frac{1}{C} \sum_{i=1}^{N} \sum_{j=1(i<j)}^{N} \frac{[d_{ij}^* - d_{ij}(m)]^2}{d_{ij}^*},$$

where $C = \sum_{i=1}^{N} \sum_{j=1(i<j)}^{N} d_{ij}^*$ and $d_{ij}(m) = \sqrt{\sum_{k=1}^{D} [y_{ik}(m) - y_{jk}(m)]^2}.$

Step 4: The steepest descent method is used to minimize the mapping error, and the relationship between m^{th} and $(m+1)^{th}$ iterations is as follows:

$$y_{pq}(m+1) = y_{pq}(m) - (MF)\Delta_{pq}(m),$$

where MF is a magic factor, which varies from 0.0 to 1.0, and

$$\Delta_{pq}(m) = \frac{\frac{\partial E(m)}{\partial y_{pq}(m)}}{\frac{\partial^2 E(m)}{\partial y_{pq}^2(m)}}$$

Now,

$$\frac{\partial E(m)}{\partial y_{pq}(m)} = \frac{-2}{C} \sum_{j=1,j\neq p}^{N} \left[\frac{d_{pj}^* - d_{pj}}{d_{pj}d_{pj}^*} \right] (y_{pq} - y_{jq})$$

$$\frac{\partial^2 E(m)}{\partial y_{pq}^2(m)} = \frac{-2}{C} \sum_{j=1,j\neq p}^{N} \frac{1}{d_{pj}d_{pj}^*} \left[(d_{pj}^* - d_{pj}) - \frac{(y_{pq} - y_{jq})^2}{d_{pj}} \left(1 + \frac{d_{pj}^* - d_{pj}}{d_{pj}} \right) \right].$$

2.3 Methodology

2.3.1 Data Collection

The experiment has been conducted on printed English alphabets of different fonts. Five different fonts of 26 alphabets are considered for the experiments. The five different fonts consist of Arial, Bardely, Calibri, Cambria, and Times New Roman, which are shown in Figure 2.1. These were stored as BMP file format. These five fonts are chosen due to their variations in shape, style, and orientation. As these image files of different letters are created manually, some preprocessing has been performed before feature extraction.

2.3.2 Preprocessing

Preprocessing is a technique for enhancing the image quality after accruing the image from different sources to eliminate the noise. It also helps to reduce the unnecessary data

Figure 2.1 Letters used for training the algorithm: (a) Arial, (b) Bardely, (c) Calibri, (d) Cambria, and (e) Times New Roman.

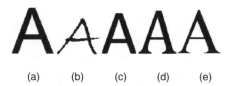

(a)   (b)   (c)   (d)   (e)

in the image of interest, which is incorporated by human inaccuracy while creating or capturing the test image. In this study, for simplicity, the printed letters are created from Microsoft Word, so no image enhancing is required, but one may use the enhancing technique if the image is scanned or captured from a newspaper or handwritten script. Here, the preprocessing technique is used for helping the feature extraction and also to eliminate the variations created while preparing the samples. The raw data is an RGB image. As color intensity information of letters is unnecessary for recognition, it is better to convert the RGB image to a binary image. First, the RGB image is converted to a grayscale image and then to a binary image. A binary image only contains 0 and 1, where 0 signifies the black pixel and 1 denotes the white pixel. Size normalization is applied to the image to convert all images into a standard size, such as 64 × 64 contour, to deal with the variable sizes of the collected data. Then, the boundary of the image is extracted to support the feature extraction data to get better clusters. Figure 2.2 shows the output image of different stages of preprocessing. The final output (Figure 2.2c) which defines the boundary of an image, has been used for extracting the feature. Figure 2.3 shows the complete steps involved in the image preprocessing with the size variation information of the image in different steps.

2.3.3 Feature Extraction

For getting good clusters, the features should be unique for each and every character. The beauty of feature extraction lies in the fact that the feature values should be similar for the same character of different fonts and different for different characters. This makes feature extraction an important part of any classification method. In this proposal, we concentrate on the longest run feature. A longest run feature of a binary image is the

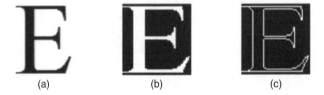

(a) (b) (c)

Figure 2.2 (a) Raw image, (b) converted binary image, and (c) boundary of the image.

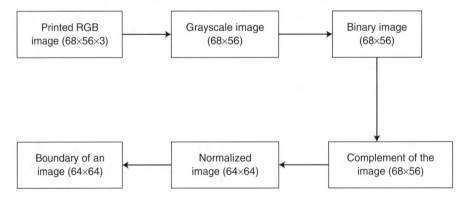

Figure 2.3 Flow diagram of preprocessing technique.

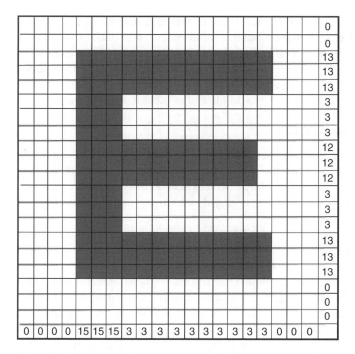

Figure 2.4 Longest run feature row-wise and column-wise for a 20 × 20 image.

sum of continuous white pixels along each row or column. As different characters have different shapes, the longest run across rows or columns gives different values. Figure 2.4 describes the concept of the longest run feature along the rows and columns.

The longest run feature gives the information on a local scale. If the size of the normalized image is 64×64, the lengths of the row-wise longest run vector across the row and column-wise longest run vector across the column are 64. The elements of the longest run vector along the rows and columns are different. The longest run vector along the row provides the information about the horizontal alignment of the image, and that across the column provides the information about the vertical alignment of the image in a miniature scale. But the problem is that the longest run values are sensitive to the shape and alignment of an image. It gives different results if the same letter of a different font is slightly thick or tilted. It also gives different results for the same letter with different fonts; for example, 'E' of Times New Roman and 'E' of Calibri font will give different values, which is not accepted. To deal with this problem, an averaged longest run vector is used for constructing the main feature vector. The longest run values for 8 or 16 consecutive rows or columns are averaged to get the averaged longest run vector. As the image contour has 64 rows and 64 columns, the length of the feature vector is 16 in the case of a 8-point averaged longest run vector or 8 in the case of a 16-point averaged longest run vector. The feature vector contains elements of the row-wise longest run vector and column-wise longest vector. Both the 8-point and 16-point averaged longest run features are tested using the three classification algorithms. For all the algorithms, the 16-point averaged longest run feature gives better recognition results compared to the 8-point averaged longest run feature in terms of recognition percentage. The main algorithm is designed using the 16-point averaged longest run vector as the main feature.

This recognition result depends on the size of the normalized image and the number of points taken for average. As in the proposal, the normalized size of the test image is 64×64; the 16-point averaged longest run feature gives better recognition results, but it may differ if the normalized image is of different size. One can use zero padding for constructing the averaged longest run vector, if the number of rows or columns of the image contour is not exactly divisible by the number of points taken for average.

2.3.4 Classification and Recognition

Both the FCM and EFC algorithms have their inherent limitations. FCM iteratively updates the clusters, and EFC determines the clusters uniquely but does not improve clusters iteratively. EFC can determine more distinct clusters, whereas FCM can determine more compact clusters. An EFCM clustering algorithm captures the advantages of both these algorithms.

Classification of huge input data using EFCM clustering has been carried out based on similarity in data. Also, the same has been done using FCM and EFC algorithms. Then, the comparison of results obtained by three algorithms is done.

Figure 2.5 shows the flowchart of developed algorithms. Initially, clustering is carried out on the input data separately using the above three algorithms. The number of clusters formed in all three algorithms is 26, as there are 26 alphabets. After forming the desired number of clusters, Euclidean distances are calculated between each test data and the cluster centers. Out of all Euclidean distances calculated, the minimum value will decide the belongingness of the test data to a predefined cluster. Now, this predefined cluster will decide the alphabet for recognition.

2.4 Results and Discussion

The experiment has been conducted on a wide range of characters to see how well the system reacts to the variations. Five different formats of 26 alphabets are considered for the experiments. Five different formats, namely Arial, Bardely, Calibri, Cambria, and Times New Roman, are used (refer to Tables A.2.1 through A.2.5 of the Appendix). These are stored as BMP file format. The shape, orientation, thickness, and style of writing of different character sets are not similar.

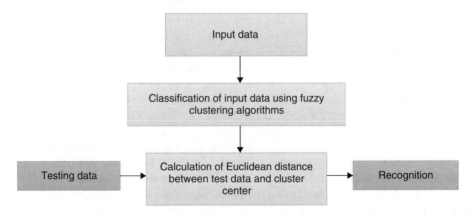

Figure 2.5 Flowchart of the proposed methodology for classification and recognition.

Compactness is an intracluster distance of the elements. The members of each cluster should be as close to each other as possible to make a cluster more compact. Variance (σ^2) is the common measure of compactness, which is given as follows:

$$\sigma^2 = \frac{1}{N} \sum_{i=1}^{N} (x_i - \mu)^2$$

Distinctness is the intercluster distance. The distance between cluster centers should be as high as possible to form clusters that can be easily distinguishable from others. Euclidean distance is the common measure of distinctness, and it is calculated as follows:

$$d_{ij} = \sqrt{\sum_{k=1}^{L} (x_{ik} - x_{jk})^2}$$

It can be observed (see Table 2.1) that the compactness and distinctness values are well balanced in EFCM, where as in the case of FCM, the value of distinctness is poor, and for EFC, the value of compactness is poor. FCM and EFCM give more compact clusters, as they update the membership values and the cluster centers iteratively. The EFCM algorithm is found to yield a good distinct cluster as the EFC algorithm does. Both EFCM and EFC algorithms determine the clusters uniquely by calculating the entropy value of each point. So, from the above points, it is clear that the EFCM has a good balance of compactness and distinctness in forming the clusters. However, the performance of these fuzzy clustering algorithms may be data dependent.

The cluster centers obtained using all the algorithms need to be visualized in order to conceive their special positions. The data of cluster centers, in this experiment, is of higher dimensions. This high-dimensional data is to be mapped to either 2D or 3D for proper visualization. Sammon's mapping technique is used to visualize the data. Here, the role of Sammon's mapping is just to visualize the higher dimensional data by mapping it into a lower dimension. Figure 2.6 shows the 2D view of the cluster centers of input data, as achieved by all the three algorithms. Note that two axes of this figure represent dimensionless numbers. The figures obtained using EFC and EFCM (refer to Figure 2.6b and 2.6c) are seen to be more distinct compared to that obtained by the FCM algorithm (refer to Figure 2.6a). The EFCM algorithm is able to form clusters, which are as compact as the FCM developed clsuters, and also their distinctness is comparable with that obtained by the EFC algorithm.

Table 2.2 displays the result of recognition using all three said algorithms. The successful recognition rate of EFCM is found to be more than that of the remaining two. It is also observed that the letters that have similarity are misidentified more. However, the recognition rate can be improved using appropriate feature extraction techniques with

Table 2.1 Output of different clustering algorithms for input data

Algorithm	Compactness	Distinctness
FCM	0.7289	5.0032
EFC	0.170	5.6281
EFCM	0.5312	5.3643

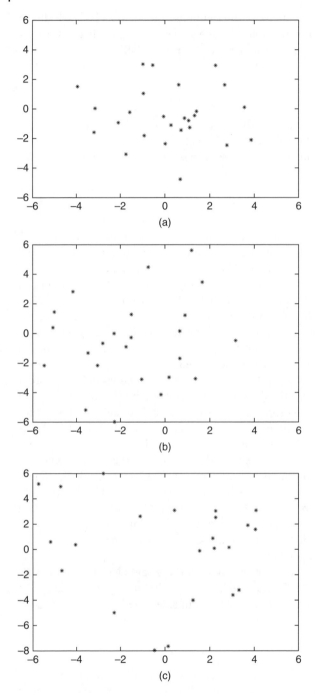

Figure 2.6 Visualization of cluster centers of input data obtained using the (a) FCM algorithm, (b) EFC algorithm, and (c) EFCM algorithm.

Table 2.2 Results of recognition of alphabets with three algorithms for Times New Roman

Input alphabets	Output of FCM	Output of EFC	Output of EFCM
A	A	A	A
B	?	B	B
C	C	?	C
D	?	D	?
E	E	E	E
F	F	F	F
G	?	?	?
H	H	?	H
I	?	I	I
J	?	J	?
K	K	K	K
L	L	L	L
M	M	M	M
N	N	N	N
O	O	?	O
P	P	?	P
Q	Q	Q	Q
R	?	R	R
S	S	?	S
T	T	T	T
U	U	U	U
V	V	V	V
W	W	W	W
X	X	?	?
Y	Y	Y	Y
Z	Z	Z	Z
% of recognition	76.92	73.07	84.61

the clustering techniques. Results of Table 2.2 have been obtained using Times New Roman font as testing data and all the remaining four fonts as training data.

Table 2.3 shows the percentage of recognition accuracy of each font by using all three algorithms. From Table 2.3, it can be observed that the EFCM has a higher percentage of recognition accuracy compared to FCM and EFC except for the Bardely font, as the shape of this font is different compared to the others. EFCM has maintained a better percentage of recognition accuracy compared to both FCM and EFC. It could be because of the balance in compactness and distinctness values, which shows the quality of clusters formed. Moreover, the results of EFCM are found to be more consistent than those of the remaining two algorithms.

The CPU time values of the FCM, EFC, and EFCM are found to be equal to 0.176206, 0.180281, and 0.20401 seconds, respectively. Out of all the algorithms, EFCM is taking

Table 2.3 Recognition accuracy with respect to each font

Test Font	FCM (%)	EFC (%)	EFCM (%)
Arial	73.08	73.07	76.92
Bardely	57.69	76.92	61.54
Calibri	84.62	73.07	84.62
Cambria	70.37	73.07	73.08
Times New Roman	76.92	73.07	84.61

more time because it is a combination of both FCM and EFC. In other words, EFCM initially creates the cluster centers using entropy values of the data points like the EFC algorithm, and then updates the created cluster centers iteratively, as it is done in FCM.

2.5 Conclusion and Future Scope of Work

EFCM is a hybrid clustering technique, and the main advantage of EFCM over FCM and EFC is that it can produce distinct as well as compact clusters, as discussed earlier in this chapter. The results shown in Table 2.2 support the statement that the compactness and distinctness values are well balanced in EFCM, whereas in the case of FCM, the value of distinctness is poor, and for EFC, the value of compactness is poor. This study is meant to show the efficiency of EFCM for English character recognition, compared to the existing clustering techniques. By observing the results shown in Tables 2.2 and 2.3 and the distributions of clusters in Figure 2.6, it is clear that the EFCM algorithm gives better cluster formation and recognition accuracy than the EFC and FCM do. The method of averaging the longest run feature gives better results than using the longest run feature itself; also, averaging the longest run feature reduces the size of the feature vector or the database. If we use only the longest run feature, the size of the feature vector for each font would be 128×26, but in our case, it is 8×26. Reduction of length of the feature vector also reduces the execution time of the proposed algorithm. Less execution time will help for the implementation of the letter recognition technique in real-time applications, like identity card detection, reading the number plates of vehicles, and so on. The English alphabet does not have much variation, unlike Chinese, Hindi, and Bengali alphabets, which makes it difficult to distinguish. It is a challenge for the researchers to obtain the same accuracy for recognizing English characters is obtained for Chinese, Bengali, and Hindi characters.

The accuracy obtained in this study with a comparatively small data set is not up to the mark. In future, attempts will be made to improve the accuracy of the result by incorporating some new features or a combination of different features. For recognition technique, we have used the Euclidean distance between the test data and the cluster center; researchers may think of a better recognition technique to improve the accuracy. The next approach will be to run the proposed EFCM-based technique on handwritten characters and to get an accuracy level obtained by the current classical and soft computing techniques. A statistical analysis like Z-test or McNemar's test will be carried out in future to get a clear idea about the differences between the mentioned clustering techniques.

References

1 Suen, C.Y., Berthod, M., and Mori, S. (1980) Automatic recognition of handprinted characters—the state of the art. *Proceedings of the IEEE*, **68** (4), 469–487.

2 Fukushima, K. (1988) Neocognitron: A hierarchical neural network capable of visual pattern recognition. *Neural networks*, **1** (2), 119–130.

3 Gan, K.W. and Lua, K.T. (1992) Chinese character classification using an adaptive resonance network. *Pattern Recognition*, **25** (8), 877–882.

4 Sharan, A. and Mitra, S. (1994) Handwritten character recognition by an adaptive fuzzy clustering algorithm, in *Fuzzy Systems, 1994. IEEE World Congress on Computational Intelligence., Proceedings of the Third IEEE Conference on*, IEEE, pp. 1820–1824.

5 O'Hair, M.A. (1990) A whole word and number reading machine based on two dimensional low frequency Fourier transforms, *Tech. Rep.*, DTIC Document.

6 Ashir, A.M. and Shehu, G.S. (2015) Adaptive clustering algorithm for optical character recognition, in *Electronics, Computers and Artificial Intelligence (ECAI), 2015 7th International Conference on*, IEEE, SG–13.

7 Ryan, M. and Hanafiah, N. (2015) An examination of character recognition on ID card using template matching approach. *Procedia Computer Science*, **59**, 520–529.

8 Hu, J. and Pavlidis, T. (1996) A hierarchical approach to efficient curvilinear object searching. *Computer vision and image understanding*, **63** (2), 208–220.

9 Ramteke, R. and Mehrotra, S. (2008) Recognition of handwritten Devanagari numerals. *International Journal of Computer Processing of Oriental Languages*, Chinese Language Computer Society & World Scientific Publishing Company.

10 Arora, S., Bhattacharjee, D., Nasipuri, M., Malik, L., Kundu, M., and Basu, D.K. (2010) Performance comparison of svm and ann for handwritten devnagari character recognition. *arXiv preprint arXiv:1006.5902*.

11 Namboodiri, A.M. and Jain, A.K. (2004) Online handwritten script recognition. *Pattern Analysis and Machine Intelligence, IEEE Transactions on*, **26** (1), 124–130.

12 Bhowmik, T.K., Parui, S.K., and Roy, U. (2008) Discriminative HMM training with ga for handwritten word recognition, in *Pattern Recognition, 2008. ICPR 2008. 19th International Conference on*, IEEE, pp. 1–4.

13 Cho, S.B. (2002) Fusion of neural networks with fuzzy logic and genetic algorithm. *Integrated Computer-Aided Engineering*, **9** (4), 363–372.

14 Baraldi, A. and Blonda, P. (1999) A survey of fuzzy clustering algorithms for pattern recognition. i. *Systems, Man, and Cybernetics, Part B: Cybernetics, IEEE Transactions on*, **29** (6), 778–785.

15 Zadeh, L.A., Abbasov, A.M., and Shahbazova, S.N. (2015) Analysis of Twitter hashtags: Fuzzy clustering approach, in *Fuzzy Information Processing Society (NAFIPS) held jointly with 2015 5th World Conference on Soft Computing (WConSC), 2015 Annual Conference of the North American*, IEEE, pp. 1–6.

16 Kaufman, L. and Rousseeuw, P.J. (2009) *Finding groups in data: an introduction to cluster analysis*, vol. 344, John Wiley & Sons.

17 Dunn, J.C. (1973) A fuzzy relative of the isodata process and its use in detecting compact well-separated clusters. *Cybernetics and Systems*, **3** (3), 32–57.

18 Bezdek, J.C. (1973) Fuzzy mathematics in pattern classification. PhD thesis, Cornell University, Ithaca, NY.

19 Keller, J.M., Gray, M.R., and Givens, J.A. (1985) A fuzzy k-nearest neighbor algorithm. *Systems, Man and Cybernetics, IEEE Transactions on*, (4), 580–585.

20 Yao, J., Dash, M., Tan, S., and Liu, H. (2000) Entropy-based fuzzy clustering and fuzzy modeling. *Fuzzy Sets and Systems*, **113** (3), 381–388.

21 Zigkolis, C.N. and Laskaris, N.A. (2009) Using conditional FCM to mine event-related brain dynamics. *Computers in Biology and Medicine*, **39** (4), 346–354.

22 Ichihashi, H., Honda, K., Notsu, A., and Hattori, T. (2007) Aggregation of standard and entropy based fuzzy c-means clustering by a modified objective function, in *Foundations of Computational Intelligence, 2007. FOCI 2007. IEEE Symposium on*, IEEE, pp. 447–453.

23 Han, S., Lee, I., and Pedrycz, W. (2009) Modified fuzzy c-means and Bayesian equalizer for nonlinear blind channel. *Applied Soft Computing*, **9** (3), 1090–1096.

24 Dey, V., Pratihar, D.K., and Datta, G.L. (2011) Genetic algorithm-tuned entropy-based fuzzy c-means algorithm for obtaining distinct and compact clusters. *Fuzzy Optimization and Decision Making*, **10** (2), 153–166.

25 Dongre, V.J. and Mankar, V.H. (2011) A review of research on Devnagari character recognition. *arXiv preprint arXiv:1101.2491*.

26 Rajashekararadhya, S. and Ranjan, P.V. (2009) A novel zone based feature extraction algorithm for handwritten numeral recognition of four Indian scripts. *Digital Technology Journal*, **2** (1).

27 Arora, S., Bhattacharjee, D., Nasipuri, M., Basu, D.K., and Kundu, M. (2010) Recognition of non-compound handwritten Devnagari characters using a combination of MLP and minimum edit distance. *arXiv preprint arXiv:1006.5908*.

28 Malik, L. and Deshpande, P. (2009) Recognition of printed Devnagari characters with regular expression in finite state models, in *International workshop on machine intelligence research*.

29 Das, N., Sarkar, R., Basu, S., Saha, P.K., Kundu, M., and Nasipuri, M. (2015) Handwritten Bangla character recognition using a soft computing paradigm embedded in two pass approach. *Pattern Recognition*, **48** (6), 2054–2071.

30 Chacko, A.M.M. and Dhanya, P. (2015) A comparative study of different feature extraction techniques for offline Malayalam character recognition, in *Computational Intelligence in Data Mining-Volume 2*, Springer, pp. 9–18.

31 Kannan, R.J. and Subramanian, S. (2015) An adaptive approach of Tamil character recognition using deep learning with big data: a survey, in *Emerging ICT for Bridging the Future-Proceedings of the 49th Annual Convention of the Computer Society of India (CSI) Volume 1*, Springer, pp. 557–567.

32 Kimura, F., Takashina, K., Tsuruoka, S., and Miyake, Y. (1987) Modified quadratic discriminant functions and the application to Chinese character recognition. *Pattern Analysis and Machine Intelligence, IEEE Transactions on*, (1), 149–153.

33 Alginahi, Y.M. (2013) A survey on Arabic character segmentation. *International Journal on Document Analysis and Recognition (IJDAR)*, **16** (2), 105–126.

34 Pratihar, D.K. (2014) *Soft computing: Fundamentals and Applications*, Narosa Publishing House, New Delhi.

Appendix

Table A.2.1 Database for Arial font

16 average row-wise longest run feature				16 average column-wise longest run feature				Letter
1.5	1	3.8125	1.4375	2.1875	2.75	2.75	2.125	A
5.625	2.875	3.125	6	6.125	1	1.125	2.6875	B
3.5	1.4375	1.5	3.3125	3.6875	1	1	1.6875	C
5.0625	1	1	5.3125	7.375	1	1.0625	3.3125	D
7.375	3.75	3.75	7.6875	5.875	1	1	2.0625	E
7.625	3.5625	3.5625	1.5625	6.3125	1	1	1.6875	F
3.25	1.375	3.8125	3.5	3.3125	1	1.375	3.5	G
1.5	3.25	3.375	1.5625	6.5625	1	1	6.5	H
2.75	1	1	2.8125	0	4.5	1	3.75	I
1.6875	1	1.625	3.6875	1.5625	1.1875	4	3.875	J
1.875	1.4375	1.375	1.75	6.1875	1.125	1.375	1.3125	K
1.6875	1	1	7.5625	7.875	1	1	1.375	L
1.75	1	1	1.5	8.9375	2.8125	2.9375	8.4375	M
1.625	1.1875	1.1875	1.6875	7.75	1.6875	1.4375	7.8125	N
3.3125	1	1	3.25	3.4375	1	1	3.3125	O
5.6875	2.875	3.25	1.4375	5.9375	1	1	1.9375	P
3.375	1	1.4375	3.6875	3.1875	1	1.25	3.5	Q
5.625	3	2.5	1.5625	6.25	1	1.4375	2.1875	R
3.6875	3.25	2.875	4	2.5	1.0625	1	2.5	S
5.875	1	1	1.4375	1.0625	4.25	4.1875	1.25	T
1.4375	1	1	3.4375	6.625	1	1.0625	6.5	U
1.4375	1	1	1.5625	2.125	2.6875	2.5625	2.375	V
1.3125	1	1	1.375	4.875	4.375	4.25	5.0625	W
1.75	1.3125	1.4375	1.75	1.625	1.625	1.625	1.5625	X
1.875	1.3125	1	1.5625	1.5	3.3125	2.9375	1.625	Y
7.125	1.5	1.4375	8.1875	1.9375	1.25	1.1875	1.9375	Z

Table A.2.2 Database for Bardely font

16 average row-wise longest run feature				16 average column-wise longest run feature				Letter
1.0625	1.4375	3.5	1.8125	2.1875	2.25	3.25	2.0625	A
4.1875	2.4375	4.0625	4.1875	1.75	3.625	1.875	1.9375	B
2.0625	1.0625	1.3125	4.25	3.1875	1.125	2	2	C
3	1.4375	1.25	4.1875	3.875	1.0625	1	2.1875	D
6	5.625	1.1875	3.75	3.5625	1	1.125	1.9375	E
4.5625	4.8125	1.0625	1.25	3.875	2	1.0625	1.75	F
3.375	1.5625	4.4375	1.1875	2	1	1.625	2.5	G
0.9375	4.125	1	1.1875	4.5625	1	2.875	3.25	H
4.625	2.375	2.625	4.3125	2.5	1.5	2.5625	1.4375	I
1	1	1.0625	3.9375	1.625	1.9375	1.375	3.1875	J
1.25	2.6875	1.75	2.8125	3.5625	1.1875	1.75	1.6875	K
0.875	1.0625	1	5.125	4.3125	1	1.25	1.6875	L
1.3125	1.8125	1.9375	1.3125	4.4375	3.125	2.125	5	M
1.3125	1.4375	2.375	1.875	3.9375	1.375	1.25	5	N
3.5	1.25	1.25	3.125	3.3125	1.25	1.4375	2.3125	O
3.9375	1.8125	3.3125	1.25	2.625	2.75	1.1875	1.625	P
3.8125	1.75	2.5	3.5	3.75	2.5625	2.3125	3.75	Q
3.875	3.1875	2	2.75	3.25	1.8125	1.75	1.8125	R
3.0625	2.125	2.5	3.25	1.8125	1.4375	1.0625	2.0625	S
4.5	1	1	1.4375	1.3125	4.5625	1.125	1.9375	T
1.1875	1.25	1.5625	2.3125	4	1.4375	3.5625	1.4375	U
1.125	1	1.0625	1.125	2.0625	3.125	2	1.9375	V
1.375	1.9375	2	1.5625	5.8125	2.5	5.4375	4.75	W
1.5	1.375	1.3125	2.1875	1.625	1.8125	2.125	1.375	X
1.75	1.25	1	1.125	1.125	3.8125	1.6875	1	Y
2.875	1.25	1.1875	3.875	1.625	1.5625	1.6875	1.625	Z

Table A.2.3 Database for Calibri font

16 average row-wise longest run feature				16 average column-wise longest run feature				Letter
1.6875	1.1875	4.375	1.5625	2.375	2.625	2.625	2.5625	A
4.875	2.5625	2.75	5.375	5.875	1	1.5625	2.8125	B
4.4375	1	1.0625	4	3.4375	1.1875	1	1.3125	C
4.875	1.125	1.0625	5	7.25	1	1.0625	3.4375	D
7	3.25	3.25	7.0625	4.3125	2.25	1	1.5625	E
6.9375	3.5	3.5	1.6875	4.4375	2.625	1	1.4375	F
3.9375	2.625	2.0625	3.75	2.9375	1.0625	1.25	3.5	G
1.375	3.1875	3.1875	1.375	6	1	1	5.9375	H
3.0625	1	1	3.125	3.9375	1	1	3.9375	I
1.9375	1	1	5	1.0625	1	4.125	4.125	J
2.125	1.8125	1.625	2.125	6.375	1.375	1.625	1.3125	K
1.875	1	1	6.8125	4.375	4.375	1	1.25	L
1.4375	1	1	1.25	8.375	2.875	3.0625	8.25	M
1.9375	1.25	1.1875	1.75	8	1.9375	1.875	8	N
3.5625	1.0625	1	3.4375	3.5625	1	1.0625	3.5	O
5.0625	1.125	4.375	1.6875	5.8125	1	1.125	2.125	P
3.4375	1	1.125	3.5	3.4375	1	2.0625	2.875	Q
4.9375	2.5625	2.1875	1.6875	6	1	1.75	2.125	R
4.75	2.3125	2.375	5.625	2.0625	1	1.0625	2.75	S
6.125	1	1	1.5	1.3125	4.375	4.3125	1.1875	T
1.5625	1	1.125	3.625	6.8125	1	1	6.875	U
1.5	1	1	1.6875	2.4375	2.6875	2.6875	2.3125	V
1.1875	1	1	1.375	4.5625	4.5	4.25	4.6875	W
1.75	1.4375	1.4375	1.875	1.5	1.875	1.9375	1.5625	X
1.8125	1.25	1	1.5625	1.5	3.375	3.4375	1.5	Y
7.125	1.25	1.3125	7.125	1.8125	1.25	1.375	1.5625	Z

Table A.2.4 Database for Cambria font

16 average row-wise longest run feature				16 average column-wise longest run feature				Letter
1.4375	1	3.5	2.375	1.5625	3	3.25	2.1875	A
5.1875	2.5	2.6875	5.375	3.9375	2.5	1	2.75	B
4.125	1.0625	1	4.3125	3.6875	1.3125	1	1.75	C
4.8125	1.0625	1.125	4.875	4.1875	4.25	1.25	3.625	D
6.8125	2.5	2.5	6.9375	4.25	2.5	1.1875	3.1875	E
6.5625	2.375	2.375	2.8125	3.9375	2.5	1.125	2.8125	F
3.875	1	2.625	3.3125	3.625	1.125	1.1875	4.125	G
2.375	4.375	1	2.4375	4.0625	2.5625	2.6875	4.0625	H
5.8125	1	1	5.8125	1.125	4	3.875	1.25	I
4.375	1	1	3.0625	0.8125	3.875	3.8125	0.75	J
2.5	1.5625	1.5	2.5	4.0625	2.6875	1.75	1.1875	K
2.8125	1	1.3125	6.3125	3.9375	4.125	1	2.3125	L
1.9375	1	1	2.5625	7.4375	3	3.0625	7.1875	M
2.3125	1.25	1.25	2.3125	7	1.625	1.6875	7.125	N
3.25	1	1.0625	3.25	3.8125	1.125	1.0625	3.875	O
5.0625	1.125	3.8125	3	4.0625	2.875	1.0625	2.125	P
3.0625	1	1.4375	3.75	3.125	1.0625	1.5625	3	Q
4.5	1.1875	2.3125	2.6875	4.125	2.875	1.875	1.9375	R
4.4375	2.125	1.9375	4.9375	2.3125	1.375	1.125	2.6875	S
5.625	1.3125	1	2.75	2	4.25	4.25	2.125	T
2.625	1	1	3.25	3.875	3.9375	1.375	6.25	U
2.5625	1.0625	1	1.4375	1.6875	3.3125	2.5625	2.4375	V
1.8125	1	1	1.375	4.5	4.375	5.4375	3.5	W
2.9375	1.25	1.4375	2.8125	1.625	1.8125	2.0625	1.3125	X
2.5625	1.25	1	2.6875	1.375	2.9375	3.0625	1.3125	Y
6.4375	1.625	1.6875	6.625	2.25	1.5625	1.5625	2.4375	Z

Table A.2.5 Database for Times New Roman font

16 average row-wise longest run feature				16 average column-wise longest run feature				Letter
1.125	1	3.5	2.3125	2.0625	2.6875	2.8125	1.75	A
4.4375	2.1875	2.25	4.75	4.0625	3.125	1.375	3.25	B
3.25	1.1875	1.0625	3.75	3.4375	1.1875	1.0625	2.625	C
4.625	1	1	4.5	4.25	4.1875	1	3.25	D
6.25	3.9375	1.375	6.25	4.3125	2.5625	1.125	3.5625	E
6.5	2.5	2.5	3.125	4.25	3.0625	1.75	3.25	F
3.0625	2.5	1	3.1875	3.125	1.1875	1.0625	5.375	G
2.5625	2.75	2.75	2.625	4.25	3.5	3.375	4.3125	H
5.1875	1	1	5.1875	1.25	4.25	4.125	1.25	I
4.0625	1	1	3	1.0625	4.5	3.9375	1.0625	J
2.8125	1.125	1.0625	3.1875	5.75	1.1875	1.6875	1.5625	K
2.875	1	1.125	6.5	4.25	4.3125	1	1.8125	L
1.6875	1	1	2.125	7.6875	2.5625	2.6875	7.4375	M
2.125	1.1875	1.125	2.5625	6.5625	1.5625	1.25	7.3125	N
2.875	1	1	2.75	3.75	1.125	1.0625	3.625	O
4.6875	1.375	2.8125	3.0625	4.375	2.6875	1.25	1.9375	P
2.9375	1	1.5625	3.5	3.0625	1.4375	1.4375	3.125	Q
3.9375	1.5625	1.9375	2.875	4.375	3	2	1.875	R
3.4375	2.4375	2.0625	3.625	3.1875	1	1	3.4375	S
5.625	1	1	2.75	1.75	4.4375	4.375	1.6875	T
2.4375	1	1	3.3125	6.625	1.125	1	6.5	U
2.5	1	1	1.0625	2.125	2.375	2.5	2.0625	V
2	1	1.0625	1.125	3.0625	4.125	4.375	3.1875	W
2.9375	1.0625	1.1875	2.9375	1.5625	1.8125	1.9375	1.4375	X
2.5625	1	1.1875	2.625	1.375	3.125	2.9375	1.75	Y
6.75	1.25	1.375	7.4375	2.5625	1.4375	1.3125	2	Z

3

A Two-Stage Approach to Handwritten Indic Script Identification

Pawan Kumar Singh, Supratim Das, Ram Sarkar, and Mita Nasipuri

Department of Computer Science and Engineering, Jadavpur University, Kolkata, West Bengal, India

3.1 Introduction

There is an enormous diversity among the scripts used in the world. Almost every country has its own language/script(s), which can be distinguished from the others in different facades. Script identification is defined as a process of recognizing the scripts, written or hand-printed, in any multilingual, multiscript environment. Script identification is stimulated by optical character recognition (OCR) research, which is used to recognize the characters written in a particular script of the underlying document. The tool OCR is generally described as the method of reading the optically scanned text by the electronic device [1]. Until the last few decades, researchers have paid almost no attention to the problem of script/language. But, as the world is getting progressively more interlinked, there is a real need for automatic script identification because the growing requirement for processing of documents in our daily life causes people to frequently face situations where the diversity of scripts/languages makes such manual text processing impractical. Also, in a multilingual country like India, where a wide variety of scripts and languages are prevalent, researchers of the OCR community have always felt the lack of a workable solution for the script identification module when they deal with multilingual documents. In general, any OCR engine is made in such a way that it can process documents written in a particular script (or language) within its knowledge. OCR in a multilingual environment can be made using either of the subsequent two choices: (1) designing a comprehensive OCR system that can identify each character of the text words of the input scripts that may prevail in the document pages, and (2) designing a script identification system to recognize different script words present in the document pages and then running individual OCR available for each script. For the first option, it is generally not feasible to recognize characters that are written in different scripts using a single OCR module. This is due to the fact that the structural properties, style, and nature of writing are deterministic features for character recognition that mostly differ from one script to another [2]. For instance, the features that are used to recognize Devanagari script might not be appropriate for recognizing

Hybrid Intelligence for Image Analysis and Understanding, First Edition.
Edited by Siddhartha Bhattacharyya, Indrajit Pan, Anirban Mukherjee, and Paramartha Dutta.
© 2017 John Wiley & Sons Ltd. Published 2017 by John Wiley & Sons Ltd.
Companion Website: www.wiley.com/go/bhattacharyya/hybridintelligence

other Indic scripts. An alternative option for identifying scripts in a multiscript environment is to apply a pool of OCRs corresponding to different scripts. The characters present in an input script document can then be identified reliably by selecting the appropriate OCR system from the said repository. Consequently, the vast majority of the OCR algorithms used in these applications are selected based upon *a priori* knowledge of the script and/or language of the documents under analysis. Unfortunately, this information may not be readily available. Here appears the issue of script identification. Almost all existing works on OCR systems make an important implicit assumption that the script type of the document to be processed is known beforehand. This assumption requires human intervention to select the appropriate OCR algorithm, limiting the possibility of completely automating the analysis process, especially in a multilingual environment. Hence, in this regard, there is a requirement to develop a pre-OCR script identification system to identify the script type of the document, so that a specific OCR tool can be selected. Automatic script identification facilitates sorting, searching, indexing, and retrieving multilingual documents. Script recognition also helps in text area identification, video indexing, and retrieval when dealing with a multiscript environment. So, the manifold applicability along with the challenges of the script identification problem are now inspiring the researchers a lot. However, research in script identification is still in its early stage, and, hence, not much literature is available in this field.

The problem of script identification is generally conceived on any of the following three levels: (1) page level, (2) text line level, and (3) word level. Performing script recognition at the word level is a much more challenging task than at the other two, higher levels. The reason for this is the information gathered from a few characters present in a word image sometimes may not be enough to identify the script truthfully. In addition to this, recognition of scripts at word level also facilitates automation of grouping words belonging to a specific script/language and also separates interlaced words pertaining to other scripts. In contrast, the computational complexity of feature extraction at page level and text line level can be sometimes too laborious and time-consuming. Therefore, script identification at word level is always an apparent choice.

The remainder of the chapter is organized as follows: Section 3.2 presents a brief review of some of the previous approaches to handwritten script identification, and some basic information related to the scripts used in the present chapter is described in Section 3.3. The proposed methodology is presented in Section 3.4. Experimental results and their analyses are given in Section 3.5. Section 3.6 concludes the work and lists some future directions.

3.2 Review of Related Work

Script identification depends on the fact that the character set of each script has unique spatial distribution and visual attributes that make it possible to distinguish it from other scripts. So, the primary task included in recognition of a script is to formulate a technique to determine these features or attributes from a given document and then classify the document's script accordingly. The reported works on script identification have acknowledged multiple approaches and features, which may be grouped into two broad categories – (1) structure-based methods and (2) visual appearance-based methods. Abundant work [1–13] has been reported in literature by means of structure-based methods. A.L. Spitz [1] proposed a technique to distinguish Han-based script (Chinese, Korean, and Japanese) from script based on Latin (German, Russian, English, and French) by examining the upward concavities of the connected components. A two-stage classifier is also designed for the classification purpose. Initially, the separation between Han-based scripts and Latin-based scripts was done by computing variances of their upward concavity distributions. Furthermore, the analysis of the distribution of optical density in the script image helped to classify within Han-based scripts. A projection profile technique to classify Korean, Chinese, Roman, Arabic, and Russian script characters was proposed by S. Wood *et al.* in [2]. The reason for applying this technique was that the projection profiles of document images were sufficient to identify different scripts. However, no suggestion was provided regarding how these projection profiles can be examined automatically for script recognition without human intervention, and also no recognition accuracy was claimed to substantiate their argument. J. Hochberg *et al.* [3] described a script identification system for recognizing six different scripts, namely, Arabic, Cyrillic, Chinese, Devanagari, Japanese, and Roman. A five-element feature set (i.e., aspect ratio, number of holes, relative horizontal centroid, relative vertical centroid, and sphericity) was extracted from all the connected components assuming 8-connectivity. A set of Fisher linear discriminants (FLDs), one FLD for each pair of script classes, was utilized for categorization. The classification of the document to their respective script classes was finally done based on where it is classified most often. An automatic script identification scheme from documents printed in Chinese, Roman, Arabic, Devanagari, and Bangla scripts was designed by U. Pal *et al.* in [4]. The features were extracted, combining statistical features and a water overflow analogy from a reservoir. B.B. Chaudhuri *et al.* in [5] reported a dual procedure based on interdependency between text-line and interline gap for recognition of six different handwritten Indic scripts, namely, Oriya, Hindi, English, Gujarati, Bangla, and Malayalam. A script identification model from a trilingual document printed in Devanagari, English, and Telugu scripts was introduced by M.C. Padma *et al.* in [6]. The classification was done at the text line level using the concept of both top and bottom profile-based features. M. Hangarge *et al.* in [7] examined texture as a tool for determining the script of handwritten document images of the three Indic scripts, namely, English, Devanagari, and Urdu. The identification of scripts was based on the observation that each script text possesses a distinct visual texture. A set of 13 spatial spread features were extracted using morphological filters. S.K. Sangame *et al.* in [8] proposed a script identification model to classify text words written in English and Kannada scripts from a handwritten bilingual document. A nine-element feature set (comprising a top-horizontal line, vertical line, bottom-component, top-holes, bottom-holes, top-down curves, bottom-up curves, right curve, and left curve) was estimated from both the scripts based on the vertical projection profiles of the word images. K. Roy *et al.* in [9] described a method for word-level identification of handwritten Oriya and Roman scripts from Indian postal documents by using the concept of a water reservoir analogy, as explained in [4]. A method for wordwise handwritten script identification for Indian postal automation was also proposed in [10] by using the run length smoothing algorithm (RLSA). A 25-element feature set based on fractal dimension, Matra/Shirorekha, a water reservoir, and other topologies was extracted to identify handwritten Bangla and English script words. A system that separated the scripts of handwritten words from a document, written in Bangla or Devanagari mixed with Roman scripts using a multilayer perceptron (MLP) classifier, was presented by R. Sarkar *et al.* in [11]. A set of eight different word-level holistic features was extracted from the word images. An intelligent feature-based technique, which recognized the scripts of handwritten words from a document page, written in Devanagari script mixed with Roman script using a MLP classifier, was reported by P.K. Singh *et al.* in [12]. A 39-element feature set was designed, of which eight features were topological (as described in [11]) and the remaining 31 were based on the convex hull of each word image. In an extended version described in [13], P.K. Singh *et al.* introduced statistical significance tests for evaluating the performances of multiple classifiers with the set of designed features, as described in [12], on numerous subsets of the data sets. A printed script identification scheme for Kashmiri, Roman, Devanagari, and Urdu document scripts at the word level was presented in [14]. The feature extraction was based on statistical features calculated using density of four mentioned scripts, and the classification was done using a tree classifier. S.M. Obaidullah *et al.* [15] proposed a two-stage approach for printed script identification at the page level for six official languages of India, namely, Bangla, Devanagari, Malayalam, Urdu, Oriya, and Roman scripts, using some abstract/mathematical, structure-based, and script-dependent features. A. Saidani *et al.* in [16] introduced structural features that were intrinsic features for word-level identification of both printed and handwritten Arabic and Latin scripts. In structure-based methods, scripts were identified based on the features extracted from either text line [4–7] or word [8–13, 16].

These above methods generally extracted features based on connected component analysis for determining the script of the text. Unfortunately, the paradox inherent in using such features was achieved only after fine segmentation the underlying document image. Consequently, the accuracy of script identification in turn is dependent on the accuracy of the intermediate steps, namely, text line and word segmentation, which are, in general, script dependent. But it is not easy to locate a common segmentation process, as different script classes show script-biased behavior. In addition to this, the presence of noise, skewness or slant, nonuniform gap between *interwords* (or *intrawords*), and other relevant degradations significantly affect the connected component analysis procedure, thus making these approaches inefficient. Due to these limitations, structure-based methods might not be a good choice to be a comprehensive methodology.

On the other hand, fine segmentation of the original document image into its corresponding text lines and words is not needed in case of visual appearance-based methods. As a result, the visual appearance-based methods make the task of script recognition much easier and less computationally expensive than the structure-based methods. But the literature survey depicts that only a few works have been proposed using visual appearance-based methods [17–19]. These methods utilize the texture-based features. These features can also be extracted from a segment of a text section of a script document image. A rotation-invariant texture feature extraction method for automatic script identification for six languages (viz., Chinese, Greek, English, Russian, Persian, and Malayalam) was described by T.N. Tan in [17]. In the initial stage, a uniform text block was produced from the input document image, and texture features were then extracted from this text block using a 16-channel Gabor filter. In order to achieve invariance to rotation, Fourier coefficients for this set of 16 channel outputs were calculated. One drawback of this method was that the text blocks extracted from the input documents did not necessarily have uniform character spacing. A. Busch *et al*. in [18] investigated the application of texture as a device in order to determine the document image script, based on the observation that different script text possesses a different visual texture. Experimentation with a number of texture features was evaluated on eight printed scripts, namely, Farsi, Chinese, Japanese, Greek, Cyrillic, Hebrew, Sanskrit, and Latin. A method for automatic identification of scripts/languages from document images was proposed by G.S. Peake *et al*. in [19]. The feature extraction is done using multiple-channel Gabor filters and gray-level co-occurrence matrices for seven printed languages: Chinese, English, Greek, Korean, Malayalam, Persian, and Russian. They extended work described in [1], where they applied some preprocessing techniques to acquire homogeneous text blocks from the input script document images. These consist of text-line localization, outsized text-line elimination, padding, and spacing normalization. Documents were also skew-corrected so that it was not necessary to generate rotation-invariant features. Script identification was then performed by using KNN (*k*-nearest neighbor) classifier. One of the critical problems encountered in using Gabor filters was the large computational cost due to the repeated filtering of input images.

Prevailing techniques for Indic script identification use the texture features such as wavelet-based co-occurrence histograms [20], Gabor filters [21, 22], wavelet packet-based features [23], histogram of oriented gradients (HOG) [24], and a combination of discrete cosine transform (DCT), moment invariants [25], among others. However, visual appearance-based methods are more suitable for a widespread approach to the problem of script identification. But hardly any researcher showed attention toward word-level handwritten script identification from Indic documents in the literature. This motivates us to propose a handwritten word-level script identification technique based on texture features. The experiment is conducted on eight popular official scripts used in India, namely, Bangla, Devanagari, Gurumukhi, Oriya, Malayalam, Telugu, Urdu, and Roman.

3.3 Properties of Scripts Used in the Present Work

India is a multilingual country where 23 constitutionally recognized languages are written using 12 major scripts. The officially recognized languages [26] are as follows: Assamese, Bengali, Bodo, Gujarati, Hindi, Marathi, Nepali, Oriya, Sindhi, Sanskrit, Punjabi, Tamil, Telugu, Kannada, Malayalam, Kashmiri, Manipuri, Konkani, Maithali, Santhali, Dogari, Urdu, and English. The 12 major modern scripts currently being used are: Bangla, Devanagari, Kannada, Gurumukhi, Gujarati, Telugu, Tamil, Malayalam, Manipuri, Oriya, Urdu, and Roman. Of these, the Urdu and Roman scripts are derived from the Persian and Latin scripts, respectively. The early Brahmi script (300 BC) is the mother of the first 10 scripts, which are also referred to as Indic scripts. Table 3.1 illustrates some vital information about eight scripts used for the present work, whereas Figure 3.1 shows a portion of handwritten document pages written in eight scripts.

Table 3.1 Important information related to scripts [26] used in the present work

Scripts	Origin	Basic character set		Writing style	Number of native speakers (millions) in india	Used to write languages
		Vowels	Consonants			
Bangla		11	39	Left to right	207	Bengali, Assamese, Manipuri, etc.
Devanagari		15	33		366	Hindi, Nepali, Marathi, Konkani, Sindhi, Sanskrit, etc.
Gurumukhi	Brahmi	12	30		57.1	Punjabi, Sanskrit, Sindhi, Braj Bhasha, Khariboli, etc.
Oriya		14	38		32.3	Oriya, Sanskrit, etc.
Malayalam		13	36		35.71	Malayalam.
Telugu		16	37		69.7	Telugu.
Roman	Latin	5	21		341	German, English, Spanish, Italian, etc.
Urdu	Persian	10	28	Right to left	60.3	Urdu, Bati, Burushaski, etc.

3.4 Proposed Work

The literature review of the script identification work has already revealed promising outcomes in recognizing the restricted number of script types in idyllic conditions. In reality, there are a number of disadvantages still prevalent in the preceding techniques, leading to difficulties in the classification of some scripts. For example, the computation of convex hull features from the word images by P.K. Singh *et al.* as described in [13] is exceedingly prone to noise and dependent on the quality of the input word image, as high variances can be observed if the word images possess these attributes. One of the main drawbacks of the technique proposed by A.L. Spitz [1] is that effective discrimination between scripts having similar character shapes, such as between Latin-based scripts and Han-based scripts, is not achieved even though such scripts can be easily visually classified by any common observers. Since texture analysis provides a global measure of the properties of a region, they are considered as reasonable alternatives for finding the solution to script identification problems without requiring analysis of each component of the script classes. Texture is a key component of human visual perception, and texture descriptors are used to measure the content of texture present in an image. Since an image is made up of pixels, texture can also be defined as an entity consisting of mutually related pixels and group of pixels. This group of pixels is called the texture primitive or texture element (*texel*) [27]. The main objective of script identification problems is to discover a connected set of pixels that satisfy a given gray-level property and that occur repetitively in a particular script constituting a textured region. Texture characteristics present in each of these regions are unique in nature. As a result, a usual way of recognizing the script document image may be based on its visual appearance

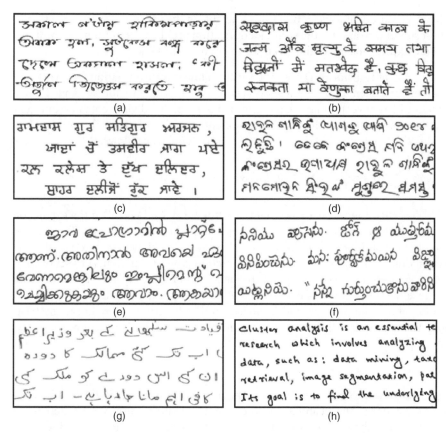

Figure 3.1 Segments of sample document pages written in: (a) Bangla, (b) Devanagari, (c) Gurumukhi, (d) Oriya, (e) Malayalam, (f) Telugu, (g) Urdu, and (h) Roman scripts, respectively.

as observed at a glimpse by a casual witness. This does not require the investigation of the individual character patterns present in the script document. The algorithms based on visual appearance extract discriminatory texture features from each script region in order to facilitate classifications of the input script patterns. This is the motive behind using texture-based features in our present work. A two-stage technique for the said purpose has been used, which is discussed in detail in the next subsection.

3.4.1 Discrete Wavelet Transform

A number of texture features has been reported in the literature. Features extracted from the Fourier transform perform poorly in practice, due to its lack of spatial localization. Gabor filters provide a better way for spatial localization; nevertheless, their effectiveness is restricted in reality for the reason that one cannot localize a spatial formation in natural textures for a single filter resolution. In comparison to the Gabor transform, the features based on wavelet transform possess a few advantages:

- Representation of textures at the most appropriate scale can be performed by varying the spatial resolution.

- As there is a variety of wavelet functions available in the literature, so the selection of the best suited wavelets for the script identification problem can be done accordingly based on texture analysis of the given scripts to be identified.

Discrete wavelet transform (DWT) is basically a linear transformation that works on a data vector having length of an integer power of two. This data vector is changed into a numerically different vector of the same length. It is a tool that divides data into different frequency components, and then each component is studied with resolution matched to its scale [27]. For an image $f(x)$, the DWT can be defined as follows:

$$f(x) = \sum_{\ell \in Z} c(\ell) \varphi_\ell(x) + \sum_{j=0}^{\infty} \sum_{k \in Z}^{n} d(j,k) W_{j,k}(x) \tag{3.1}$$

where $c(\ell)$ is called the wavelet coefficient; and the term $d(j,k)$ is called the expansion coefficient, which represents detailed information of the original image. The wavelet series converts a continuous function as a set of expansion coefficients. $\varphi_\ell(x)$ is called the scaling function. In the wavelet domain, this function is defined in terms of a set of basis functions that are finite over a range of frequencies and locations (i.e., frequency and spatially localized functions). This means that each coefficient provides some limited information about the position and frequency of the function. There are many basis functions that can satisfy the above requirement. One such basis function is given as follows:

$$\phi_{s\ell}(x) = 2^{-s/2} \phi(2^s x - \ell) \tag{3.2}$$

where ϕ is called the mother wavelet, whose magnitude is given by $2^{-s/2}$; the term s determines the scale of the wavelet and is also known as the width of the wavelet; whereas the term ℓ determines the position of the wavelet. Once a basis function is obtained, one can derive the set of basis functions by applying translation and dilation. Dilation is used to rescale the mother wavelet to any power of an integer, where the power of stretching or compressing is called scaling. By translation, it changes the offset of the wavelet for delaying it or hastening it.

The resolutions of the mother wavelet can be changed to yield a scaling function that is given as:

$$W(x) = \sum_{k=-1}^{N-2} (-1)^k C_{k+1} \phi(2x + k) \tag{3.3}$$

where C_k is called the wavelet coefficient. The necessary conditions for a wavelet coefficient are the following:

1. The sum of odd- and even-numbered coefficients should be one.
2. The product of adjacent odd or even coefficients should be zero.
3. The sum of the products of the coefficients and their complex conjugates should be two.

The values of the wavelet coefficients can be obtained using the formula:

$$c(\ell) = \int f(x) \varphi_\ell(x) \, dx \tag{3.4}$$

Similarly, the values of the expansion coefficients can be calculated as:

$$d(j,k) = \int f(x) W_{j,k}(x)\, dx \tag{3.5}$$

3.4.1.1 Haar Wavelet Transform

Haar used a square pulse as a wavelet to approximate a function [27]. The basis vectors should be orthogonal. These orthogonal basis functions are scale-varying wavelets. Mathematically, Haar transform is defined as:

$$\phi_j^i = \phi(2^j x - i) \quad where, i = 0, 1, 2, 3, \dots, 2^j - 1 \tag{3.6}$$

The mother wavelet function $\phi(x)$ of the Haar wavelet is defined as:

$$\phi(x) = \begin{cases} 1 & for\ 0 \le x \le 1/2 \\ -1 & for\ 1/2 \le x \le 1 \\ 0 & otherwise \end{cases} \tag{3.7}$$

The implementation of a wavelet is also possible by using a set of finite impulse response filters such as low-pass and high-pass filters, which are generally used in pairs. Wavelet transforms are implemented by quadrature mirror filters. There are two kinds of filters: g (high-pass) and h (low-pass). A pair of such filters is used to divide the frequency into subbands. This procedure is repeated recursively till the lowest frequency band of the given input image is reached. Since the application of two filters to an image increases its size, down-sampling is done to make sure that the size of the image does not increase, thus ensuring that the result is of the same size as the original image. Here, LL represents the approximation coefficients and can be subjected to the next level of decomposition. HH, HL, and LH represent the detail coefficients. This is called one-level decomposition. Similarly, the decomposition can be continued till it is practically possible. The representation of the levels of the image is called a *wavelet decomposition tree*. A two-level decomposition of a wavelet transformation is shown in Figure 3.2. The DWT is performed first for all rows and then for all columns of an image. Down-sampling is represented by the symbol $\downarrow 2$. In the proposed work, the g and h filters for a Haar wavelet are expressed as follows:

$$g = \frac{1}{\sqrt{2}}[1, 1] \tag{3.8}$$

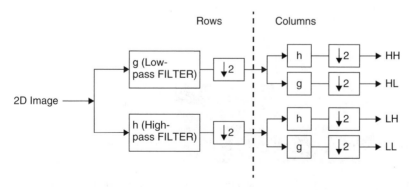

Figure 3.2 Wavelet decomposition tree for a 2D image using both low-pass and high-pass filters.

$$h = \frac{1}{\sqrt{2}}[1, -1] \tag{3.9}$$

The proposed algorithm for performing 2D wavelets is as follows:

1. Read the given image $f(x, y)$.
2. Apply g to $f(x, y)$, and down-sample the result by 2 along x. This gives the image T1.
3. Apply h to $f(x, y)$, and down-sample the result by 2 along x. This gives the image T2.
4. Apply g to T1, and down-sample the result by 2 along x. This gives the image LL.
5. Apply h to $f(x, y)$, and down-sample the result by 2 along x. This gives the image LH.
6. Apply g to T2, and down-sample the result by 2 along x. This gives the image HL.
7. Apply h to T2, and down-sample the result by 2 along x. This gives the image HH.
8. Display the image.
9. Exit.

The application of the Haar wavelet to a sample word image written in Bangla script is shown in Figure 3.3. Depending on the requirement, we can discard any of the high- or low-frequency components. It is generally seen that the low-frequency component is the smoother representation of the original image, whereas the high-frequency component represents a sharp transition like edge, line, and so on. So, wavelet analysis can also be useful for edge detection purposes. Unlike the DCT, the wavelet transform is not Fourier-based, and, therefore, wavelets perform a better job of handling discontinuities in data. Here, we have used discrete Haar wavelet transform. The Haar wavelet is computed by calculating the sums and differences of adjacent elements in a given image. The 2D Haar wavelet transforms data into four bands: LL, HL, LH, and HH. It can decompose the original image into different components in the frequency domain, that is the average component (LL) and three detail components (LH, HL, and HH), as shown in Figure 3.3. In order to preserve the most representative information and to avoid unnecessary computations by performing different image-smoothing operations, only the resultant LL component of the word image has been taken into account for

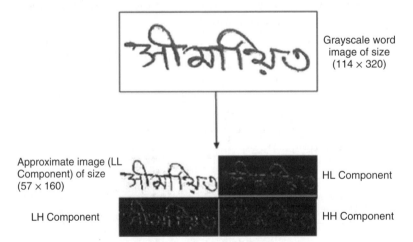

Figure 3.3 Pictorial description of the LL, HL, LH, and HH components after applying Haar wavelet transform to the original grayscale word image written in Bangla script.

the proposed work. It may also be noted that the LL component of the word image is smaller than the original word image (due to down-sampling). For handwritten word images, smoothing is an important task as data may be noisy. So, the preprocessing task, which is generally an important phase in any image-processing domain, is avoided by performing Haar wavelet transform.

3.4.2   Radon Transform (RT)

Radon transform (RT) for 2D function is named after Johann Radon. RT was first invented in 1917 by Radon, who also implemented this transform for 2D space [28]. Later, it was also generalized for higher dimensional space. RT is generally applied in tomography, where an image is created with the help of projection data related with cross-sectional scans of an object. Let an unknown density correspond to a function f, and the output of a tomographic scan is achieved in the form of projection data, which is represented by the RT. Hence, the original density can be reconstructed from the projection data through the inverse of the RT. Thus, it leads to the mathematical foundation for tomographic reconstruction, which is also known as image reconstruction. Since the RT of a Dirac delta function is a distribution supported on the graph of a sine wave, so the RT data is sometimes also referred to as a *sinogram*. As a result, the RT of a number of small objects can be expressed graphically as a number of blurred sine waves having different amplitudes and phases. The RT is useful in barcode scanners, computed axial tomography (CAT) scans, reflection seismology, and electron microscopy of macromolecular assemblies like viruses and protein complexes, and in searching for solutions of hyperbolic partial differential equations [29]. It is mainly used for line detection applications in computer vision, image processing, and seismology. It can extract line parameters from a 2D image that has embedded lines. If an image contains lines, then for each line, the RT creates a peak positioned at the corresponding line parameters. This property of the RT can detect a line present at any angle. Although this transform is mainly used to detect lines, generalized RT was proposed for the detection of shapes, which include arbitrary curves such as circles, hyperbola, and so on. The authors in [30] used RT for skew angle detection in Bangla script. In our present work, features are extracted from the resultant word images, attained after the application of Haar wavelet transform on the original word images. Here, in the second stage, only the LL component of the word images undergoes RT.

The RT is computed as the projections of an image matrix taken along specified directions. A projection of a 2D function $f(x, y)$ is defined as a set of line integrals. The Radon function calculates the line integrals from multiple sources along parallel paths or rays originating from a source in a certain direction. The beams are spaced 1 pixel unit apart. The source and sensor have a rotation angle θ about the center of the image. Now, this process is repeated for a given set of angles, usually $\theta \in [0°, 180°)$. It is obvious that angle 180° and angle 0° have the same result. Each ray has a distance x' from the center of the image for a particular angle. The x'-axis and each ray are perpendicular to each other [28].

To represent an image, the Radon function takes multiple, parallel-beam projections of the image from different angles by rotating the source of the parallel beam around the center of the image. Figure 3.4 shows a single projection at a specified rotation angle on a handwritten Bangla word image.

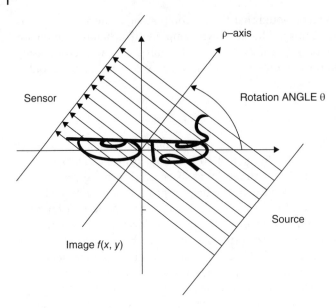

Figure 3.4 Illustration of a single projection at a specified rotation angle θ on a handwritten Bangla word image.

RT can also be defined as the projection of the image intensity along a radial line at a specific angle. The radial coordinates are the values along the x'-axis, which is oriented at $\theta°$ counterclockwise from the x-axis. Both the axes originate from the center of the image. For example, the line integral of $f(x, y)$ in the vertical direction is the projection of $f(x, y)$ onto the x-axis, and the line integral in the horizontal direction is the projection of $f(x, y)$ onto the y-axis. Figure 3.5 shows the horizontal and vertical projections for a simple 2D function.

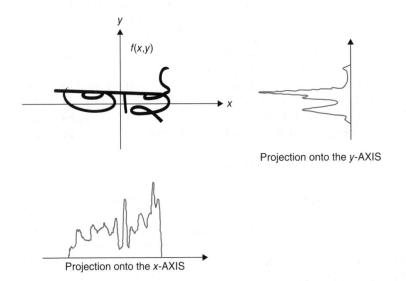

Figure 3.5 Horizontal and vertical projections of a word image written in Bangla script.

Projection can be computed along any angle θ by using the general equation [29] of the RT, as given here:

$$\Re(x',\theta) = \int_{-\infty}^{\infty} \int_{-\infty}^{\infty} f(x,y) \quad \delta(x\cos\theta + y\sin\theta - x') \quad dx\,dy \qquad (3.10)$$

where $x' = x\cos\theta + y\sin\theta$, θ is the angle of incidence of the beams; and δ is a dirac delta function (i.e., $\delta(b) = 1$ if $b = 0$ and otherwise $\delta(b) = 0$). Figure 3.6 illustrates the geometry of the RT.

RT has very interesting properties [31, 32] relating to the application of affine transformations. RT of any translated, rotated, or scaled image can be computed by knowing the transformation of the original image and the parameters of the affine transformation applied to it. RT of handwritten word images written in eight handwritten scripts – Bangla, Devanagari, Gurumukhi, Oriya, Malayalam, Telugu, Urdu, and Roman – are shown in Figure 3.7.

The output of the RT is the radon-transformed matrix where each column represents the RT at a particular angle θ. It is obvious that the sum of values of each column of the radon-transformed matrix will be the same. For simplicity, the word images are rotated at 12 different orientations ($\theta = 0°, 15°, 30°, 45°, 60°, 75°, \ldots, 165°$), and some informative statistical measures have been estimated at each of these orientations. It has been observed that the values present in the° column (i.e., for each rotation angle) in the radon-transformed matrix are not symmetrically distributed. So, for each column, column-wise skewness is computed that is taken as feature value. In general, skewness is the measure of symmetry. It is used because column-wise data distribution is off-centered. An example of symmetrical distribution is normal distribution that has no skewness. Skewness mainly occurs due to the existence of extremely large and small values in a data set. If the tail of the distribution points toward the positive x-axis, then it is

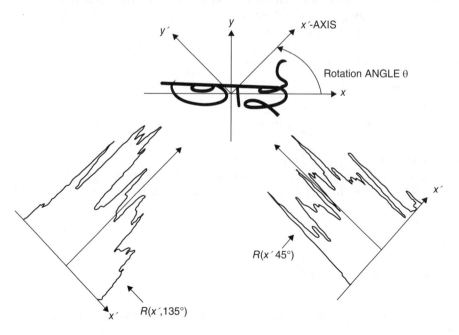

Figure 3.6 Pictorial description of the geometry of the radon transformation.

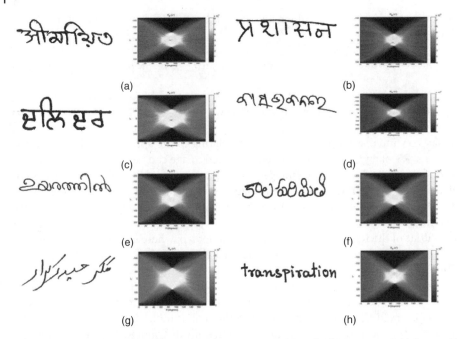

Figure 3.7 Illustration of RT of the word images written in (a) Bangla, (b) Devanagari, (c) Gurumukhi, (d) Oriya, (e) Malayalam, (f) Telugu, (g) Urdu, and (h) Roman scripts, respectively.

called positively skewed distribution. Here, unlike with the normal distribution, mean, median, and mode are scattered. For positively skewed distribution, median is on the right side to the mode and mean is on the right side to the median. Similarly, for negatively skewed distribution, the tail of the distribution points toward the negative x-axis. For more illustration, refer to Figure 3.8.

Skewness of a distribution is measured using the coefficient of skewness. Four well-known coefficients of the skewness are:

1. *Pearson coefficient of skewness*: It is also known as the Pearson mode skewness, which is defined as:

$$S_{k_p} = \frac{(mean - mode)}{\sigma} = \frac{3(mean - median)}{\sigma} \tag{3.11}$$

where σ is known as the standard deviation.

As from the above discussion, it is understood that for positively skewed distribution, mean is greater than median and standard deviation is always a positive term;

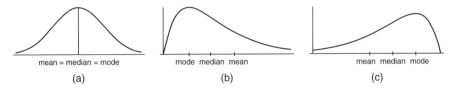

Figure 3.8 Graph for: (a) normal distribution, (b) positively skewed distribution, and (c) negatively skewed distribution.

thus, this coefficient (i.e., S_{k_p}) is greater than zero for positively skewed distribution. Similarly, the coefficient is less than zero for negatively skewed distribution and zero for symmetric distribution.

2. *Quartile coefficient of skewness*: In this method, the distribution is divided into four parts, where Q_1, Q_2, and Q_3 are the 1st, 2nd, and 3rd quartiles, respectively. Q_2 is also the median of the distribution, as it is known that median divides the distribution into two equal halves. Quartile coefficient of skewness is defined as:

$$S_{k_q} = \frac{(Q_3 - median) - (median - Q_1)}{(Q_3 - Q_1)} \tag{3.12}$$

For positively skewed distribution, $(Q_3 - median)$ is greater than $(median - Q_1)$; thus, this coefficient (i.e., S_{k_q}) is greater than zero for positively skewed distribution. Similarly, this coefficient is less than zero for negatively skewed distribution, and coefficient is zero for symmetric distribution.

3. *Percentile coefficient of skewness*: In this method, the distribution is divided into 100 parts, where P_{10} is the 10th percentile and P_{90} is the 90th percentile. Percentile coefficient of skewness is defined as:

$$S_{k_{pp}} = \frac{(P_{90} - median) - (median - P_{10})}{(P_{90} - P_{10})} \tag{3.13}$$

For positively skewed distribution, $(P_{90} - median)$ is greater than $(median - P_{10})$; thus, this coefficient (i.e., $S_{k_{pp}}$) is greater than zero for positively skewed distribution. Similarly, this coefficient is less than zero for negatively skewed distribution, and coefficient is zero for symmetric distribution.

4. *Moment coefficient of skewness*: The moment coefficient of skewness of a data set is defined as:

$$g_1 = \frac{m_3}{m_2^{3/2}} \tag{3.14}$$

where

$$m_3 = \sum_{i-1}^{n} (x - \bar{x})^3 / n \tag{3.15}$$

and

$$m_2 = \sum_{i=1}^{n} (x - \bar{x})^2 / n \tag{3.16}$$

and where $\bar{x}$ is the mean, n is the sample size, m_3 is the third moment, and m_2 is the variance of the data set.

For positively skewed distribution, this coefficient is greater than zero. Similarly, this coefficient is less than zero for negatively skewed distribution and zero for symmetric distribution.

We have used the moment coefficient of skewness in the selected columns of the radon-transformed matrix. Next, the peakedness of the distribution has been considered by calculating kurtosis. So, for each column (i.e., for each rotation angle) in the radon-transformed matrix, row-wise kurtosis has been computed and taken as a feature value. Basically, kurtosis measures the way that values are bundled across the center of the distribution. According to the value of kurtosis coefficient, the distribution can have three types of shape as shown in Figure 3.9. A distribution that is less peaked than normal is called platykurtic. If a distribution is identical to the normal distribution, then it is called mesokurtic. If a distribution is more peaked than normal, then it is called leptokurtic. Two well-known coefficients of kurtosis are as follows:

1. *Percentile coefficient of kurtosis*: This measure of kurtosis is based on both quartiles and percentiles and is defined as:

$$k = \frac{Q}{P_{90} - P_{10}} \tag{3.17}$$

where $Q = \frac{1}{2}(Q_3 - Q_1)$ is the semi-interquartile range. If $k = 0.263$, then distribution is said to be normal or mesokurtic. If $k < 0.263$ and $k > 0.263$, then distribution is said to be platykurtic and leptourtic, respectively.

2. *Moment coefficient of kurtosis*: This measure of kurtosis uses the fourth moment about the mean and is expressed in dimensionless form as:

$$a_4 = \frac{m_4}{\sigma^4} = \frac{m_4}{m_2^2} \tag{3.18}$$

where m_4, the fourth moment about the mean, is defined as:

$$and \quad m_4 = \sum_{i=1}^{n} (x - \bar{x})^4 / n \tag{3.19}$$

If the value of a_4 is less than or greater than 3, then the distribution is known as platykurtic or leptourtic. If the value of a is equal to 3, then the distribution is known as mesokurtic.

For the present work, the moment coefficient of kurtosis is chosen to measure the kurtosis in selected columns of the radon-transformed matrix. A binarized word image (foreground pixel taken as 0, and background pixel taken as 1) is inverted, and radon transformation is performed in the same way with a 15° interval gap. This is done intentionally, as radon transform for an image $f(x, y)$ at a specified rotation angle θ is a projection that computes the line integral. So, this inversion should be reflected in a radon transform matrix. As in this case, foreground pixels are set to 0 so background pixels are contributing, while calculating the line integral in radon transform and column-wise standard deviation is computed at 12 orientations. Standard deviation is the most commonly used measure of dispersion, which is a measure of spread of data about the mean. Here, skewness and kurtosis calculations are not included because, considering only background pixels (which are large in number compared to foreground pixels), all the line integrals along the parallel beam sum to a large number that makes the calculation of skewness less important. Similarly, most results show that column-wise data distribution is platykurtic in nature, so the calculation of column-wise kurtosis is excluded in the present work. The pictorial representation of the feature extraction procedure from the RT is also shown in Figure 3.10. The proposed algorithm for the present work is discussed here:

- *Step 1*: Convert the RGB word image into a grayscale image.
- *Step 2*: Perform a discrete Haar wavelet transform, and select the upper left zone (LL) component.

Figure 3.9 Illustration of (a) mesokurtic, (b) leptourtic, and (c) platykurtic curves.

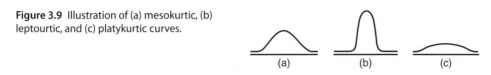

(a) (b) (c)

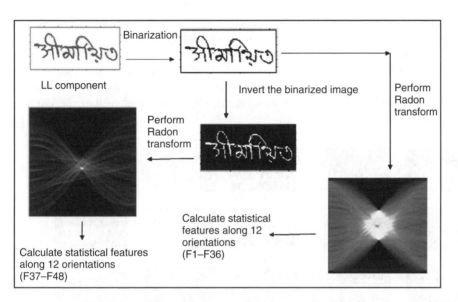

Figure 3.10 Pictorial representation of the two-stage approach of the proposed script identification technique.

- *Step 3*: Binarize the LL component using Otsu's global thresholding approach.
- *Step 4*: Apply radon transform to the binarized LL component for angle 0° to 180°. Output of this step is called a radon-transformed matrix.
- *Step 5*: Extract features from this matrix by calculating column-wise standard deviation, kurtosis, and skewness along 12 different orientations ($\theta = 0°, 15°, 30°, 45°, 60°, 75°,\ldots, 165°$).
- *Step 6:* Perform Step 3 by considering the foreground pixel as 0 and background as 1.
- *Step 7:* For this radon-transformed matrix, compute column-wise standard deviation along 12 orientations, as in Step 5.
- *Step 8:* Exit.

3.5 Experimental Results and Discussion

For the evaluation of the current script identification technique, a data set of a total of 16,000 handwritten words has been collected. Here, each of the eight scripts (Bangla, Devanagari, Gurumukhi, Oriya, Malayalam, Telugu, Urdu, and Roman) contains about 2000 words. 100 different native writers belonging to different age groups and educational levels assisted us in creating the database. The writers were asked to use black or blue ink pens for writing 50 to 100 words inside an A4-sized sheet. The materials of the contents were collected from different types of sources (viz., class notes of students of different age groups, articles of a popular daily newspaper, etc.). This is done to ensure that the probability of getting dissimilar words becomes higher. A few word samples written in the above-mentioned eight scripts from our data set are shown in Figure 3.11. The handwritten sheets are then scanned using a flatbed scanner with 300 dpi grayscale image resolution. The original gray-tone word images are converted into two-tone images (0 and 1) by Otsu's global thresholding approach [33], where the label 1 represents the object and 0 represents the background. A total of 9600 words (1200 words per script) have been used for the training purpose, whereas the remaining 6400 words (800 words per script) have been used for the testing purpose. To realize the effectiveness of the proposed approach, our comprehensive experimental

Figure 3.11 Sample word images of eight handwritten scripts from our database written in: (a) Bangla, (b) Devanagari, (c) Gurumukhi, (d) Oriya, (e) Malayalam, (f) Telugu, (g) Urdu, and (h) Roman, respectively.

tests, which are discussed in the subsequent subsections, can be summarized as follows:

1. The proposed approach is tested using multiple classifiers on multiple data sets. These data sets are randomly selected from the test data sets, and, based on the consequences of the statistical significance tests, the best classifier for the approach is chosen.
2. The statistical performance analysis of the best classifier with respect to different parameters is evaluated.
3. The individual performance of DWT and Radon transform is observed and compared with the two-stage approach. Also, a comparison of the current approach is done with some other conventional methods previously proposed.
4. Finally, the reasons for misclassification of some of the word images are briefly analyzed.

3.5.1   Evaluation of the Present Technique

The designed feature set has been individually applied to seven well-known classifiers namely, naive Bayes, Bayes net, MLP, support vector machine (SVM), random forest, bagging, and multiclass classifier. The script identification performances of the present technique on each of these classifiers, and their corresponding success rates achieved at a 95% confidence level, are shown in Table 3.2. The model building times required for the above-mentioned classifiers are shown in Figure 3.12. All the experiments are implemented in MATLAB 2010 under a Windows XP environment on an Intel Core 2 Duo 2.4 GHz processor with 2 GB of RAM. The accuracy rate, used as an assessment criterion for measuring the recognition performance of the proposed system, is expressed as follows:

$$Accuracy\ rate(\%) = \frac{No.\ of\ correctly\ classified\ text\ words}{Total\ no.\ of\ text\ words}\ X\ 100(\%) \qquad (3.20)$$

It can be observed from Table 3.2 that the highest recognition accuracy of the proposed script identification system is found to be 97.69% using SVM classifier. Since that SVM classifier is found to be suitable for the current problem, a brief discussion about this classifier will be noteworthy.

SVM classifier effectively maps pattern vectors to a high-dimensional feature space where a "best" separating hyperplane (the *maximal margin* hyperplane) is constructed. To construct an optimal hyperplane, SVM employs an iterative training algorithm that minimizes an error function, which can be defined as follows:

Table 3.2 Success rates of the proposed script identification technique using seven well-known classifiers

	Classifiers						
	Naive bayes	Bayes net	MLP	SVM	Random forest	Bagging	Multiclass classifier
Success rate (%)	87.62	85.07	95.1	97.69	90.82	89.79	91.55
95% confidence score (%)	91.88	89.49	96.39	99.65	93.94	90.3	95.23

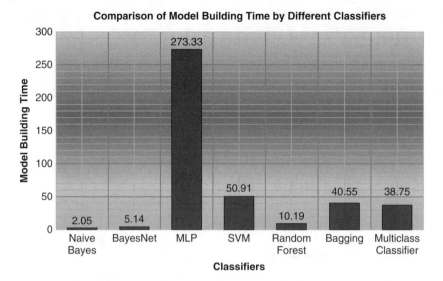

Figure 3.12 Graphical comparison of model building time (seconds) required by seven different classifiers.

Find $W \in \mathbb{R}^m$ and $b \in \mathbb{R}$ to minimize $1/2\ W^T W + C \sum_{i=1}^{N} \xi_i$ subject to constraints $y_i(W^T \phi(x_i) + b) \geq 1 - \xi_i$ and $\xi_i \geq 0$ (where $i = 1, 2, \dots, N$), where C is the capacity constant; W is the vector of coefficients; b is a constant; and ξ_i represents parameters for handling nonseparable data (inputs). The index i labels the N training cases. Note that $y \in \pm 1$ represents the class labels, and x_i represents the independent variables. The kernel ϕ is used to transform data from the input (independent) to the feature space. It should be noted that the larger the C, the more the error is penalized. Thus, C should be chosen with care to avoid overfitting. A number of kernels can be used in SVM models. These include linear, polynomial, radial basis function (RBF), and sigmoid kernels. For the present work, we have applied a polynomial kernel, which can be written as:

$$K(X_i, X_j) = (\gamma X_i . X_j + C)^d \tag{3.21}$$

where $K(X_i, X_j) = \phi(X_i) \cdot \phi(X_j)$ is the kernel function that represents a dot product of input data points mapped into the higher dimensional feature space by transformation ϕ.

3.5.1.1 Statistical Significance Tests

The statistical significance of the experimental setup has also been measured as an essential part for validating the performance of the multiple classifiers using multiple data sets. To do so, we have performed a safe and robust nonparametric Friedman test [13] with the corresponding post-hoc tests. For the experimentation, the number of data sets (N) and the number of classifiers (k) are set as 12 and 7, respectively. These data sets are chosen randomly from the test set. The performances of the classifiers on different data sets are shown in Table 3.3. On the basis of these performances, the classifiers are then ranked for each data set separately; the best performing algorithm gets the rank 1, the second best gets rank 2, and so on (see Table 3.3). In case of ties, average ranks are assigned.

Table 3.3 Recognition accuracies of seven classifiers and their corresponding ranks on 12 different data sets (ranks in parentheses are used for performing the Friedman test)

Data sets	Naive bayes (%)	Bayes net (%)	MLP (%)	SVM (%)	Random forest (%)	Bagging (%)	Multiclass classifier %
1	90(5)	87(7)	95(2)	98(1)	88(6)	91(4)	93(3)
2	88(6.5)	88(6.5)	89(5)	96(1)	91(3)	90(4)	93(2)
3	93(3)	92(4)	97(1)	95(2)	90(6.5)	90(6.5)	91(5)
4	91(5.5)	92(4)	91(5.5)	97(1)	93(3)	90(7)	95(2)
5	90(6.5)	90(6.5)	95(2.5)	96(1)	93(4)	91(5)	95(2.5)
6	96(2.5)	96(5.5)	93(4)	98(1)	90(6.5)	91(5)	90(6.5)
7	93(2.5)	93(2.5)	91(6)	95(1)	91(6)	92(4)	91(6)
8	88(7)	89(6)	96(1.5)	96(1.5)	92(4)	90(5)	95(3)
9	95(2)	93(4)	91(5.5)	97(1)	90(7)	91(5.5)	94(3)
10	92(2)	89(4)	88(5)	94(1)	90(3)	86(6.5)	86(6.5)
11	90(4.5)	89(6)	88(7)	95(1)	93(2)	91(3)	90(4.5)
12	89(6.5)	89(6.5)	92(4)	96(1)	94(2.5)	90(5)	94(2.5)
Mean Rank	$R_1 = 4.458$	$R_2 = 4.958$	$R_3 = 4.083$	$R_4 = 1.125$	$R_5 = 4.458$	$R_6 = 5.042$	$R_7 = 3.875$

Let r_j^i be the rank of the jth classifier on i^{th} data set. Then, the mean of the ranks of the jth classifier over all the N data sets will be computed as:

$$R_j = \frac{1}{N} \sum_{i=1}^{N} r_j^i \qquad (3.22)$$

The null hypothesis states that all the classifiers are equivalent, and so their ranks R_j should be equal. To justify it, the Friedman statistic [13] is computed as follows:

$$\chi_F^2 = \frac{12N}{k(k+1)} \left[\sum_j R_j^2 - \frac{k(k+1)^2}{4} \right] \qquad (3.23)$$

Under the current experimentation, this statistic is distributed according to χ_F^2 with $k - 1(= 6)$ degrees of freedom. Using Equation (3.23), the value of χ_F^2 is calculated as 27.52. From the table of critical values (see any standard statistical book), the value of χ_F^2 with 6 degrees of freedom is 18.548 for $\alpha=0.05$ (where α is known as the level of significance). It can be seen that the computed χ_F^2 differs significantly from the standard χ_F^2. So the null hypothesis is rejected.

Iman *et al.* [13] have derived a better statistic using the following formula:

$$F_F = \frac{(N-1)\chi_F^2}{N(k-1) - \chi_F^2} \qquad (3.24)$$

where F_F is distributed according to the F-distribution with $k - 1(=6)$ and $(k - 1)(N - 1)$ (=66) degrees of freedom. Using Equation (3.22), the value of F_F is calculated as 6.806. The critical value of $F(6, 66)$ for $\alpha =0.05$ is 2.1829 (see any standard statistical book), which shows a significant difference between the standard and calculated values of F_F. Thus, both Friedman and Iman *et al.* statistics reject the null hypothesis.

As the null hypothesis is rejected, a post-hoc test known as the Nemenyi test [13] is carried out for pairwise comparisons of the best and worst performing classifiers. The performances of two classifiers are significantly different if the corresponding average ranks differ by at least the critical difference (CD), which is expressed as:

$$CD = q_\alpha \sqrt{\frac{k(k+1)}{6N}} \qquad (3.25)$$

For Nemenyi's test, the value of $q_{0.05}$ for seven classifiers is 2.949 (see Table 5a of [34]). So, the CD is calculated as $2.949\sqrt{\frac{7.8}{6.12}}$, that is, 2.6 using Equation (3.23). Since the difference between mean ranks of the best and worst classifier is much greater than the CD, we can conclude that there is a significant difference between the performing ability of the classifiers. For comparing all the classifiers with a *control classifier* (here, SVM), we have applied the Bonferroni–Dunn test [13]. For this test, CD is calculated using the same Equation (3.23). But here, the value of $q_{0.05}$ for seven classifiers is 2.638 (see Table 5b of [34]). So, the CD for the Bonferroni–Dunn test is calculated as $2.638\sqrt{\frac{7.8}{6.12}}$ (i.e., 2.32). As the difference between the mean ranks of any classifier and SVM is always greater than CD, so the chosen control classifier performs significantly better than other classifiers. A graphical representation of the above-mentioned post-hoc tests for comparison of seven different classifiers is shown in Figure (3.13).

After performing the above-mentioned statistical significance tests on the 12 data sets using seven classifiers, it can be concluded that SVM outperforms all other classifiers.

3.5.2 Statistical Performance Analysis of SVM Classifier

Detailed error analysis of SVM classifier with respect to different statistical parameters is also performed in the present work. These parameters are as follows: kappa statistics, mean absolute error (MAE), root mean square error (RMSE), true positive rate (TPR), false positive rate (FPR), precision, recall, F-measure, Matthews correlation coefficient (MCC), and area under the ROC curve (AUC). Table 3.4 provides a statistical performance analysis with respect to the said seven parameters for each of the scripts used in the present work. The above-mentioned parameters are defined here in detail:

1. *Kappa statistics*: Cohen's kappa indicates the agreement between two raters who each classify N items into C mutually exclusive categories. It specifies the proportion of agreement beyond that expected by chance (i.e., the achieved beyond-chance agreement as a proportion of the possible beyond-chance agreement). It takes the form:

$$k = \frac{observed\ agreement - chance\ agreement}{1 - chance\ agreement} \qquad (3.26)$$

   In terms of symbols, this can be expressed as:

$$k = \frac{P_o - P_c}{1 - P_c} \qquad (3.27)$$

   where, P_o is the proportion of observed agreements; and P_c is the proportion of agreements expected by chance.

Pariwise Comparison of Classifiers for Nemenyi's Test

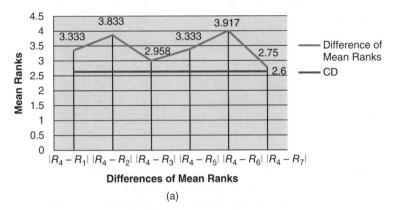

(a)

Comparison of Classifiers with SVM for Bonferroni–Dunn's Test

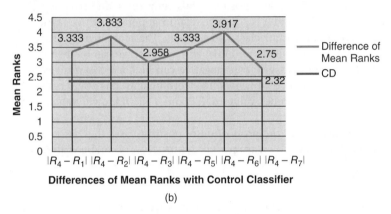

(b)

Figure 3.13 Comparison of multiple classifiers for: (a) Nemenyi's test and (b) Bonferroni–Dunn's test.

2. *Mean absolute error (MAE):* This is a model evaluation metric used with a regression model. The MAE of a model with respect to a test set is the mean of the absolute values of the individual prediction errors over all instances in the test set. Each prediction error is the difference between the true value and the predicted value for the instance. It is defined as:

$$MAE = \frac{\sum_{i=1}^{n} abs(y_i - \lambda(x_i))}{n} \tag{3.28}$$

where y_i is the true target value for test instance x_i; $\lambda(x_i)$ is the predicted target value for test instance x_i; and n is the number of test instances.

3. *Root mean square error (RMSE):* This is also a model evaluation metric used with a regression model. The RMSE of a model with respect to a test set is the square of the values of the individual prediction errors over all instances in the test set. Each prediction error is square of the difference between the true value and the predicted

		Predicted	
		Positive	Negative
	Positive	TP	FP
Actual	Negative	FN	TN

Figure 3.14 Confusion matrix for a classification rule.

value for the instance. It is defined as:

$$RMSE = \frac{\sum_{i=1}^{n}(y_i - \lambda(x_i))^2}{n} \tag{3.29}$$

The performance of a classification rule with respect to predictive accuracy can be summarized by a confusion matrix. Let us take a classification problem where there are only two classes to be predicted, referred to as the "positive" class and the "negative" class. In this case, the confusion matrix will be a 2×2 matrix as illustrated in Figure 3.14.

Let the number of true positives, false positives, false negatives, and true negatives be symbolized by TP, FP, FN, and TN, respectively. In general, *positive* stands for identified and *negative* stands for misclassified samples. Therefore, true positive can be the number of correctly identified word images, false positive can be the number of incorrectly identified word images, true negative can be correctly misclassified, and false negative can be incorrectly misclassified word images. It is obvious that the classification rule is better for higher values of TP and TN, and lower values of FP and FN.

4. *TP rate (TPR)*: This is defined as the proportion of positive cases that are correctly identified as positive. It is also known as *recall* or *sensitivity* and can be written as:

$$TPR = \frac{TP}{TP + FN} \tag{3.30}$$

5. *FP rate (FPR)*: This is defined as the proportion of negative cases that are incorrectly classified as positive. It is also known as *specificity* and can be written as:

$$FPR = \frac{FP}{FP + TN} \tag{3.31}$$

6. *Precision*: This is defined as the proportion of the predicted positive cases that are correct. It is also known as *consistency* or *confidence*, and can be written as:

$$Precision = \frac{TP}{TP + FP} \tag{3.32}$$

7. *F-measure*: This measure has been widely employed in information retrieval and is defined as the harmonic mean of recall and precision.

$$F - Measure = \frac{2 \cdot recall \cdot precision}{recall + precision} \tag{3.33}$$

8. *Matthews correlation coefficient (MCC)*: The MCC is used to measure the quality of binary (two-class) classification in machine learning. It considers all the values of TP, FP, FN, and TN and is generally viewed as a balanced measure that can be used even if the classes are of very different sizes. The MCC is mathematically defined as a correlation coefficient between the observed and predicted binary classifications;

Table 3.4 Statistical performance measures along with their respective means (shaded in gray) achieved by the proposed technique for eight handwritten scripts

	Kappa statistic	MAE	RMSE	TP rate	FP rate	Precision	Recall	F-measure	MCC	AUC
Bangla				0.973	0.001	0.994	0.973	0.983	0.981	0.986
Devanagari				0.983	0.004	0.970	0.983	0.977	0.973	0.989
Gurumukhi				0.952	0.006	0.957	0.952	0.954	0.948	0.973
Oriya				0.994	0.000	0.999	0.994	0.996	0.996	0.997
Malayalam	0.9736	0.0058	0.076	0.974	0.008	0.945	0.974	0.959	0.953	0.983
Telugu				0.977	0.004	0.971	0.977	0.974	0.970	0.986
Roman				0.962	0.002	0.983	0.962	0.972	0.968	0.980
Urdu				1.000	0.000	1.000	1.000	1.000	1.000	1.000
Mean				**0.977**	**0.003**	**0.977**	**0.977**	**0.977**	**0.974**	**0.987**

it returns a value between −1 and +1. A coefficient of +1 signifies a perfect prediction, 0 denotes no better than random prediction, and −1 specifies total disagreement between prediction and observation. The MCC can be directly calculated from the confusion matrix using the formula:

$$MCC = \frac{TP \times TN - FP \times FN}{\sqrt{(TP + FP)(TP + FN)(TN + FP)(TN + FN)}} \tag{3.34}$$

9. *Area under the ROC curve (AUC):* A receiver operating characteristic (ROC), or ROC curve, is a graphical plot that illustrates the performance of a binary classification system as its discrimination threshold is varied. The curve is created by plotting the TPR against the FPR at various threshold settings. A classification rule no better than chance is reflected by an ROC curve that follows the diagonal of the square, from the lower left corner to the top right corner. A classification rule that is worse than chance would produce an ROC curve that lay below the diagonal–but in this case, performance superior to chance could be obtained by inverting the labels of the class predictions. The AUC is then simply the area under the ROC curve. The AUC of a ROC curve is a way to reduce ROC performance to a single value representing expected performance. The performance of SVM classifier, represented as a point on the ROC curve for the present work, is shown in Figure 3.15.

3.5.3 Comparison with Other Related Works

For the justification of the two-stage approach used in the present work, both the DWT and RT have been individually applied on the present database. For DWT, the statistical features have been computed in two ways: (1) considering only the LL component and (2) considering the entire four components of the original word image. Similarly, for RT, feature extraction is done both with and without considering the first stage of our proposed approach (i.e., DWT). This means that in the first case, the RT is applied directly on the original word images, and in the second case the transformation is applied only on the LL component of the word images. In short, the statistical features are extracted by considering the following four cases:

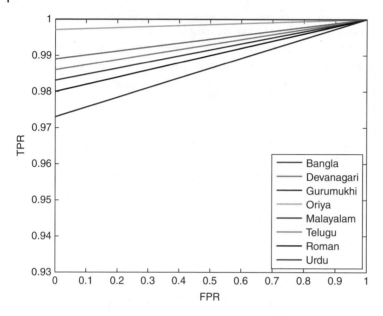

Figure 3.15 Graph showing the performance of SVM classifier on the ROC curve for eight handwritten scripts.

- *Case I*: Computation taking all four components from the original word image obtained after DWT.
- *Case II*: Computation taking only the LL component of the word image obtained after DWT.
- *Case III*: Computation after performing RT from the original word image.
- *Case IV*: Computation after performing RT from the LL component of the word image.

A comparison of the results considering all of these cases is shown in Table 3.5. Based on the values of statistical performance parameters defined above, it can be concluded that the fourth case (i.e., our proposed approach) has surpassed all the remaining cases except for the execution time. Here, the execution time is calculated as the time taken from the extraction of statistical features to the computation of the respective parameters. It can be noted that the execution time for the fourth case is the maximum, but still it takes only about 4 minutes of additional time than the best case (Case II).

Table 3.5 Comparison of statistical performance parameters for the four cases (best case is styled in bold)

	Kappa statistic	MAE	RMSE	TP rate	FP rate	Precision	Recall	F-measure	MCC	AUC	Execution time (secs)
Case I	0.8703	0.0533	0.1537	0.889	0.019	0.888	0.889	0888	0.870	0.979	450.720
Case II	0.8835	0.0426	0.1449	0.900	0.017	0.900	0.900	0.900	0.883	0.981	**398.060**
Case III	0.9185	0.0316	0.1257	0.930	0.012	0.931	0.930	0.930	0.919	0.985	505.132
Case IV	**0.9736**	**0.0058**	**0.076**	**0.977**	**0.003**	**0.977**	**0.977**	**0.977**	**0.974**	**0.987**	612.697

Table 3.6 Comparison of the present script identification result with state-of-the art methods

Researchers	Feature set	Feature dimension	Type of *indic* scripts used	Execution time (sec)	Success rate (%)
Sarkar *et al.* [11]	Holistic features	8		955.803	85.02
Singh *et al.* [13]	Topological and Convex hull based Features	39		1076.575	91.47
Bashir *et al.* [14]	Density based statistical features	4		649.518	76.23
Obaidullah *et al.* [15]	Abstract/ mathematical, structure based and script dependent features	41		725.962	90.14
Hiremath *et al.* [20]	Wavelet based co-occurrence histogram	32	Bangla, Devanagari, Gurumukhi, Oriya,	801.244	92.55
Ma *et al.* [21]	Gabor filter based Features	32	Malayalam, Telugu, Urdu and Roman.	651.046	93.81
Chaudhuri *et al.* [22]	Gabor filters	30		665.95	94.36
Padma *et al.* [23]	Wavelet packet based Features	44		682.507	91.54
Singh *et al.* [24]	Shape based and HOG features	87		861.207	93.43
Singh *et al.* [25]	DCT and Moment Invariants	92		759.860	94.25
Propoased Method	**DWT and Radon Transform**	**48**		**612.697**	**97.69**

For comparison of the present work with some recent works, the feature set as described in [11, 13–15] and [20–25] has been implemented and evaluated on the present script data set. From the experiment, it was noted that the result with the current feature set not only gives higher identification accuracies but also is very fast compared to other methods. So, it may be concluded from the analysis that the proposed technique outperforms the previous ones. Comparisons in terms of the accuracy and the computational complexity obtained by different state-of-the-art feature sets with the proposed feature set are detailed in Table 3.6.

3.5.4 Error Analysis

Sample handwritten word images that are successfully classified by the present technique are shown in Figure 3.16. It is observed from the confusion matrix that all the word images written in Urdu script have been classified quite successfully. In certain cases, some small words containing fewer than three characters (see Figure 3.17a) are misclassified by the present script recognition technique. The possible reason for this might be that a significant number of discriminatory feature values is not found to distinguish them from the others. Also, the presence of skewness in some of the word images (see

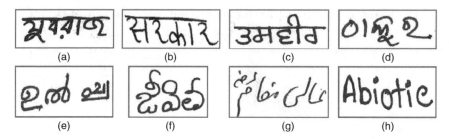

Figure 3.16 Samples of successfully classified handwritten word images written in: (a) Bangla, (b) Devanagari, (c) Gurumukhi, (d) Oriya, (e) Malayalam, (f) Telugu, (g) Urdu, and (h) Roman scripts, respectively.

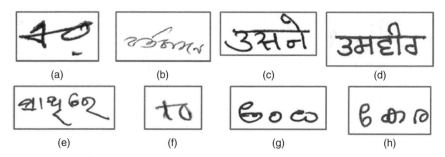

Figure 3.17 Sample handwritten word images misclassified by the present technique due to the presence of: (a) a significantly smaller number of characters constituting the word; (b) skewness; (c–e) structural similarity in Devanagari (misclassified as Gurumukhi), Gurumukhi (misclassified as Devanagari), and Oriya (misclassified as Bangla); (f) Matra-like structure in Roman script; and (g–h) abrupt spaces in Malayalam and Telugu scripts, respectively.

Figure 3.17b) and structural similarity among the characters of different scripts, such as Devanagari and Gurumukhi (Figure 3.17c–d) and Bangla and Oriya (Figure 3.17e), lead to wrong identification. Even discontinuities in Matra in certain scripts like Bangla and Devanagari and the existence of Matra-like structure in Roman script (Figure 3.17f) misclassify them among each other. The reasons for misclassification of Malayalam and Telugu scripts are mainly due to the existence of nonuniform spaces in between characters of a single word (Figure 3.17g–h).

3.6 Conclusion

Multiscript identification from handwritten documents is a challenging research area in the document image analysis domain. There is an increasing concern for designing methods that can handle documents written in more than a single script or language. Motivated by the demands of growing world trade, it can be expected that multiscript OCR systems will become more widespread in both academic and commercial worlds. Identification of the script or language of the documents, therefore, becomes a crucial need in the automation of multilingual text recognition.

Handwritten script identification is a very significant task in any multiscript and multilingual OCR development. In this chapter, a two-stage approach for word-level handwritten script identification from eight popular scripts in India has been presented. The proposed scheme is based on texture-based features. At present, 48 distinct features have been extracted and tested using multiple classifiers. Finally, an accuracy of 97.69% has been achieved using the SVM classifier. Experiments on handwritten word images show excellent script identification results that also exceed other state-of-the-art techniques. This guarantees the suitability of the scheme to script recognition problem. In addition to this, the scheme also works well for single-character words. This scheme is applicable for different writing styles of handwriting, which ensures flexibility of the present technique.

In future, this technique can also be extended to identify other scripts in any multiscript environment. More data samples considering all official Indic scripts need to be collected in near future for the comprehensive evaluation of the proposed methodology. In short, the proposed methodology could be used as a common word-level script identification module for the development of any multiscript OCR system.

Acknowledgments

The authors are thankful to the Center for Microprocessor Application for Training Education and Research (CMATER) and Project on Storage Retrieval and Understanding of Video for Multimedia (SRUVM) of the Computer Science and Engineering Department, Jadavpur University, for providing infrastructure facilities during progress of the work.

References

1 Spitz, A.L. (1997) Determination of script and language content of document images. *IEEE Transactions on Pattern Analysis and Machine Intelligence*, **19** (3), 235–245.

2 Wood, S.L., Yao, X., Krishnamurthy, K., and Dang, L. (1995) Language identification for printed text independent of segmentation. *IEEE of International Conference on Image Processing*, pp. 428–431.

3 Hochberg, J., Kerns, L., and Thomas, P.K.T. (1997) Automatic script identification from images using cluster based templates. *IEEE Transactions on Pattern Analysis and Machine Intelligence*, **19** (2), 176–181.

4 Pal, U. and Chaudhuri, B.B. (2002) Identification of different script lines from multi-script documents. *Image and Vision Computing*, **20** (13–14), 945–954.

5 Chaudhuri, B.B. and Bera, S. (2009) Handwritten text line identification in Indian scripts. *10th International Conference on Document Analysis and Recognition (ICDAR)*.

6 Padma, M.C. and Vijaya, P.A. (2009) Identification of Telugu, Devnagari and English scripts using discriminating features. *International Journal of Computer Science and Information Technology (IJCSIT)*, **1** (2).

7 Hangarge, M. and Dhandra, B.V. (2010) Offline handwritten script identification in document images. *International Journal of Computer Applications (IJCA)*, **4** (6).

8 Sangame, S.K., Ramteke, R.J., Andure, S., and Gundge, Y.V. (2012) Script identification of text words from a bilingual document using voting techniques. *World Journal of Science and Technology*, **2** (5), 114–119.

9 Roy, K. and Pal, U. (2006) Word-wise handwritten script separation for Indian postal automation. *International Workshop on Frontiers in Handwriting Recognition, La Baule*, pp. 521–526.

10 Roy, K., Pal, U., and Chaudhuri, B.B. (2005) Neural network based word-wise handwritten script identification system for Indian postal automation. *International Conference on Intelligent Sensing and Information Processing, Chennai*, pp. 581–586.

11 Sarkar, R., Das, N., Basu, S., Kundu, M., Nasipuri, M., and Basu, D.K. (2010) Word level script identification from Bangla and Devnagari handwritten texts mixed with Roman scripts. *Journal of Computing*, **2**, 103–108.

12 Singh, P.K., Sarkar, R., Das, N., Basu, S., and Nasipuri, M. (2013) Identification of Devnagari and Roman script from multiscript handwritten documents. *5th International Conference on PReMI, LNCS 8251*, pp. 509–514.

13 Singh, P.K., Sarkar, R., Das, N., Basu, S., and Nasipuri, M. (2014) Statistical comparison of classifiers for script identification from multi-script handwritten documents. *International Journal of Applied Pattern Recognition (IJAPR)*, **1** (2), 152–172.

14 Bashir, R. and Quadri, S.M.K. (2015) Density based script identification of a multilingual document image. *International Journal of Image, Graphics and Signal Processing*, **2**, 8–14.

15 Obaidullah, S.M., Roy, K., and Das, N. (2013) Comparison of different classifiers for script identification from handwritten document. *IEEE International Conference on Signal Processing, Computing and Control (ISPCC)*, pp. 1–6.

16 Saidani, A., Echi, A.K., and Belaid, A. (2013) Identification of machine-printed and handwritten words in Arabic and Latin scripts. *12th IEEE International Conference on Document Analysis and Recognition (ICDAR)*, pp. 798–802.

17 Tan, T.N. (1998) Rotation invariant texture features and their use in automatic script identification. *IEEE Transaction on Pattern Analysis and Machine Intelligence*, **20** (7), 751–756.

18 Busch, A., Boles, W.W., and Sridharan, S. (2005) Texture for script identification. *IEEE Transaction on Pattern Analysis and Machine Intelligence*, **27** (11), 1720–1732.

19 Peake, G.S. and Tan, T.N. (1998) Script and language identification from document images. *Lecture Notes in Computer Science: Asian Conference Computer Vision, LNCS-1352*, pp. 97–104.

20 Hiremath, P.S. and Shivashankar, S. (2008) Wavelet based co-occurrence histogram features for texture classification with an application to script identification in a document image. *Pattern Recognition Letters 29*, pp. 1182–1189.

21 Ma, H. and Doermann, D. (2004) Word level script identification on scanned document images. *SPIE Conference on Document Recognition and Retrieval, San Jose, CA*, pp. 124–135.

22 Chaudhuri, S. and Gulati, R.M. (2016) Script identification using Gabor feature and SVM classifier. *7th International Conference on Communication, Computing and Virtualization, Procedia Computer Science*, pp. 85–82.

23 Padma, M.C. and Vijaya, P.A. (2010) Global approach for script identification using wavelet packet based features. *International Journal of Signal Processing, Image Processing and Pattern Recogntion*, **20** (3).

24 Singh, P.K., Mondal, A., Bhowmik, S., Sarkar, R., and Nasipuri, M. (2014) Word-level script identification from multi-script handwritten documents. *3rd International Conference on Frontiers in Intelligent Computing Theory and Applications (FICTA)*, pp. 551–558.

25 Singh, P.K., Khan, A., Sarkar, R., and Nasipuri, M. (2014) A texture based approach to word-level script identification from multi-script handwritten documents. *6th IEEE International Conference on Computational Intelligence and Communication Networks (ICCICN)*, pp. 228–232.

26 Languages of India. Available from http:// www.newworldencyclopedia.org/entry/ Languages_of_India

27 Gonzalez, R.C. and Woods, R.E. (1992) *Digital Image Processing*, vol. 1, Prentice-Hall, Noida, India.

28 Deans, S.R. (1983) *The Radon transform and some of its applications*, John Wiley and Sons, New York.

29 Lei, Y., Zheng, L., and Huang, J. (2014) Geometric invariant features in the Radon transform domain for near-duplicate image detection, *Pattern Recognition*, **47**, pp. 3630–3640.

30 Habib, S.M.M., Noor, N.A., and Khan, M. (2006) Skew angle detection of Bangla script using radon transform. *9th International Conference on Computer and Information Technology*, pp. 105–109.

31 Hoang, T.V. and Tabbone, S. (2012) Invariant pattern recognition using the RFM descriptor. *Pattern Recognition*, **45** (1), 271–284.

32 Lei, Y., Wang, Y., and Huang, J. (2011) Robust image hash in radon transform domain for authentication. *Signal Process.: Image Commun.*, **26** (6), 280–288.

33 Ostu, N. (1978) A thresholding selection method from gray-level histogram. *IEEE Trans. Systems Man Cybernet. SMC-8*, pp. 62–66.

34 Demsar, J. (2006) Statistical comparisons of classifiers over multiple data sets. *Journal of Machine Learning Research*, **7**, 1–30.

4

Feature Extraction and Segmentation Techniques in a Static Hand Gesture Recognition System

Subhamoy Chatterjee, Piyush Bhandari, and Mahesh Kumar Kolekar

Indian Institute of Technology Patna, Bihta, Patna, Bihar, India

4.1 Introduction

Today, human–computer interaction (HCI) and human alternative and augmentative communication (HAAC) are the two most rapidly growing fields of research. It's even a challenging task to control computers and intelligent machines without any physical interaction. The huge development of intelligent and fast computing has encouraged us to think about HCI by means of some intelligent computing and algorithms.

Gestures are the most popular means for HCI by some intelligent computing and algorithms. A gesture is a form of nonverbal communication in which visible body parts express certain meaningful information. Gestures include the movement of hands, legs, face, head, or any other body parts, or meaningful poses of hands, legs, or face. Static and dynamic hand gestures are the most popularly used gestures and have a widespread application in HCI, robot control, sign language communication, and surveillance [1]. Hand gestures have the capability to express some meaningful information to interact with the surroundings. For an example, waving the hand in the air to express "Goodbye" or showing a "V" sign to express victory are some popular hand gestures that have been used by human beings for a long period.

There are mainly two types of gestures: static gestures [2] that include some poses of body, fingers, and palms; and dynamic gestures [3] like movement of hands or any other body parts. However, static hand gestures are more popularly used in HCI compared to dynamic hand gestures because they are easy for human beings to imitate. Sometimes, static hand gestures act as a specific transition stage in a dynamic hand gesture recognition system, too. For that reason, HCI by means of some well-known hand gestures is currently becoming popular. The most popular techniques used in static hand gesture recognition systems include glove-based techniques [4] and vision-based techniques [2]. Glove-based techniques are not an automatic choice for a gesture recognition system as the glove and its associated cables make the system very cumbersome and hamper the free movement of hands. In addition to it, gloves and sensors are very expensive. On the other hand, vision-based techniques require only one or more cameras to capture the gesture images. Hence, vision-based techniques are very easy to provide HCI.

Hybrid Intelligence for Image Analysis and Understanding, First Edition.
Edited by Siddhartha Bhattacharyya, Indrajit Pan, Anirban Mukherjee, and Paramartha Dutta.
© 2017 John Wiley & Sons Ltd. Published 2017 by John Wiley & Sons Ltd.
Companion Website: www.wiley.com/go/bhattacharyya/hybridintelligence

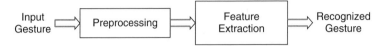

Figure 4.1 Operational flowchart of proposed static hand gesture recognition system.

As shown in Figure 4.1, a vision-based static hand gesture recognition approach consists of three steps: preprocessing, feature extraction, and classification. In the pre processing step, captured hand gesture images are segmented from the background, and the region of interest (ROI) is extracted by a proper skin color region segmentation algorithm [2]. Some morphological operations [5] are also performed in the preprocessing step to obtain a smooth, closed, and complete contour or shape of the gestures. Many researchers have employed various rotation [6] and illumination normalization [7] techniques. Skin color–based ROI segmentation is a vital issue in hand gesture–based HCI systems. In Section 4.2, various gesture segmentation algorithms and our proposed illumination and skin color complexion invariant segmentation algorithm are discussed in detail.

To find a proper feature descriptor that is capable of representing the shapes or contours of hand gesture images is another important aspect in a static hand gesture recognition system. Feature extraction is one of the most important steps in static hand gesture recognition algorithms. Feature extraction techniques employed in static hand gesture recognition are basically classified into two classes: contour-based features [6] and silhouette-based features [2]. The feature descriptors used in the literature are localized contour sequences [6], Krawtchouk moments [2], Tchebichef moments [8], geometric moments [2], Zernike moments [2], and Gabor wavelets [9]. In user-independent gesture recognition, none of these features have shown satisfactory gesture recognition accuracy. To improve gesture recognition accuracy, researchers have proposed methods like parallel feature fusion [8], serial feature fusion [8], and multifeature fusion [10]. None of these proposed feature fusion methods are satisfactory choices in user-independent gesture recognition, however, as they are not efficient in real time. All the feature extraction and feature-level fusion techniques used in static hand gesture recognition algorithms will be discussed in Sections 4.3 and 4.4.

In user-independent gesture recognition, some misclassification occurs because of different hand shapes, aspect ratios, and styles of gesticulation by the users. In addition, a static hand gesture sign language database contains some gestures that are of similar shapes [11]. For an example in American Sign Language, gestures 7, 8, and 9 have similar shapes, and these gestures cause a deterioration in the overall gesture recognition rate. Misclassification of similar-shape gestures is a frequently occurring problem in static hand gesture recognition systems. Some recent trends in classifying similar-shape gestures [12–14] have been reported in Section 4.5. In [14], researchers have employed the F-ratio-based weighted feature extraction technique for similar-shape character recognition. The F-ratio-based weighted feature technique has shown a satisfactory improvement in recognition of digits 7, 8, and 9 of American Sign Language digits. Researchers have also proposed techniques based on zoning methods [15–17] for similar-shape gesture classification. Various zoning methods and their effectiveness in user-independent gesture recognition are discussed in Section 4.4.

4.2 Segmentation Techniques

The main objective of gesture segmentation is to extract the ROI from the background of the image. A flowchart of the hand region extraction method is given in Figure 4.2. In this section, we will discuss various gesture segmentation algorithms, rotation and illumination normalization techniques, and morphological operations.

4.2.1 Otsu Method for Gesture Segmentation

The Otsu method, [6] named after Nobuyunki Otsu, is vividly used in image processing for segmentation of gray images. Otsu's algorithm automatically performs clustering-based image thresholding and converts the gray-level image into a binary image. The algorithm assumes that the image contains two classes of pixels following bimodal histogram (foreground pixels and background pixels); it then calculates the optimum threshold separating the two classes so that their combined spread or intraclass variance is minimum and their interclass variance is maximum. The Otsu algorithm is described by the following steps:

Step 1: Select an initial threshold value T for segmentation of gray-level image $f(x, y)$.
Step 2: Using this threshold T, the binary image is segmented into two classes.
Step 3: Determine the optimum threshold T^*. T^* is the value of T for which the ratio of between-class variance and total variance is maximized. The optimum threshold value T^* converts the gray-level histogram to a bimodal histogram, as shown in Figure 4.3.

In other words, the algorithm treats the segmentation of a grayscale image as a binary classification problem. Using a threshold T, the L gray levels image is segmented in two classes, $\Omega_0 = 1, 2, \ldots T$ and $\Omega_1 = T + 1, T + 2, \ldots L$. The optimal threshold T^* is determined as that value of T for which the ratio of between-class variance σ_b^2 and total class variance σ_T^2 is maximum. If the number of pixels at the ith gray level is n_i and the total number of pixels is N, then for a given T, the between-class variance and the total class

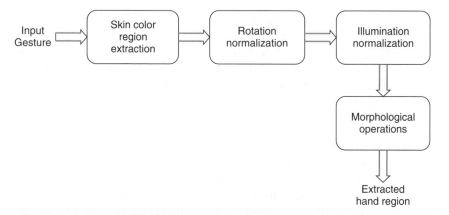

Figure 4.2 Flowchart of hand region extraction method.

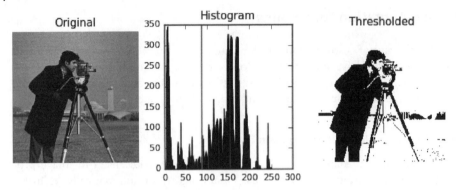

Figure 4.3 Otsu method for thresholding:(a) gray-level image, (b) bimodal histogram, and (c) segmented binary image.

variance are defined as follows:

$$\sigma_b^2 = \omega_0(\mu_0 - \mu_T)^2 + \omega_1(\mu_1 - \mu_T)^2 \tag{4.1}$$

$$\sigma_T^2 = \sum_{i=1}^{L} (i - \mu_T)^2 P_i \tag{4.2}$$

where $\omega_0 = \sum_{i=1}^{T} P_i\omega_i$, $\mu_0 = \sum_{i=T+1}^{L} iP_i/\omega_1$, and $\mu_T = \sum_{i=1}^{L} iP_i$.

Typically, the entire histogram is scanned to find the optimum threshold T.

Otsu's segmentation has been widely used in various literature [6, 17, 18]. The main drawback of the Otsu method is that it assumes that the image histogram is bimodal. So, it fails if the number of classes is variable or the two classes have different sizes. In complex background color images, the number of clusters usually varies and the cluster sizes are different. For that reason, it can't perform skin color segmentation in color images and complex background images. Various skin color-based segmentation methods [2, 8, 19] have been proposed by the researchers for segmentation of color and complex background database images.

4.2.2 Color Space–Based Models for Hand Gesture Segmentation

Most of the researchers have proposed color space–based [19] skin color region segmentation methods for static hand gesture recognition because of their simplicity and effectiveness. Most of the color space models use some primary colors like red, green, or blue to represent the images. Various color space models like RGB [20], YCbCr [8], YIQ [21], and HSV [22] color spaces have been employed for skin color region detection in static hand gesture recognition algorithms. A brief study on the skin color region detection using these color space models is given in this section.

4.2.2.1 RGB Color Space–Based Segmentation

RGB color space [20] is the most popular color space model in image processing and computer vision because of its similarity with the human visual system. Red, green, and blue: these three primary colors are used in RGB color space to represent an image. Although it is very similar to the human visual system, RGB color space models are very sensitive to illumination changes.

The normalized RGB color space models [8] have been proposed to make the color space model illumination invariant. The normalization techniques are given here:

$$R_{normalized} = Red/(Red + Green + Blue) \tag{4.3}$$

$$G_{normalized} = Green/(Red + Green + Blue) \tag{4.4}$$

$$B_{normalized} = Blue/(Red + Green + Blue) \tag{4.5}$$

Here, the sum of the normalized values of the red, green, and blue components are unity.

$$R_{normalized} + G_{normalized} + B_{normalized} = 1 \tag{4.6}$$

The normalized RGB color space model is very useful in illumination-invariant skin color detection and is widely used in static hand gesture recognition algorithms. In [23], researchers discovered threshold values of R, G, and B components for Asian and European skin color, which are given here:

$$R > 95, G > 40, B > 20 \tag{4.7}$$

$$max(R, G, B) - min(R, G, B) > 15 \tag{4.8}$$

$$|R - G| > 15, R > G, R > B \tag{4.9}$$

4.2.2.2 HSI Color Space–Based Segmentation

The HSI color space [23] consists of hue, saturation, and intensity components of the RGB color space points. It is basically a combination of HSV and HSL color space. RGB color space models represent color components in Cartesian coordinates, whereas HSI color space models represent colors in cylindrical coordinates. For that reason, it is perceptually more reliable than the RGB color space models. The threshold values for skin color region segmentation in the HSI model are as follows:

$$0.23 < S < 0.68, 0 < H < 50, 0 < I < 64 \tag{4.10}$$

4.2.2.3 YCbCr Color Space–Based Segmentation

YCbCr color space [24] is widely used in image processing and computer vision as it represents statistically independent color components: Y, Cb, and Cr, where Y is the luminance component of the color and Cb and Cr are the chrominance between blue-yellow and red-yellow components. YCbCr color space has mainly two advantages: it shows uniform clustering of color components, and it has a uniform separation between chrominance and luminance.

The threshold values for Asian and European skin color detection [8] are:

$$Y > 80, 85 < C_b < 135, \ 135 < C_r < 180 \tag{4.11}$$

4.2.2.4 YIQ Color Space–Based Segmentation

YIQ color space is derived from YCbCr color space. Here, Y corresponds to luminance value, I corresponds to intensity value, and Q corresponds to chrominance value. Researchers have proposed hand gesture segmentation using the YIQ color space model [2] and specified threshold values for skin color regions for intensity (I) and amplitude (θ).

$$30 \leq I \leq 100, \ 105 \leq \theta \leq 150 \tag{4.12}$$

where θ is the amplitude of the Chroma cue vector.

$$\theta = tan^{-1}(V/U) \tag{4.13}$$

where U and V are the orthogonal vectors of YUV space.

4.2.3   Robust Skin Color Region Detection Using *K*-Means Clustering and Mahalanobish Distance

Color space–based methods struggle to exhibit proper ROI segmentation with illumination changes and different skin color complexion of users. Most of the times, different illumination levels and skin color complexions change the predefined threshold values of different color space models. A skin color region detection method using K-means clustering and Mahalanobish distance has been proposed, which is robust to illumination and skin color complexion changes.

In this proposed algorithm, we have used hand gesture images of uniform background for semisupervised learning about skin color regions and used this *a priori* knowledge to find out the skin color regions of the complex background images. Uniform background hand gesture images have only two clusters: foreground and background. For that reason, these images are very effective to find out the skin color region through K-means clustering [25]. We found Mahalanobish distance [26] between the skin color region centroids and the complex background images and then again perform K-means clustering for extracting the foreground region using proposed Algorithm 4.1.

Algorithm 4.1   Hand gesture segmentation using K-means clustering and Mahalanobish distance

---

Input: Hand gesture images of complex and uniform background databases
Output: Segmented region of interest (ROI)

1. Convert the RGB images of the uniform background database to YCbCr and reshape it.
2. Perform modified K-means clustering with k=2.
   $[a_1, b_1] = mkmeans(image_1, 2)$
3. As the background is uniform, the hand region has the minimum area. Let *Hand* be the minimum area that corresponds to all uniform background images.
4. Repeat step 2 for *Hand* with cluster size $(k) = 2$.
5. Convert the RGB color images of the complex database to normalized RGB color images.
6. Repeat step 1.
7. Find the Mahalanobish distance (d) between cluster centers of *Hand* and *YCbCr* images of complex databases.
8. Repeat step 2 with cluster size (k)=2 for d. $[a_3, b_3] = mkmeans(d, 2)$.
9. **if** $b = 2$, **then**
10. | return Foreground;
11. **else**
12. | return Background;
13. **end**

4.2.3.1 Rotation Normalization

A real-time efficient static hand gesture database must contain images captured in different angles to make it rotation variant, as shown in Figure 4.4. All the hand gesture images in our complex background database are captured in different angles so that the hand gesture images are rotation variant. To make the gesture recognition algorithm rotation invariant, we have employed a rotation-invariant algorithm based on the following steps.

Algorithm 4.2 Rotation normalization technique

  Input: Segmented hand gesture images
  Output: Rotation normalized hand gestures
1. Find the direction of first principal axes of a segmented hand gesture.
2. Find the rotation angle between the first principal axes and the vertical axes.
3. Rotate it by an angle equal to the rotation angle.

4.2.3.2 Illumination Normalization

Illumination changes result in improper ROI segmentation as well as gesture misclassification. To improve the overall gesture recognition accuracy, we have implemented some illumination-invariant algorithms. Our illumination-invariant method consists of three steps: (1) Power law transform [6], (2) RGB to normalized RGB [8], and (3) homomorphic filtering [6], as shown in Figure 4.5.

4.2.3.3 Morphological Filtering

To remove the undesirable background noises and to get the proper shape of hand regions, some morphological operations are performed on the segmented binary hand gesture images. In this work, a sequence of four morphological operations – (1) erosion,

Figure 4.4 Hand gesture images captured in different angles.

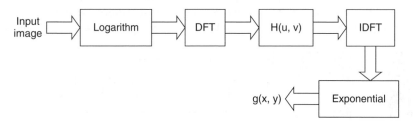

Figure 4.5 Block diagram of homomorphic filtering.

Figure 4.6 (a) RGB color image, (b) segmented image, and (c) ROI after morphological operation.

(2) dilation, (3) opening, and (4) closing – is performed in order to get the proper binary hand silhouette [10], as shown in Figure 4.6.

4.3 Feature Extraction Techniques

Both contour-based [5, 24] and shape-based features [2, 8, 9] are used to represent the shape information of binary silhouettes in static hand gesture recognition. Contour-based features represent images based on some values calculated on the boundary pixels of the binary hand silhouettes. In contrast, shape-based features represent the shape of binary hand silhouettes. The main problem associated with the feature extraction part is to discriminate similar-shape gestures. Gestures of similar shapes have almost the same boundary profiles, and for that reason contour-based feature extraction techniques misclassify gestures with similar boundary profiles. On the other hand, shape-based features are calculated on the whole shape of any image. So, these kind of features are the most appropriate features for similar-shape gesture classification.

In this section, we will discuss both shape-based and contour-based features to extract desired information from the segmented hand regions. Nonorthogonal moments like geometric moments [2] and orthogonal moments like Tchebichef [8] and Krawtchouk moments [2] are used as shape-based features. All these moments are discrete. Contour signature [24] and localized contour sequence (LCS) [6] are used as contour features. We have designed an artificial neural network (ANN)-based [27] classifier for gesture classification. The basic steps of feature extraction in a Krawtchouk static hand gesture recognition system are shown in Figure 4.7. First, a Krawtchouk segmented hand gesture is cropped so that the ROI contains the maximum area in the bounding box. Then, the corresponding feature matrices are calculated.

4.3.1 Theory of Moment Features

Moment features [2] are wildly used in image processing and pattern recognition as these features can represent both the globalized and localized properties of an image

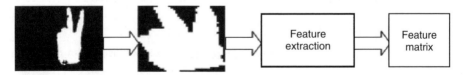

Figure 4.7 Basic steps of feature extraction.

shape. Both continuous and discrete moment features are used by the researchers in static hand gesture recognition. Discrete moment features are more popular than the continuous moment features because continuous moments result in some discretization error, which affects the gesture recognition accuracy. Moment features can also be classified as orthogonal and nonorthogonal features. Here, in this work, we have used both orthogonal and nonorthogonal moment features.

For an image $I(x, y)$ of size $A \times B$, a geometric moment [2], Krawtchouk moment [2], and Tchebichef moment [8] are explained here:

1. **Geometric moment:** The $(a + b)$ order geometric moment is given as:

$$GM_{ab} = \sum_{x=0}^{A-1} \sum_{y=o}^{B-1} x^a y^b I(x, y) \tag{4.14}$$

where $a = 0, 1, 2, 3 \dots A$, $b = 0, 1, 2 \dots B$.
The corresponding central moment of order $(a + b)$ is defined as:

$$\mu_{ab} = \sum_{x=0}^{A-1} \sum_{y=0}^{B-1} (x - \bar{x})^a (y - \bar{y})^b I(x, y) \tag{4.15}$$

where $\bar{x} = GM_{10}/GM_{00}$ and $\bar{y} = GM_{01}/GM_{00}$.
Now, the normalized central moment is defined as:

$$\eta_{ab} = \mu_{ab}/\mu_{00}^\gamma \tag{4.16}$$

where γ is the scaling normalization factor and is given by $\gamma = (a + b)/2 + 1$.

2. **Krawtchouk moment:** Krawtchouk moments are discrete orthogonal moments derived from the Krawtchouk polynomials. The nth-order Krawtchouk polynomial at a discrete point x with $(0 < p_1 < 1, 0 < p_2 < 1)$ is defined in terms of hyper geometric function [2] as:

$$k_1(x, p, n) = F(-n, -x, -N, 1/n) \tag{4.17}$$

By definition,

$$F(a, b, c, z) = \sum_{v=0}^{n} a_v b_v z^v / (c_v v!) \tag{4.18}$$

where a_v is the Pochhammer function and is given by the following function:

$$a_v = a(a - 1) \dots (a + v - 1) \tag{4.19}$$

From the Krawtchouk polynomials, we can get a weight function that is used to normalize feature values. The weight function of the Krawtchouk polynomial is given as:

$$W(x, p, n) = \frac{N!}{(x!(N - x)!)} p_1^x (1 - p_1)^{N-x} \tag{4.20}$$

Krawtchouk polynomials are normalized by dividing with weight function as given here:

$$k_1^1(x, p_1, N) = k_1(x, p_1, N) \sqrt{W(x, p_1, N)/\rho(x, p_1, N)} \tag{4.21}$$

and

$$\rho(n, p_1, N) = \frac{(-1)^n (1 - p_1)^n n!}{p_1^n (-N)_n} \tag{4.22}$$

The constants p_1 and p_2 are called shift parameters. Normally, these are set to 0.5 to make the ROI centralized. The Krawtchouk moment of order (n+m) is given by the following equation:

$$Q_{nm} = \sum_{x=0}^{N-1} \sum_{y=0}^{M-1} I(x, y) k_1^1(x, p_1, n) k_2^1(y, p_2, m) \tag{4.23}$$

3. **Tchebichef moment:** Tchebichef moments [8] are same as Krawtchouk moments, as they are also derived from orthogonal basis functions. The 1D Tchebichef polynomial at a discrete point x is defined as:

$$t_n(x) = (1 - N)_n 3_{F_2}(-n, -x, 1 + n, 1, 1 - N, 1) \tag{4.24}$$

where $3_{F_2}(.)$ is a hypergeometric function and is defined by the following equation:

$$3_{F_2}(a_1, a_2, a_3; b_1, b_2; z) = \frac{(a_1)_v (a_2)_v (a_3)_v z^v}{(b_1)_v (b_2)_v v!} \tag{4.25}$$

where a_v is the Pochhammer function and is given as:

$$a_v = a(a - 1) \ldots (a + v - 1) \tag{4.26}$$

The Tchebichef moment of order $(n + m)$ is given as:

$$T_{nm} = \frac{1}{\sqrt{\rho(n, N)\rho(m, M)}} \sum_{x=0}^{N-1} \sum_{y=0}^{M-1} t_n(x) t_m(y) I(x, y) \tag{4.27}$$

where $\rho(n, N)$ is the normalization constant and is given by the following equation:

$$\rho(n, N) = (2n)! \frac{(N + n)!}{(2n + 1)!(N - n - 1)!} \tag{4.28}$$

4.3.2 Contour-Based Features

The shape of contours of static hand gestures is a very important property, which could be used to distinguish hand gestures. In this section, we will discuss mainly two contour-based features, contour signature and localized contour sequence. Both of these features are popularly used by the researchers in static hand gesture recognition.

1. **Localized contour sequence:** The LCS has been confirmed by various researchers as a very efficient representation of contours [6]. As LCS is not bounded by the shape complexity of the gestures, many researchers have used it as an efficient feature descriptor of static hand gesture recognition. The Canny edge detection algorithm is used for edge detection of hand gesture images in LCS feature descriptors.

LCS uses a contour-tracking algorithm in either a clockwise or anticlockwise direction. The contour pixels are numbered sequentially from the topmost left pixel. A contour sequence is calculated on each pixel (x_i, y_i) by the following algorithm:

A. Create a window W.

B. Find out the intersection points (a_i, b_i) of the window (W) and the contour. Find the chord (C) connecting the bottom-left and bottom-right points of the contour.

C. Calculate the ith sample $h(i)$ by computing the perpendicular Eucledian distance from the top-left point of the contour to the chord.

D. For N number of contour points, we will get N number of contour sequence $h(i)$.

F. Find the average number of boundary pixels in contours. As the number of contour points are different, $h(i)$ should be sampled to the average value.

The LCS feature set is position and size invariant, and for that reason it is a robust feature descriptor in static hand gesture recognition.

2. **Contour signature:** Contour signature is another powerful feature descriptor used by the researchers [24]. Contour signature finds out the amplitude and directions of the boundary points from the contour centroids. A contour signature is calculated by the following algorithm.

A. Find out the edges of the gesture images by the Canny edge detection method.

B. Find out the centroids (C_x, C_y) of the contours.

C. Find the amplitude (A_i) and phase (θ_i) of the boundary points (x_i, y_i) from the centroids. Let C_x and C_x be the centroids of the contour of the hand gesture image. Let the ith boundary point be (x_i, y_i). Then the amplitude (A_i) of the i_{th} boundary point is given by:

$$A_i = \sqrt{(x_i^2 + y_i^2)} \qquad (4.29)$$

The phase (θ_i) is given by the following equation:

$$\theta_i = tan^{-1}\left(\frac{y_i}{x_i}\right) \qquad (4.30)$$

D. Find out the feature vector F_i by concatenating A_i and θ_i.

$$F_i = [A_i\theta_i] \qquad (4.31)$$

E. Contours of different gestures have different numbers of boundary pixels. So, the length of feature vector F_i also varies with images. To overcome this problem, all the feature vectors are sampled to a value N, where N is the average value of feature lengths.

Contour signature is affected by shape complexity and contour noises. For that reason, in user-independent gesture recognition, Contour signature doesn't show satisfactory gesture recognition accuracy.

4.4 State of the Art of Static Hand Gesture Recognition Techniques

In user-independent gesture recognition, as the gesture images of different users are used for training and testing purposes, some misclassification occurs because of different hand shapes, aspect ratios, and styles of gesticulation by the users. In this section, we will introduce some feature enhancement techniques to improve user-independent gesture recognition.

4.4.1 Zoning Methods

A zoning method (Z_M) [15] can be considered as a partition of feature set (F) into M subset, named zones $(Z_M) = [z_1 z_2 \ldots z_M]$, where $z_1 z_2 \ldots z_M$ are the zones or smaller sub-spaces of feature vectors, each one providing zonal information of patterns. The main goal of zoning is to extract useful information from the subregions of the images to recognize similar-shape images. Zoning topologies can be classified into two main categories: static [17] and adaptive [28]. Static topologies are designed without using *a priori* information on feature distribution in pattern classes. Conversely, adaptive zoning topologies can be considered as the results of the optimization procedure for zoning design. As the adaptive zoning topologies use *a priori* information on feature distribution and classification, these methods are not real-time efficient. Hence, various static zoning topologies are discussed in this chapter. The taxonomy of zoning topologies is shown in Figure 4.8.

Static zoning topologies can be classified into two broad classes: uniform zoning [29] and nonuniform zoning [30]. In uniform zoning, gesture images are divided into grids of the same sizes. We have used 2×2, 3×2, 2×3, 1×4, and 4×1 regular grids for uniform zoning. In nonuniform zoning, gesture images are divided into unequal grids based on some partition criteria. We have partitioned gesture images into zones by centroid-based partition. We have used 4×1 and 5×1 grids in nonuniform zoning.

4.4.2 F-Ratio-Based Weighted Feature Extraction

F-ratio-based weighted feature extraction [14] is a very useful technique for similar-shape gesture recognition. To improve the user-independent gesture recognition, F-ratio-based weighted feature extraction will be discussed in this section. F-ratio is a statistical measure that is defined by the ratio of the between-class variance and within-class variance. The F-ratio is calculated statistically from the features of the similar-shape gestures, and it is multiplied with the feature vectors for feature enhancement. The weighted feature vector includes more useful information to discriminate the similar classes. The F-ratio (F_i) is defined by:

$$F_i = \frac{S_{bi}^2}{S_{wi}^2} \tag{4.32}$$

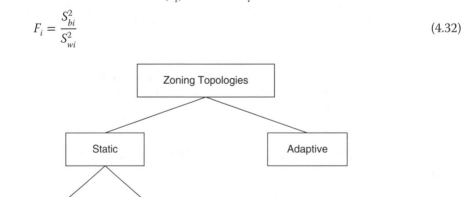

Figure 4.8 Zoning topology.

where S_{bi}^2 and S_{wi}^2 are between-class and within-class variance. They are defined by the following equation with class number l, class size L, and *a priori* probability $P(\omega_l)$ of class ω_l.

$$S_{wi}^2 = \sum_{l=1}^{L} P(\omega_l)E(x_i - m_{li})^2/\omega_l \tag{4.33}$$

$$S_{bi}^2 = \sum_{l=1}^{L} P(\omega_l)(m_{li} - m_{0i})^2 \tag{4.34}$$

where $m_{li} = E(x_i/\omega_l)$ and $m_{0i} = E(x_i) = \sum_{l=1}^{L} P(\omega_l)m_{li}$. Here, the function E calculates the expectation value of its argument.

Though the F-ratio-based weighted feature extraction technique results in a significant improvement of user-independent gesture recognition accuracy, it is not real-time efficient as the similar-shape gestures are determined manually.

4.4.3 Feature Fusion Techniques

In feature fusion, two different feature sets are combined by a proper rule or algorithm. Various feature fusion techniques are reported in literature [8, 10, 31]. In this section, we will discuss two feature fusion methods, MLE-based hidden feature fusion, and serial feature fusion and parallel feature fusion.

1. **MLE-based hidden feature fusion:** In this method, it is assumed that there must be a hidden feature (f_h) of two different feature sets f_1 and f_2. It is assumed that the original two features are obtained by two different transformations [31], as given here:

$$f_1 = Af_h + \epsilon_1, \quad f_2 = Bf_h + \epsilon_2 \tag{4.35}$$

where ϵ_1 and ϵ_2 are independent noises from Gaussian distribution $N(0, \sigma I)$. A and B are constants.
Let f_1 and f_2 be represented by:

$$f_1 = [f_1^1, f_1^2, f_1^3 \dots\dots f_1^n] \tag{4.36}$$

and

$$f_2 = [f_2^1, f_2^2, f_2^3 \dots\dots f_2^n] \tag{4.37}$$

Then the log likelihood for a hidden feature is given by:

$$F_h = [f_h^1, f_h^2, f_h^3 \dots\dots f_h^n] \tag{4.38}$$

The linear transformation of F_h, A, and B is given by:

$$L(F_h, a, b) = -\frac{1}{2\sigma^2} \sum_{i=1}^{n} \|f_1 - Af_h^1\|_2^2 - \frac{1}{2\sigma^2} \sum_{i=1}^{n} \|f_2 - Bf_h^1\|_2^2 + c \tag{4.39}$$

where c is constant. This equation is closely related to the problem of dictionary learning. So, we can say that it is a joint dictionary-learning problem in which A and B are the dictionaries of two features f_1 and f_1, respectively.

2. **Serial feature fusion and parallel feature fusion:** In [8], researchers have introduced two new feature fusion strategies, which enhance the classification accuracy significantly. Let X and Y be two features that consist of n and m numbers of feature vectors x_i and y_i. Here, $x_i \epsilon X$ and $y_i \epsilon Y$. Then the serially combined feature is defined as $F_s = [x_i \; y_i]$. The dimension of the resultant feature is $(n + m)$. In case of parallel feature fusion, the resultant feature is expressed by a supervector $F_p = [x_i + y_i]$. If the dimensions of X and Y are not the same, then the lower dimension vector is up-sampled to the dimension of an upper dimensional feature. Thus, the size of the resultant parallel feature becomes m if $m > n$ or n if $n > m$.

For an example, let $X = [x_1 \; x_2]$ and $Y = [y_1 \; y_2 \; y_3]$. Then, to find the parallel feature fusion of X and Y, we have to make X and Y the same length by up-sampling X to a length of 3. Now, $X = [x_1 \; x_2 \; x_3]$. Parallel combination is defined in a superplane as $F_p = [x_i + y_i]$.

Numerical unbalance is the main problem in feature-level fusion. Let a feature vector $X = [0.5 \; 0.9]$, and another is $Y = [15 \; 14]$. Then, after parallel combination, attributes of Y will be more than X. It means almost all the significance of X will be lost. To overcome this numerical unbalance condition, features have to be normalized first. Normalization of features is commonly done by dividing its maximum value: $\overline{X} = X/max(X)$ and $\overline{Y} = Y/max(Y)$.

Even after normalization, numerical imbalance may occur because of difference in feature sizes. To avoid this, the lower dimensional one is multiplied by a normalization constant. We have empirically selected this normalization constant as $v = n^2/m^2$ assuming $n > m$. Then the serial feature fusion and parallel fused feature vectors are given by:

$$F_s = [x \; vy] \tag{4.40}$$

and

$$F_p = [x_i + vy_i] \tag{4.41}$$

4.5 Results and Discussion

Extracting ROI from the captured hand gesture color images is a challenging task in static hand gesture images. We have used images captured in complex backgrounds, varring the illumination level, skin color complexion, and rotation angle. Most of the color space–based skin color segmentation algorithm fails to extract ROI from the complex background color images. Hence, we have emphasized rotation and illumination normalization in the preprocessing part. Our proposed skin color segmentation algorithm has shown robustness in varring the illumination level and rotation angle.

Different styles of gesticulation by the users and different hand shapes and aspect ratios are the most significant problems in user-independent static hand gesture recognition. So, to choose a proper feature descriptor that could overcome all the aforementioned problems is of paramount importance in user-independent static hand gesture recognition. In American Sign Language digits, gestures 7, 8, and 9 have almost similar shape and affect the gesture recognition accuracy seriously. Most of the moment and contour-based features misclassify these similar-shape gestures. To overcome this similar-shape gesture misclassification, F-ratio-based feature enhancement technique, serial feature fusion and parallel feature fusion, MLE-based hidden feature fusion, and zoning methods have been employed. Improvement in gesture recognition accuracy by these techniques will be discussed in this section.

4.5.1 Segmentation Result

1. **Database:** We have used two hand gesture databases. In case of the first database, a uniform black background is used behind the user to avoid background noises as shown in Figure 4.9. The second one is taken in a complex background as shown in Figure 4.10. In both the databases, the forearm region is separated by wrapping a black cloth. In the second database, the hand region is restricted to have maximum area compared to other regions.

   A Logitech c120 webcam has been used to capture the hand gesture images. The resolution of the grabbed image is 320×240 for both the databases. All images are taken in various angles and in different light conditions to make our gesture recognition algorithm rotation and illumination invariant. We have used the uniform background database only for semisupervised learning purposes. The second database is mainly used for testing and training purposes. The data set consists of 1500 gestures of 10 classes, with 15 samples from each class of 10 users. We have conducted two experiments to evaluate the gesture recognition performance by our proposed feature descriptors. In the first experiment, gesture recognition performance is evaluated in the user-dependent condition. 1000 gestures of 10 classes for all the users are used for training purposes, and 500 gestures of 10 classes for all the users are used for testing purposes. In the second experiment, the data set is equally divided into training and testing data sets of 750 gestures of 10 classes for five different users to make the system user independent.

2. **Experimental results:** We have used two different skin color segmentation methods: YCbCr color model–based skin color segmentation, and K-means clustering and Mahalanobish distance–based skin color segmentation. Skin color region detection

Figure 4.9 Uniform Background Database: (a) digit 1, (b) digit 2, and (c) digit 3.

Figure 4.10 Complex Background Database: (a) digit 0, (b) digit 1, and (c) digit 2.

in complex backgrounds and varying illumination conditions is a difficult task for researchers. Most of the researchers have employed color space–based skin color segmentation for static hand gesture recognition. Color space–based skin color segmentation methods are not robust for skin color detection, because in varying illumination conditions and in complex backgrounds, threshold values for the color space models also vary. The result of YCbCr color model–based segmentation is shown in Figure 4.11. It can be concluded from the segmentation result that the extracted foreground is quite different from the desired response and may result in misclassification in gesture recognition by any classifier.

We have proposed a skin color detection process using semisupervised learning based on K-means clustering and Mahalanobish distance, which has shown robustness in varying illumination conditions and complex backgrounds as shown in Figure 4.12. Our proposed illumination normalization technique has also shown effectiveness in varying illumination conditions.

4.5.2 Feature Extraction Result

We have used orthogonal and nonorthogonal moment features as the shape-based feature descriptors and contour signature and localized contour sequence (LCS) as contour-based feature descriptors. Moment features are calculated on the pixel values of images. For that reason, we have resized and cropped our original images into a 40×40 size so that the hand region becomes the maximum region. We have empirically selected the order of geometric moment as 49 (n=m=7), and the Krawtchouk, and Tchebichef moments as 64 (m=8=n=8). We have empirically selected a translation parameter p as 0.5 of the Krawtchouk moment to make the feature scale invariant. On the other hand, the geometric moment has an order of 49, so the size of the feature is 49. contour-based features like contour signature and LCS are calculated on boundary pixels of images.

Figure 4.11 (a) Input gesture, (b) YCbCr segmented image, and (c) extracted hand region after morphological operation.

Figure 4.12 (a) input gesture, (b) segmented image by our proposed method, and (c) extracted hand region after morphological operation.

1. **Classification performance of the proposed feature descriptors:** Both probabilistic classifiers like the hidden Markov model (HMM) [32] and Bayesian belief network [33] and statistical classifiers like support vector machines (SVMs) [34] have been used for static hand gesture classification. In this work, we have implemented a feedforward multilayer ANN classifier. Performance of the proposed feature descriptors in terms of classification accuracy in both user-dependent and user-independent conditions are shown in Table 4.1. It shows that the Krawtchouk moment is the best in terms of classification accuracy. In user-independent conditions, neither these moment features nor the contour-based features have shown satisfactory classification accuracy. The geometric moment shows the worst performance in both user-dependent and user-independent conditions. Though both the contour-based features have shown significant gesture recognition accuracy in the user-dependent condition, they are unable to perform well in user-independent gesture recognition. The Krawtchouk moment, Tchebichef moment, contour signature, and LCS features have shown more than 95% accuracy in the user-dependent condition. In the user-independent condition, only the Krawtchouk moment has shown significant gesture recognition accuracy, as shown in Table 4.1.

2. **Why the Krawtchouk moment is performing better:** To classify similar-shape gestures as shown in Figure 4.13, the localized properties of images should be included with the global feature descriptor. The appropriate selection of shifting parameters p_1 and p_2 of the Krawtchouk moment feature descriptor enables to extract significant local properties along with global features. The parameters p_1 and p_2 shift the support of the moment horizontally and vertically, respectively. In the globalized Krawtchouk moment, features p_1 and p_2 are set to 0.5, so that the support of the moment feature is centralized. The confusion matrix of the Krawtchouk moment is shown in Table 4.2. From Table 4.2, it is clear that most of the misclassification occurs for the similar-shape gestures 7, 8, and 9. To classify similar-shape gestures 7, 8, and 9 as shown in Figure 4.12, some localized property of Krawtchouk moments should be introduced, which has been done by introducing zoning methods.

3. **Why other features are not performing well:** In the cases of the geometric and Tchebichef moments, mismatches occur more than with Krawtchouk moments, as shown in Table 4.3 and Table 4.4. This is because the geometric moment is a global feature, and it only represents the statistical attributes of the shape. On the other

Table 4.1 User-dependent and user-independent classification results

Features	Classification accuracy (user-dependent condition) (%)	Classification accuracy (user-independent condition) (%)
Krawtchouk moment	99.27	91.53
Tchebichef moment	95.56	82.67
Geometric moment	88.7	76.2
Contour signature	98.5	78.8
LCS	98.8	81.2

Table 4.2 Confusion matrix of the Krawtchouk moment for user-independent condition

Class	0	1	2	3	4	5	6	7	8	9
0	138	0	10	0	1	1	0	0	0	0
1	0	150	0	0	0	0	0	0	0	0
2	0	0	150	0	0	0	0	0	0	0
3	0	0	0	132	0	1	4	0	0	13
4	0	0	0	0	150	0	0	0	0	0
5	14	1	0	6	12	112	0	0	0	5
6	0	0	0	1	0	0	148	0	1	0
7	0	0	0	0	0	0	0	132	1	17
8	0	0	1	0	1	0	1	10	133	4
9	0	0	0	0	4	1	4	0	13	128

Table 4.3 Confusion matrix of the Tchebichef moment for the user-independent condition

Class	0	1	2	3	4	5	6	7	8	9
0	131	0	6	2	0	0	2	0	1	8
1	0	150	0	0	0	0	0	0	0	0
2	0	1	147	0	0	0	0	0	1	1
3	2	0	0	145	0	3	0	0	0	0
4	0	0	1	0	120	0	15	14	0	0
5	0	0	0	20	0	124	0	0	1	5
6	0	0	1	0	10	0	136	0	3	0
7	0	9	6	0	3	0	12	106	10	4
8	0	2	15	0	23	0	7	30	58	15
9	7	0	1	0	0	1	1	2	15	123

hand, although the Tchebichef moment is orthogonal, it doesn't show satisfactory classification performance in the user-independent condition. In the case of the geometric moment, gesture 7 is misclassified as 8 and 9; gesture 8 is misclassified as 6, 7, and 9; and gesture 9 is misclassified as 6 and 7, as shown in Table 4.4. In the case of the Tchebichef moment, gesture 8 has been misclassified as 7 and 9, and gesture 9 has been misclassified as 7 and 8, as shown in Table 4.3. In the case of the Krawtchouk moment, mismatch occurs less than with the geometric and Tchebichef moments, as shown in Table 4.2.

4. **Classification results of feature enhancement, zoning, and feature fusion techniques:** To overcome this similar-shape gesture mismatch problem, we have introduced some feature enhancement techniques, feature fusion and zoning methods as described in Section 4.4. Here we will discuss the improvements in gesture recognition by these feature enhancement methods. Table 4.5 has shown the classification results of Krawtchouk moment features by zoning methods. By using 4×1 nonuniform zoning methods, gesture recognition accuracy has been improved significantly,

Table 4.4 Confusion matrix of the geometric moment for the user-independent condition

Class	0	1	2	3	4	5	6	7	8	9
0	150	0	0	0	0	0	0	0	0	0
1	0	148	0	2	0	0	0	0	0	0
2	0	0	150	0	0	0	0	0	0	0
3	3	2	0	142	0	3	0	0	0	0
4	10	10	0	0	112	1	10	12	5	0
5	20	0	20	0	0	93	7	2	8	0
6	6	0	10	0	3	0	124	0	6	1
7	1	0	7	0	10	14	0	79	7	32
8	0	0	6	0	6	8	21	42	45	22
9	1	0	5	0	0	2	15	27	0	100

Table 4.5 Classification result of Krawtchouk moment zonal features

Zoning methods	Accuracy (%) for Krawtchouk moment
2 × 2 uniform grid	92.5
3 × 2 uniform grid	93.1
2 × 3 uniform grid	92.8
4 × 1 uniform grid	94.7
5 × 1 uniform grid	94.33
4 × 1 nonuniform grid	95.8
5 × 1 nonuniform grid	95.4

and it shows 95.8% classification accuracy for Krawtchouk moments. Table 4.6 has shown classification results of various feature-level fusion and MLE-based hidden feature fusion of moment features. Parallel fusion of Krawtchouk and Tchebichef moments has shown the best gesture recognition accuracy of 95.33%, and the hidden feature of Krawtchouk and Tchebichef moments has shown the best gesture recognition accuracy of 94.5% among the hidden features. Gesture classification results by *F*-ratio-based enhanced features have been discussed in Table 4.7. From Table 4.5 to Table 4.7, it is clear that our proposed methods have shown a great improvement in user-independent gesture classification accuracy. Among the proposed methods, 41 centroid-based nonuniform Krawtchouk zonal features have shown the best gesture recognition accuracy of 95.8%.

4.6 Conclusion

In this chapter, we have compared various static hand gesture image segmentation and feature extraction techniques. The problems associated with real-time

Figure 4.13 ASL similar-shape gestures: (a) 7, (b) 8, and (c) 9.

Table 4.6 Classification results of serial, parallel, and MLE-based hidden feature fusion

Fused features	Accuracy (%)
KM+TM serial fused feature	93.53
KM+TM parallel fused feature	95.33
KM+GM serial fused feature	94.93
KM+GM parallel fused feature	94.2
MLE-based hidden features	
Krawtchouk+Tchebichef	94.5
Krawtchouk+Geometric	92.7
Geometric+Tchebichef	84.7

Table 4.7 Classification result of *F*-ratio-based enhanced features

F-ratio-based enhanced features	Accuracy (%)
Krawtchouk moment	91.88
Tchebichef moment	84.56
Geometric moment	79.45
Contour signature	80.5
LCS	83.5

user-independent static hand gesture recognition, like illumination variance, rotation variance, and similar-shape gesture misclassification, have been addressed. In varying illumination conditions and skin color complexions, threshold values for skin color regions also change, which results in deterioration of gesture segmentation. To overcome this problem, we have proposed a skin color segmentation method based on K-means clustering and Mahalanobish distance, and it shown a great improvement in static hand gesture image segmentation. We have also proposed the following shape- and contour-based feature extraction techniques: geometric moments, Krawtchouk moments, Tchebichef moments, contour signature, and LCS. Their performance in both user-independent and user-dependent gesture recognition has been reported. Due to different gesticulation styles of users and similar-shape gesture images, these well-known feature descriptors exhibit some misclassification in user-independent gesture recognition. To overcome this similar-shape gesture mismatch problem, we have proposed the following feature enhancement techniques: *F*-ratio-based feature enhancement, serial feature fusion and parallel feature fusion, MLE-based hidden feature fusion, and zoning methods. Experimental results show that centroid partition-based 4×1 nonuniform zoning of Krawtchouk moments has the highest gesture classification accuracy of 95.8%.

4.6.1 Future Work

Our database contains images with small rotation and illumination variations. In future, we want to perform gesture recognition with more illumination and rotation angle variations so that we can test the performance of proposed algorithms in real-time situations. In future, we will test with more variation in skin color complexion data. This work has been tested only for American Sign Language digits. We would like to test proposed gesture recognition algorithms with other data sets, such as ASL alphabets, to verify the capability of similar-shape gesture classification. In real-time scenarios, hand gestures might be occluded with face or any other body parts. In future, we will work on such occluded hand gesture images. We are also interested in using gesture recognition for video surveillance applications such as abnormal activity recognition.

Acknowledgment

The authors would like to acknowledge the support of NIT Rourkela for capturing the American Sign Language static hand gesture digit database.

References

1 Himanshu Rai, Maheshkumar H Kolekar, Neelabh Keshav, and JK Mukherjee. Trajectory based unusual human movement identification for video surveillance system. *Progress in Systems Engineering*, pages 789–794, 2015.

2 S Padam Priyal and Prabin Kumar Bora. A robust static hand gesture recognition system using geometry based normalizations and Krawtchouk moments. *Pattern Recognition*, **46** (8): 2202–2219, 2013.

3 Ruiduo Yang, Sudeep Sarkar, and Barbara Loeding. Handling movement epenthesis and hand segmentation ambiguities in continuous sign language recognition using nested dynamic programming. *IEEE Transactions on Pattern Analysis and Machine Intelligence*, **32** (3): 462–477, 2010.

4 David J Sturman and David Zeltzer. A survey of glove-based input. *Computer Graphics and Applications*, **14** (1): 30–39, 1994.

5 Rafael C Gonzalez, Richard E Woods, and Steven L Eddins. Digital image processing using MATLAB. 2002.

6 Dipak Kumar Ghosh and Samit Ari. A static hand gesture recognition algorithm using k-mean based radial basis function neural network. In *International Conference on Information, Communications and Signal Processing*, pages 1–5–. IEEE, 2011.

7 Sauvik Das Gupta, Souvik Kundu, Rick Pandey, Rahul Ghosh, Rajesh Bag, and Abhishek Mallik. Hand gesture recognition and classification by discriminant and principal component analysis using machine learning techniques. *Hand*, **1** (9), 2012.

8 Subhamoy Chatterjee, Dipak Kumar Ghosh, and Samit Ari. Static hand gesture recognition based on fusion of moments. In *Intelligent Computing, Communication and Devices*, pages 429–434. Springer, 2015.

9 Deng-Yuan Huang, Wu-Chih Hu, and Sung-Hsiang Chang. Gabor filter-based hand-pose angle estimation for hand gesture recognition under varying illumination. *Expert Systems with Applications*, **38** (5): 6031–6042, 2011.

10 Liu Yun, Zhang Lifeng, and Zhang Shujun. A hand gesture recognition method based on multi-feature fusion and template matching. *Procedia Engineering*, **29**: 1678–1684, 2012.

11 Dipak Kumar Ghosh and Samit Ari. On an algorithm for vision-based hand gesture recognition. *Signal, Image and Video Processing*, pages 1–8, 2015.

12 Kinjal Basu, Radhika Nangia, and Umapada Pal. Recognition of similar shaped handwritten characters using logistic regression. In *IAPR International Workshop on Document Analysis Systems*, pages 200–204. IEEE, 2012.

13 Sukalpa Chanda, Umapada Pal, and Katrin Franke. Similar shaped part-based character recognition using g-surf. In *International Conference on Hybrid Intelligent Systems*, pages 179–184. IEEE, 2012.

14 Tetsushi Wakabayashi, Umapada Pal, Fumitaka Kimura, and Yasuji Miyake. F-ratio based weighted feature extraction for similar shape character recognition. In *International Conference on Document Analysis and Recognition*, pages 196–200. IEEE, 2009.

15 Donato Impedovo and Giuseppe Pirlo. Zoning methods for handwritten character recognition: a survey. *Pattern Recognition*, **47** (3): 969–981, 2014.

16 ZC Li, HJ Li, CY Suen, HQ Wang, and SY Liao. Recognition of handwritten characters by parts with multiple orientations. *Mathematical and computer modelling*, **35** (3): 441–479, 2002.

17 Glenn Baptista and KM Kulkarni. A high accuracy algorithm for recognition of handwritten numerals. *Pattern Recognition*, **21** (4): 287–291, 1988.

18 MP Devi, T Latha, and C Helen Sulochana. Iterative thresholding based image segmentation using 2D improved Otsu algorithm. In *Global Conference on Communication Technologies*, pages 145–149. IEEE, 2015.

19 Khamar Basha Shaik, P Ganesan, V Kalist, BS Sathish, and J Merlin Mary Jenitha. Comparative study of skin color detection and segmentation in HSV and YCbCr color space. *Procedia Computer Science*, **57**: 41–48, 2015.

20 Harpreet Kaur Saini and Onkar Chand. Skin segmentation using RGB color model and implementation of switching conditions. *Skin*, **3** (1): 1781–1787, 2013.

21 Xiaolong Teng, Bian Wu, Weiwei Yu, and Chongqing Liu. A hand gesture recognition system based on local linear embedding. *Journal of Visual Languages & Computing*, **16** (5): 442–454, 2005.

22 P Ganesan, V Rajini, BS Sathish, and Khamar Basha Shaik. HSV color space based segmentation of region of interest in satellite images. In *International Conference on Control, Instrumentation, Communication and Computational Technologies*, pages 101–105. IEEE, 2014.

23 Zaher Hamid Al-Tairi, Rahmita Wirza OK Rahmat, M Iqbal Saripan, and Puteri Suhaiza Sulaiman. Skin segmentation using YUV and RGB color spaces. *JIPS*, **10** (2): 283–299, 2014.

24 P Peixoto, J Goncalves, and H Araujo. Real-time gesture recognition system based on contour signatures. In *International Conference on Pattern Recognition*, volume 1, pages 447–450. IEEE, 2002.

25 Maheshkumar H. Kolekar, Somnath Sengupta, and Gunasekaran Seetharaman. Semantic concept mining based on hierarchical event detection for soccer video indexing. *Journal of Multimedia*, pages 298–312, 2009.

26 Suli Zhang and Xin Pan. A novel text classification based on Mahalanobis distance. In *International Conference on Computer Research and Development*, volume 3, pages 156–158, 2011.

27 Simon Haykin. *Neural network*. MacMillan, New York, 1994.

28 Simone BK Aires, Cinthia Oa Freitas, Flávio Bortolozzi, and Robert Sabourin. Perceptual zoning for handwritten character recognition. In *Conference of the International Graphonomics Society, Włochy*, 2005.

29 Brijesh Verma, Jenny Lu, Moumita Ghosh, and Ranadhir Ghosh. A feature extraction technique for online handwriting recognition. In *International Joint Conference on Neural Networks*, volume 2, pages 1337–1341. IEEE, 2004.

30 Jaehwa Park, Venu Govindaraju, and Sargur N Srihari. OCR in a hierarchical feature space. *IEEE Transactions on Pattern Analysis and Machine Intelligence*, **22** (4): 400–407, 2000.

31 Jun Cheng, Can Xie, Wei Bian, and Dacheng Tao. Feature fusion for 3D hand gesture recognition by learning a shared hidden space. *Pattern Recognition Letters*, **33** (4): 476–484, 2012.

32 Maheshkumar H Kolekar and S Sengupta. Hidden Markov model based video indexing with discrete cosine transform as a likelihood function. pages 157–159, 2004.

33 Maheshkumar H Kolekar and S Sengupta. Bayesian network-based customized highlight generation for broadcast soccer videos. *IEEE Transactions on Broadcasting*, **61** (2): 195–209, 2015.

34 Maheshkumar H Kolekar and D. P. Dash. A nonlinear feature based epileptic seizure detection using least square support vector machine classifier. pages 1–6, 2015.

5

SVM Combination for an Enhanced Prediction of Writers' Soft Biometrics

Nesrine Bouadjenek, Hassiba Nemmour, and Youcef Chibani

Laboratoire d'Ingénierie des Systèmes Intelligents et Communicants (LISIC), Faculty of Electronics and Computer Sciences, University of Sciences and Technology Houari Boumediene (USTHB), Algiers, Algeria

5.1   Introduction

Nowadays, handwriting recognition has been addressed by many researchers for different purposes such as digit recognition, indexation in historical documents, writer's identification, or soft-biometrics prediction. Any behavioral or physical characteristic that gives some information concerning the identity of someone, but does not offer decent proof to exactly verify the identity, may be referred to as a soft-biometric attribute, such as ethnicity, handedness, gender, age range, height, weight, eye color, scars, marks, tattoos, and so on [1]. In forensic analysis, incorporating soft-biometric attributes could enhance the identification process provided by traditional biometric identifiers like fingerprint, face, iris, and voice. Over the past years, most of the published studies in the literature have dealt with predictions of soft biometrics from face images [2–4] or from speech signals [5–8]. From a medical perspective, numerous studies tried to clarify how gender can manage human behavior. Specially, the gender impact has been proved in Alzheimer's disease [9], asthma in childhood and adolescence [10], mental health [11], as well as crimes and violence [12, 13]. Thus, researchers in handwriting recognition were confronted with a simple question: are gender and other soft biometrics affecting the handwriting style? Beech and Mackintosh showed that prenatal sex hormones can affect a woman's handwriting when investigating the relationship between sex hormones and the handwriting style in [14]. In [15], authors investigated gender and age influence in handwriting performance in children and adolescents, where the results showed that in the first grades of primary school, females have better performances than males. In some earlier psychological investigations, differences between male's and female's handwriting were inspected [16, 17]. On the other hand, handedness prevalence was investigated in various studies, among them its impact on deaf individuals [18], the relationship between handedness and implicit or explicit self-esteem [19], or highlighting the relationship between handedness and language dominance [20, 21]. As to age influence over the handwriting performance, it was inspected in [22–25]; significant differences were found between age ranges.

Hybrid Intelligence for Image Analysis and Understanding, First Edition.
Edited by Siddhartha Bhattacharyya, Indrajit Pan, Anirban Mukherjee, and Paramartha Dutta.
© 2017 John Wiley & Sons Ltd. Published 2017 by John Wiley & Sons Ltd.
Companion Website: www.wiley.com/go/bhattacharyya/hybridintelligence

Automatic soft biometrics prediction formulates a modern topic of interest in the handwritten document analysis sphere. However, only a few studies are reported in the literature that deal with gender, handedness, age range, and nationality prediction. The first work was introduced by Cha *et al.* [26] in 2001. Thereafter, some works have followed.

A prediction system consists of two essential points, which are feature generation and classification where it is important to wisely choose valuable methods in each step to attain good performance. In earlier works on soft biometrics prediction, diverse classifiers like SVM, neural networks, and random forests were utilized, whereas the feature generation step was based on standard curvature, direction, and edge features. In most handwriting recognition tasks, SVMs seem to be the rational choice as they usually defeat other classifiers, like neural networks and hidden Markov models (HMM) [27–30]. Indeed, SVMs use the spirit of the structural risk minimization, which answers two major issues of statistical learning theory: overfitting and controlling the model complexity [31]. Also, their training process can easily handle large amounts of data without claiming for dimensionality reduction even when training examples contain errors. Besides, in our earlier works, gender, handedness, and age range prediction findings revealed in [32–34] affirm that by using one SVM classifier associated to a proper feature, the combination of several systems is beaten if they operate with powerless descriptors. Hence, a genuine approach to produce an efficient prediction is to ally robust features and SVMs.

In this work, three prediction systems are developed using SVM classifiers associated to various gradient and textural features: rotation-invariant uniform local binary patterns, a histogram of oriented gradients (HOG), and gradient local binary patterns (GLBPs). Since distinctive features profit from opposed aspects of characterization, we investigate classifier combination in order to improve the prediction accuracy. In this respect, the fuzzy integral, which has enjoyed strong success in several applications of land cover classification [35], digit recognition [36], face recognition [37], as well as combining document object locators [38], is used to produce a robust soft biometrics prediction. Presently, Sugeno's fuzzy integral and its modified form, fuzzy min-max, are used. Two standard Arabic and English data sets are used to prove the effectiveness of using a fuzzy integral operator for improving soft biometrics prediction.

The rest of this chapter is organized as follows: Section 5.2 presents the related background. Section 5.3 introduces the proposed system and methodologies. Experiments are presented and discussed in Section 5.4 and Section 5.5, respectively. The main conclusions are given in Section 5.6.

5.2   Soft Biometrics and Handwriting Over Time

After a certain meditation on the impact of soft biometrics on handwriting, researchers began to investigate the automatization process of predicting gender, handedness, age range, or ethnicity from handwritten text. The first work on writers' soft biometrics predictions was published by Cha *et al.* [26]. The aim was to classify the US population into various sub categories, such as white/male/age group 15–24 and white/female/age group 45–64, based on handwritten images of uppercase letters. Various features such as pen pressure, writing movement, and stroke formation were used with artificial neural networks. Experiments reveal a performance of 70.2% and 59.5% for gender and handedness prediction, respectively. Next, boosting techniques were employed to achieve higher performance where accuracies reached 77.5%, 86.6%, and 74.4% for gender, age, and handedness classification [39]. Thereafter, a research group on computer vision and artificial intelligence at Bern University developed the IAM handwriting data set, which is devoted to writer identification as well as gender and handedness prediction [40, 41]. Authors utilized a set of 29 on-line and off-line features through a combination of support vector machines and Gaussian mixture models (GMMs). Off-line features are based on conventional structural traits such as ascenders, descenders, and the number of points above or below the corpus line. Reported gender prediction accuracy is about 67.57% after classifier combination. The handedness prediction using GMM classifier is about 84.66%. In [32–34], similar gender and handedness prediction experiments were conducted on the IAM dataset by using more effective off-line descriptors, which were pixel density, pixel distribution, local binary patterns (LBPs), HOG, and GLBPs associated to SVM classifiers; the prediction accuracies were about 76% and 100%, respectively. Later, authors proposed a fuzzy min-max algorithm to combine SVM outputs trained with pixel density, pixel distribution, and GLBP to predict writers' gender, handedness, and age range [42]. The proposed algorithm considerably improved the prediction accuracy.

Furthermore, Al Maadeed *et al.* [43] employed a *k*-nearest neighbors (KNN) algorithm for handedness detection from off-line handwriting. A set of direction, curvature, tortuosity, and edge-based features was used. The experiments were conducted on the QUWI data set collected at Qatar University by asking 1017 writers to reproduce two texts in both English and Arabic languages [44]. The prediction accuracy was about 70%. Then, the same features were used for gender, age range, and nationality prediction with random forest and kernel discriminant analysis classifiers [45]. The best prediction scores were about 70% for gender prediction, 60.62% for age range prediction, and less than 50% for nationality classification. After that, in [46], authors proposed a fuzzy conceptual reduction approach for handedness prediction in order to select only the most charaterizing features among those mentioned earlier. The reported precision is over 83.43% with a reduction of 31.1% of the feature vector dimensionality. Not long after that, using the QUWI data set and another set of Arabic and French handwritten text, Siddiqi *et al.* [47] investigated gender classification using curvature, fractal, and textural features. The classification was based on neural networks and SVM classifiers. This same data set was used by Ibrahim *et al.* [48] for gender identification, comparing the performance of local and global features that are wavelet domain LBP, gradient features, and some features provided in the ICDAR 2013 competition on gender prediction. The classification task was achieved using SVM classifiers. The results showed that local gradient features outperform all other global features with a precision of 94.7%. Unfortunately, the QUWI data set is not publicly available to perform comparison.

The inspection of all previous works reveals that predicting writers' soft biometrics is a very complicated task, since the classification scores are commonly around 75%. In this work, individual predictors based on robust data features are developed and subsequently combined to provide effective prediction.

5.3   Soft Biometrics Prediction System

The aim of soft biometrics prediction systems is to classify writers into distinctive classes such as "male" or "female" for gender prediction, "left hand" or "right hand" for handedness prediction, and various age ranges in the case of age prediction. As to any handwriting recognition system, feature generation and classification steps are needed. As depicted in Figure 5.1, features are locally computed by applying equi-spaced grids over text images. For each descriptor, the different histogram cells are concatenated to form the feature vector of the whole image. Then, SVM outputs are combined through fuzzy integrals to aggregate a robust decision for soft biometrics prediction.

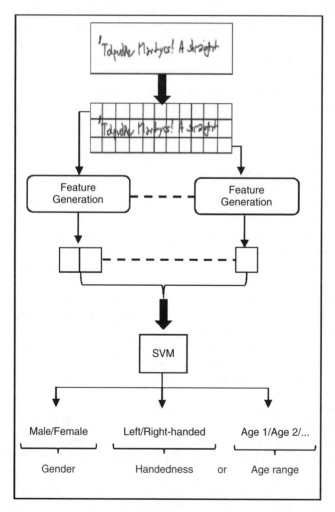

Figure 5.1 Soft-biometrics prediction system.

5.3.1 Feature Extraction

5.3.1.1 Local Binary Patterns

LBP was introduced by Ojala *et al.*, in [49] as a simple and efficient texture operator that has been used in image processing, and it has been applied successfully to face recognition [50], signature verification [51], and texture analysis [52]. An LBP operator is utilized to perform statistical and structural analysis of textural patterns [53]. They label the pixels of an image by thresholding the neighborhood of each pixel (x, y) according to the value of the central pixel taken as a threshold value, then consider the result as a binary number. Explicitly, the LBP code is obtained by comparing the gray-level value of the central pixel with neighboring gray levels. It takes 1 when the value of the central pixel is bigger than the neighboring pixel. Otherwise, it takes 0. The decimal value of the LBP code for P neighbors situated on a circle of radius R is computed as follows:

$$LBP_{P,R}(x, y) = \sum_{p=0}^{P} S(g_p - g_c)2^p \tag{5.1}$$

with:

$$S(l) = \begin{cases} 1 & l \geq 0 \\ 0 & l < 0 \end{cases}$$

g_c: gray value of the central pixel.
g_p: gray value of the pth neighbor.

Commonly, the neighbors are taken by considering a circular neighborhood; consequently, the *pth* neighbor does not belong to a pixel. Therefore, the adequate gray-level value is computed by interpolation as [49]:

$$g_p = I\left(x + Rsin\frac{2\pi p}{P}, \ y - Rcos\frac{2\pi p}{P}\right) \tag{5.2}$$

Then, image features are obtained by considering the LBP histogram, whose length is equal to 2^P.

Further extension to the original operator is the so-called rotation-invariant uniform LBP (LBP^{riu}), as reported in [53, 54], which allows invariance with respect to rotation. This extension was inspired by the fact that some binary patterns occur more commonly in texture images than others. A local binary pattern is called uniform if the binary pattern contains at most two bitwise transitions from 0 to 1 or vice versa when the bit pattern is considered circularly. It is defined as follows:

$$LBP_{P,R}^{riu2}(x, y) = \begin{cases} \displaystyle\sum_{p=0}^{P-1} S(g_p - g_c) & U(x, y) \leq 2 \\ P + 1 & \textit{otherwise} \end{cases} \tag{5.3}$$

with: $U(x, y) = \displaystyle\sum_{p=1}^{P} |s(g_p - g_c) - s(g_{p-1} - g_c)|$, and $g_p = g_0$. Moreover, the $LBP_{(P,R)}^{riu2}$ reduces the size of the LBP histogram to $(P + 2)$ [53].

5.3.1.2 Histogram of Oriented Gradients

The HOG feature was introduced for human detection by Dalal and Triggs [55]. HOG is meant to characterize the local object apperance and the shape of objects based on the distribution of local intensity gradients or edge directions. This descriptor has exhibited high performance in various applications, such as face recognition [56] and handwritten signature verification [57]. Concretely, HOG is computed by dividing the image into small connected regions called cells, and for the pixels within each cell, a histogram of gradient directions is accumulated, as reported in Algorithm 5.1. Note that orientations are selected according to the Freeman code. Besides, for each cell, the HOG histogram is normalized to scale in the range [0, 1].

Algorithm 5.1 HOG computation

Within each cell, the HOG feature is calculated by conforming to these steps:

1. For each pixel (x,y), horizontal and vertical gradient information are computed as:

$$g_x(x, y) = (x + 1, y) - (x - 1, y)$$
$$g_y(x, y) = (x, y + 1) - (x, y - 1) \tag{5.4}$$

2. The gradient magnitude and phase are obtained by using the following equations:

$$M(x, y) = \sqrt{g_x(x, y)^2 + g_y(x, y)^2}$$
$$\varphi(x, y) = arctan\left(\frac{g_x(x,y)}{g_y(x,y)}\right) \tag{5.5}$$

3. Establish the histogram of gradients by accumulating magnitudes according to their orientations.

Figure 5.2 presents an example of HOG descriptor calculation over a handwritten text image. This figure shows the HOG phase and magnitude as well as the concatenated histograms. In this example, the considered image was divided into 3×3 cells. From the histogram analysis, it is easy to see that HOG does not take the same magnitudes within the different image regions. This outcome indicates that a global histogram calculated over the entire image cannot highlight subtle handwriting characteristics.

5.3.1.3 Gradient Local Binary Patterns

GLBPs were recently introduced for human detection by Jiang *et al.* [58]. In a GLBP, gradient information and texture information are combined. Its principle idea consists of exploiting uniform LBPs to compute the histogram of oriented gradients. Presently, we investigate its efficiency for handwritten text characterization.

Thus, for a given cell, a GLBP table is settled as explained in Algorithm 5.2. First, the LBP code is calculated, then neighbor pixels with the same binary value stick together to get several "1" areas and "0" areas so that only uniform patterns are considered. Recall that, when the "1" area and "0" area appear only once, the pattern is called uniform. The size of the GLBP table is defined by all possible angle and width values. Precisely, there are eight possible Freeman directions or angle values, while the number "1" in the uniform patterns can vary from 1 to 7. This yields a 7×8 GLBP table in which gradient values are accumulated. At last, the GLBP table constituting the histogram is exploded into a vector of 56 elements. The flow of calculation is shown in Figure 5.3.

(a) Original handwriting text

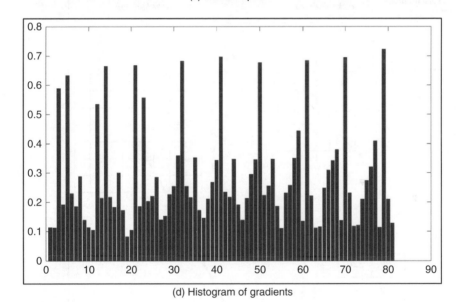

(b) Gradient magnitude

(c) Gradient phase

(d) Histogram of gradients

Figure 5.2 HOG feature calculated on a handwritten text image.

5.3.2 Classification

The soft biometrics prediction step is based on SVMs, which are machine learning tools used to solve real word classification problems such as image classification and text analysis. SVMs are binary classifiers, given a set of training examples, each labeled for belonging to one of two classes; the SVM training algorithm builds a model that assigns new examples into one class or the other, by creating an optimal linear separating hyperplane between two classes [59]. Explicitly, let $(k_n, c_n) \epsilon R^M \times \{\pm 1\}$ a set of training samples so that M corresponds to data dimension $\{n = 1, \ldots, N_c\}$, and N_c is the number of samples per a class c. SVM training selects the function f, which maximizes the margin

Algorithm 5.2 How to compute GLBP for a given cell

For each pixel:

1. Calculate the LBP code and consider only uniform patterns.
2. Generate the width and angle values such that:
   – The width corresponds to the number "1" in the LBP code.
   – The angle is the Freeman direction of the middle pixel within the "1" area in the LBP code.
3. Calculate the gradient value on the 1 to 0 (or 0 to 1) transitions in the LBP code using the value of the original pixel and the value of its neighbors.
4. Width and angle values are used for mapping the position within the GLBP table in which gradient values are accumulated.

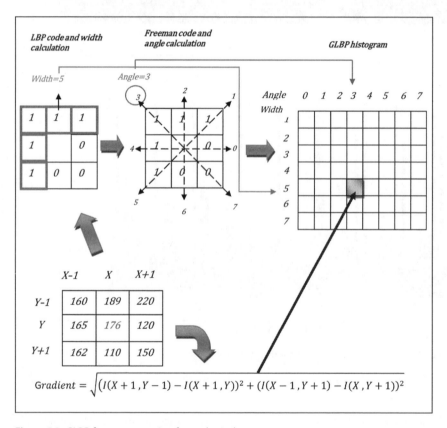

Figure 5.3 GLBP feature extraction for each pixel.

between the two classes by minimizing an upper bound on the generalization error [31]. Then, data are classified according to:

$$f(k) = sign\left(\sum_{i=1}^{SV} \beta_i q_i K(k_i, k) + b\right) \tag{5.6}$$

where b is the bias, q_i is the lass label, β_i is the Lagrange multipliers, and SV is the number of support vectors

As most data sets are not linearly separable in their given form, explicit mapping is needed to get linear learning algorithms to learn a nonlinear function or decision boundary. Kernels provide a way around this problem by providing a way to calculate the dot product of two vectors without explicitly mapping them to a specific higher dimension. There are four popular SVM kernels in the literature [31], namely, linear, polynomial, quadratic, and radial basis function (RBF), but the latter is the most promoted for pattern recognition. This kernel is described as follows:

$$K(k_i, k) = exp\left(-\frac{1}{2\sigma^2}\|k_i - k\|^2\right) \tag{5.7}$$

where σ: user-defined parameter.

5.3.3 Fuzzy Integrals–Based Combination Classifier

The prediction performance relies essentially on the feature extraction that assists SVM to discriminate between writers. Indeed, distinctive features profit from discriminative power, providing more complementary information to the other feature, which leads to eminent diversity between classifiers [51]. Hence, the combination of such predictors can improve the prediction accuracy. Recall that classifier combination was introduced for soft biometrics prediction in [41], by using classical max, min, and average rules. The investigation was conducted for gender prediction by combining GMMs. The prediction accuracy was improved from 64.25% to 67.57%. Presently, the intention is to combine different SVM decisions in order to aggregate more accurate predictions using different fuzzy integral operators. The main reason behind the use of fuzzy logic operators is that they allow modeling *a priori* knowledge about individual predictors' pertinence through fuzzy measure operators. Recall that fuzzy measure is a set of function A that certifies the following properties [35]:

$- g(\Phi) = 0$
$- g(Z) = 1$
$- g(Z_i) \leq g(Z_j)$ if $Z_i \subset Z_j$

5.3.3.1 g_λ Fuzzy Measure

The so-called g_λ fuzzy measure satisfies the following properties [35]: Let $Z = \{Z_i\}_{i=1:N}$ formulate the set of classifiers (SVM), while $g(Z_i)$ refers to their performances. Because of the nature of fuzzy measures, Sugeno stated that the fuzzy measure for the union of two classifiers does not correspond to the sum of individual fuzzy measures. To overcome this limitation, he proposed the λ fuzzy measure that expresses the degree of interaction between two classifiers Z_i and Z_j as:

$$g(Z_i \cup Z_j) = g(Z_i) + g(Z_j) + \lambda g(Z_i)g(Z_j) \tag{5.8}$$

For each class, λ is the unique nonzero root of equation (5.9) that belongs in the interval $[-1, \ldots, +\infty[$.

$$\lambda + 1 = \prod_{i=1}^{N}(1 + \lambda g(Z_i^{\pm})) \tag{5.9}$$

SVM outputs are converted into membership degrees in the two classes of interest by adjusting the membership model proposed in [35]. Specifically, the decision of each SVM Z_i is converted into membership degrees $h^+(Z_i)$ and $h^-(Z_i)$ in both positive and negative classes. Recall that, in theory, SVM outputs are defined by values that are at least greater than 1 for the positive class and at max equal -1 for the negative class. Thus, if the absolute value of the SVM decision is greater than 1, the sample is said to entirely belong to one of the classes, and its membership degree for the respective class equals 1 and that of the other class is 0. On the contrary, if the SVM decision belongs in the separating margin $]-1, +1[$, the decision is confused and the sample can belong to each class according to complementary membership degrees as shown in Algorithm 5.3. Then, the set of SVM is rearranged such that the following relation holds: $h(Z_1) \geq \dots \geq h(Z_N) \geq 0$. We obtain an ascending sequence of SVM $A_i = Z_1, \dots, Z_i$, whose fuzzy measures are constructed as:

$$g(A_1) = g(Z_1) \tag{5.10}$$

$$g(A_i) = g(A_{i-1} \cup Z_i)$$

$$g(A_i) = g(A_{i-1}) + g(Z_i) + \lambda g(A_{i-1})g(Z_i) \tag{5.11}$$

It is important to stress that equation (5.11) allows to construct the fuzzy measures in order to provide the weight of a single SVM classifier. Though there is no proper rule to follow for the attribution of $g(Z_i)$ values, in fact, they can be subjectively assigned by an expert or computed from the training data [35]. Presently, $g(Z_i)$ of the SVM predictor in both negative and positive classes is derived from the training accuracy. Let (t_i^+, t_i^-) be the training accuracy of the SVM Z_i in the two classes. These accuracies are handled through a weighted soft-max function [42], such that:

$$g(Z_i^\pm) = \alpha \frac{1 + exp(t_i^\pm)}{\sum_{i=1}^{N}[1 + exp(t_i^\pm)]} \tag{5.12}$$

where N is the number of trained SVMs, while the weight parameter α scales in the range $]0.1, 1]$ to control the importance assigned to the fuzzy measure. It is experimentally set to allow the best training accuracy. Thereafter, Sugeno's fuzzy integral and the fuzzy min-max algorithm are computed to aggregate the final prediction.

Algorithm 5.3 Fuzzy membership degrees calculation

If $Z_i \geq 1$ then: $\begin{cases} h^+(Z_i) = Z_i \\ h^-(Z_i) = 0 \end{cases}$

Else:
{

If $Z_i \leq -1$ then: $\begin{cases} h^+(Z_i) = 0 \\ h^-(Z_i) = Z_i \end{cases}$

Else:

$\begin{cases} h^+(Z_i) = (1 + Z_i)/2 \\ h^-(Z_i) = (1 - Z_i)/2 \end{cases}$

}

5.3.3.2   Sugeno's Fuzzy Integral

Sugeno's integral I_S, of a function $h: Z \rightarrow [0, 1]$ with respect to a fuzzy measure g over Z, is computed for each class $\pm$ by:

$$I_S^{\pm} = max[min(h^{\pm}(Z_i), g^{\pm}(A_i)]_{i=1:N} \tag{5.13}$$

5.3.3.3   Fuzzy Min-Max

The fuzzy min-max algorithm of a function $h: Z \rightarrow [0, 1]$ with respect to a fuzzy measure g over Z takes advantage from the flexibility and effectiveness of fuzzy logic operators and is computed for each class $\pm$ as [42]:

$$Fuzzy\ Min - Max^{\pm} = min[max(h^{\pm}(Z_i), g^{\pm}(A_i)]_{i=1:N} \tag{5.14}$$

5.4   Experimental Evaluation

In order the evaluate the proposed methods, two corpuses extracted from two public data sets were used. The samples were collected in a multiscript unconstrained writing environment. These data sets, namely, IAM and KHATT, contain handwritten sentences in English and Arabic languages, respectively.

5.4.1   Data Sets

5.4.1.1   IAM Data Set

The IAM On-Line Handwriting Database was developed by a research group on computer vision and artificial intelligence at Bern University in Switzerland.[1] The database was collected by the contribution of more than 200 writers, each of them participating with eight English texts constituting an average of seven lines. The data acquisition was carried out using a whiteboard. The database is presented in a form of handwritten sentences, where each sentence is indexed according to the writer's gender, handedness, age, and educational level. The number of collected samples for the prediction of soft biometrics is performed according to the availability of data in each subcategory within the database. Figure 5.4 presents some samples from this data set.

Presently, according to the first work published on automatic gender and handedness prediction using IAM data set [40, 41], a first corpus (IAM-1) was collected. This corpus is selected by taking only one sample from each writer. For gender prediction, 75 samples

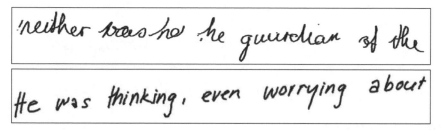

Figure 5.4 IAM data set samples.

---

1 http://www.iam.unibe.ch/fki

per class were randomly chosen and divided into 40 samples for training, 10 samples for validation, and 25 samples for testing. As to handedness prediction, unfortunately the database contains only 20 samples for the left-handed class. So for both classes, 15 samples were assigned to the training stage, while the remaining 5 samples were utilized to test the prediction performance. In a second step, the investigation was extended to a larger data set (IAM-2) to get a deeper analysis on gender and age classification. Precisely, 165 samples per class are collected for gender prediction. On the other hand, two main age ranges are available, 25–34 years and 35–56 years, to perform age prediction, where 84 samples were selected for each class. As for most of the classification and data-mining tasks [60] as well as the soft biometrics state of the art [45], two-thirds of samples were used for the training step, while the remaining one-third were used for testing the system.

5.4.1.2 KHATT Data Set

Not that long ago, Mahmoud *et al.* [61, 62] published a new Arabic data set, namely, KHATT (KFUPM Handwritten Arabic TexT), designed to serve research in Arabic handwritten text recognition, Arabic writer identification and verification, forms analysis, and segmentation of paragraphs, lines, words, subwords, and characters.[2] The database is composed of 1000 forms that cover all the shapes of Arabic characters, filled by 1000 writers from 18 different countries. Each writer has participated with one handwritten form segmented into paragraphs, then into handwritten text lines. This data set was collected by considering gender, qualification, handedness, age category, and nationality, but up to now it was not used for soft biometrics prediction. Figure 5.5 presents some examples from this data set.

To perform gender, handedness, and age range prediction, three corpuses were randomly selected. Specifically, 90 training samples and 45 test samples were collected per class for both gender and age range applications. Note that two age ranges are considered, which are "16 to 25 years." and "26 to 50 years". The handedness corpus is composed of 84 samples for both right-hand and left-hand classes, divided as 56 training samples and 28 test samples.

5.4.2 Experimental Setting

An ideal implemention of SVM classifiers requires an optimal parameters selection. First, four popular kernels available in the literature, namely, linear, polynomial, quadratic, and RBF, were tested. We investigate the best choice of SVM kernel before proceeding to

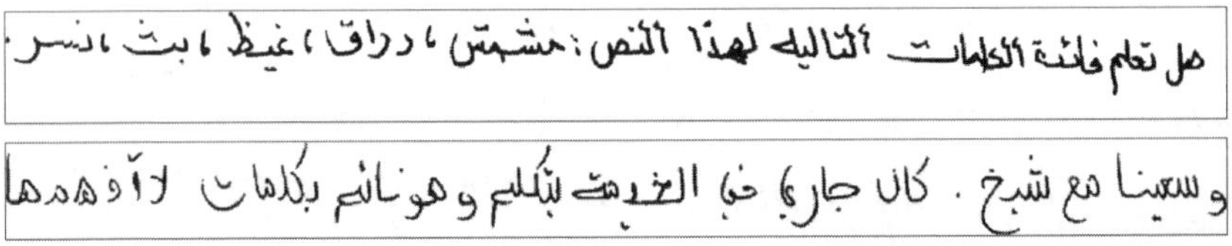

Figure 5.5 KHATT data set samples.

---

2 http://khatt.ideas2serve.net

Table 5.1 Influence of SVM kernels for gender prediction on the IAM-1 corpus (%)

Linear	Polynomial	Quandratic	RBF
66.00	62.00	62.00	72.00

the experiments. Tests were performed using the pixel density feature, which is the ratio between the number of pixels that belong to the text and the cell's size. The regularization parameter was varied in the interval [0.01:10:200]. The degree value of the polynomial kernel was varied from 1 to 3, while the RBF kernel sigma was varied in the range [2:2:80]. Then, the couple giving the best training performance was selected in each test. Results obtained in terms of accuracy for gender prediction on the IAM-1 corpus are given in Table 5.1. As can be seen, the RBF kernel provides the best accuracy (i.e., 72%), exceeding other kernels with at least 6%.

After establishing that the RBF kernel is the best choice for us, experiments on the IAM-1 corpus were conducted on the global image to optimize the performance of the LBP by choosing the type of the operator, the number of considered neighbors P, as well as the radius of the neighborhood R. The obtained results are shown in Figure 5.6. As can be seen, the LBP^{riu} allows an improvement of at least 2% over the classical operator. Also, $LBP^{riu}_{8,1}$ provides the best accuracy compared to the other configurations. Hence, this configuration is utilized for the rest of the experiments. Moreover, as claimed in [51], the appropriate grid size depends on the feature and the database that are used. Therefore, several grid size configurations were tested to find the optimal one by allowing the best training accuracy. For $LBP^{riu}_{8,1}$ and HOG features, the number of cells was experimentally varied from 1×1 cell (which corresponds to the whole image) to 5×10 cells and from 2×2 to 6×6, respectively. For GLBP, which returns a 56-element feature vector in each cell, the grid size was varied from 1×1 to 4×8 cells. Findings revealed that the number of grid cells has a meaningful impact over the prediction performance. Figure 5.7 plots the most relevant outcomes obtained using the different features. It is easy to see that

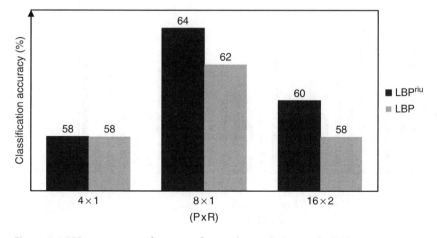

Figure 5.6 LBP operators; performances for gender prediction on the IAM-1 corpus.

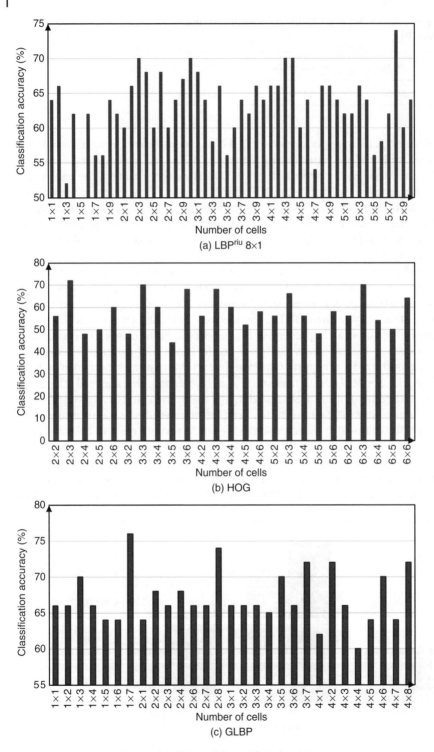

Figure 5.7 Influence of the grid size for gender prediction on the IAM-1 corpus.

the best performance of $LBP_{8,1}^{riu}$ is about 74% using a grid of 5×8 cells and allowing a gain of 10% over the $LBP_{8,1}^{riu}$ computed on the global image. As to the HOG feature, the best accuracy (i.e., 72%) is obtained using a grid of 2×3 cells, while GLBP precision (76%) is obtained with a grid of 1×7 cells. From this analysis, one can say that the grid size has indeed an influential effect on the prediction performance. Analogous experimentation was carried out for each application, and since similar behavior has been noticed, only gender prediction results using the IAM-1 data set are given.

5.4.3 Gender Prediction Results

Table 5.2 summarizes gender prediction results. As can be seen, GLBP exceeds $LBP_{riu}^{8,1}$ and HOG for the IAM-1 and IAM-2 corpuses. On the contrary, for the KHATT corpus, $LBP_{riu}^{8,1}$ gives the best performance (i.e., 75.56%), beating GLBP precision with more than 1% . Furthermore, the performance of fuzzy combination rules is evaluated comparatively to classical mean and max rules. As reported in Table 5.3, the proposed fuzzy min-max combination outperforms Sugeno's integral. Compared to individual systems, the advanced combination rule allowed a gain of 6%, 1.19%, and 7.77% for the IAM-1, IAM-2, and KHATT corpuses, respectively.

5.4.4 Handedness Prediction Results

Similarly to gender prediction experiments, the first step was initially focused on setting the parameters mentioned in this chapter. Table 5.4 reports the best results in terms of prediction accuracy. For this task, recall that the IAM-1 corpus contains only 20 samples

Table 5.2 Results of individual systems for gender prediction (%)

Corpus	$LBP_{riu}^{8,1}$	HOG	GLBP
IAM-1	74.00	72.00	**76.00**
IAM-2	72.72	74.45	**75.45**
KHATT	**75.56**	68.89	74.44

Table 5.3 Results of combination systems for gender prediction (%)

Corpus	Max	Mean	I_s	Fuzzy min-max
IAM-1	76.00	76.00	78.00	**82.00**
IAM-2	71.11	73.33	75.55	**76.64**
KHATT	77.78	70.00	**83.33**	83.33

Table 5.4 Results of handedness prediction for individual systems (%)

Corpus	$LBP_{riu}^{8,1}$	HOG	GLBP
IAM-1	**100.00**	**100.00**	**100.00**
KHATT	75.00	**83.93**	78.57

Table 5.5 Results of combination systems for handedness prediction (%)

Corpus	Max	Mean	I_s	Fuzzy min-max
KHATT	76.79	76.79	82.14	**83.93**

Table 5.6 Results of handedness prediction for individual systems (%)

Corpus	$LBP^{8,1}_{riu}$	HOG	GLBP
IAM-2	69.64	**73.21**	69.64
KHATT	**70.00**	67.78	**70.00**

Table 5.7 Results of combination systems for age prediction (%)

Corpus	Max	Mean	I_s	Fuzzy min-max
IAM-2	71.42	64.29	73.21	**75.79**
KHATT	62.22	77.78	**78.89**	75.56

for each class; for this reason, the prediction scores are high and reach 100% with all features. As to the KHATT corpus, the best precision (i.e., 83.93%) is obtained using HOG features. Furthermore, the combination step was limited to the KHATT data set since for IAM-1, the optimal performance was provided by all individual systems. Results of the different combination rules are reported in Table 5.5. As can be seen, a fuzzy min-max combination provides the best precision (i.e., 83.93%), but they did not provide any gain comparatively to the individual system trained on HOG features.

5.4.5 Age Prediction Results

Table 5.6 gives the results of age range prediction, while those obtained for the combination step are reported in Table 5.7. Recall that, for this experiment, two different age ranges were considered in the IAM-2 and KHATT corpuses. For the IAM-2 corpus, the HOG feature outperforms the others with more than 3.5%, while for the KHATT corpus, the best precision (i.e., 70.00%) is obtained using both $LBP^{8,1}_{riu}$ and GLBP features. Also, by combining these systems, the prediction accuracies are improved to more than 75.79% with the fuzzy min-max for the IAM-2 corpus. On the contrary, Sugeno's integral gives the best improvement for the KHATT corpus with a gain that exceeds 8.5% over individual systems.

5.5 Discussion and Performance Comparison

This chapter focused on developing a soft biometrics prediction system using a fuzzy integral–based combination method to ensure robust gender, handedness, and age range prediction from handwriting analysis. In fact, as reported in Table 5.8, only a few research works are available in the literature and are unfortunately using private data

Table 5.8 State-of-the-art results

Soft-biometrics	Reference	Data set	# Training data	# Test data	Features	Classifier	Precision rate (%)
Gender	[63]	English + Urdu	30	—	Human performance		67.84
	[45]	QUWI	—	—	Direction + curvature + tortuosity + chain code	Kernel discriminant analysis	73.70
	[47]	QUWI Arabic	300	100	Slant + curvature+ LBP	Neural network	71.00
	[47]	QUWI English	300	100	Slant + curvature	SVM	70.00
	[47]	MSHD French	42	42	Slant + curvature	SVM	68.25
	[47]	MSHD Arabic	42	42	LBP	SVM	74.20
	[48]	QUWI	282		Local gradient-based features	SVM	94.7
	[41]	IAM-1	80	50	Off-line+on-line	GMM	67.57
	[41]	IAM-1	24		Human performance		63.88
	Proposed	IAM-1	80	50	Fuzzy min-max($LBP_{riu}^{8.1}$ + HOG + GLBP)		82.00
		IAM-2	220	110	Fuzzy min-max($LBP_{riu}^{8.1}$ + HOG + GLBP)		77.78
		KHATT	180	90	Fuzzy min-max($LBP_{riu}^{8.1}$ + HOG + GLBP)		84.44
Handedness	[43]	QUWI	—	—	Direction + curvature + tortuosity + chain code	KNN	70.00
	[46]	QUWI	121		Fuzzy conceptual reduction	KNN	83.43%
	[40]	IAM-1	30	10	Off-line+on-line	GMM	86.64
	[40]	IAM-1	20		Human performance		62.00
	[34]b	IAM-1	30	10	GLBP	SVM	100.00
	Proposed	KHATT	112	56	Fuzzy min-max($LBP_{riu}^{8.1}$ + HOG + GLBP)		83.93
Age	[45]	QUWI	—	—	Direction + curvature + tortuosity	Random forest	62.40
	Proposed	IAM-2	112	56	Fuzzy min-max($LBP_{riu}^{8.1}$ + HOG + GLBP)		76.78
		KHATT	180	90	Fuzzy min-max($LBP_{riu}^{8.1}$ + HOG + GLBP)		78.89

sets, which does not promote a decent comparison. However, from methods analysis, the superiority of fuzzy integral–based prediction systems can easily be deduced. This fact can be explained by the incorporation of LBP features that capture the local structure information, along with HOG features that extract gradient information according to a given orientation, and GLBP features that combine texture and gradient information.

– From all experiments, we can say that for soft biometrics prediction, uniform LBP operators perform better than the classical LBP, specially with the $(P = 8, R = 1)$ configuration, which allowed at least a gain of 2%. Furthermore, computing the proposed features locally by applying equispaced grids over text images allows to get local information of the image content. The obtained findings highlighted the impact of the grid size on the reliability of the feature characterization. Precisely, for $LBP_{riu}^{8,1}$, the local calculation allowed a considerable gain of 10%. Prediction accuracies given by individual systems vary between 52% and 76%. From the inspection of all data set results, one can say that the three SVM predictors give satisfying performances, but there is no descriptor that provides the best discriminative power for SVM. According to the theory, such differences can be explored through a combination framework to improve performances.

– Arabic is written cursively and from the right to the left, where letters are normally connected to the baseline with horizontal strokes, compared to English handwriting where in general each character is connected to the next character with diagonal strokes and is written from the left to the right. Also, Arabic language has its specific diacritical marking to represent vowels such as dumma ('), hamza (ع), or chadda (ω). Regardless of all these differences between the two languages, results on both data sets are typically in the same range.

– Fuzzy integral–based combination results reveal promising results on both the IAM and KHATT databases in term of accuracy. Indeed, the proposed combination rules outperform individual prediction systems where the improvement varies from 1 to 8% for the different data sets. Also, they give comparable results in most cases, because they share the same computation concept.

5.6 Conclusion

This chapter proposed the use of fuzzy integrals as a combination classifier method, to improve the prediction of soft biometrics traits from handwriting. Firstly, three SVM predictors paired to three distinct data features were developed to achieve writers' gender, handedness, and age range predictions. Following that, SVM decisions were combined to enhance the prediction performance. Exhaustive experiments carried out on two different English and Arabic handwritten corpuses indicated that the aforementioned combination methods enhence the prediction accuracy. Besides, the proposed approach has proven itself compared to the different combination rules as well as the state of the art. As the results have shown, the combination process handles SVM outputs supplied with relevant information, which indicates that any kind of data features can be used. To enhance the results more, we plan in a future work to investigate new classifiers such as artificial immune recognition systems (AIRSs) and convolutional neural networks (CNNs). Also, it would be interesting to develop a multiclass prediction system of various soft biometrics traits.

References

1 Karthik Nandakumar and Anil K. Jain. *Soft Biometrics*, pages 1235–1239. Springer US, Boston, MA, 2009.

2 Yasmina Andreu, Pedro Garca-a-Sevilla, and R. A. Mollineda. Face gender classification: a statistical study when neutral and distorted faces are combined for training and testing purposes. *Image and Vision Computing*, **32** (1):27–36, 2014.

3 Di Huang, Huaxiong Ding, Chen Wang, Yunhong Wang, Guangpeng Zhang, and Liming Chen. Local circular patterns for multi-modal facial gender and ethnicity classification. *Image and Vision Computing*, **32** (12):1181–1193, 2014.

4 Grigory Antipov, Sid-Ahmed Berrani, and Jean-Luc Dugelay. Minimalistic CNN-based ensemble model for gender prediction from face images. *Pattern Recognition Letters*, **70**:59–65, 2016.

5 R. Djemili, H. Bourouba, and M.C.A. Korba. A speech signal based gender identification system using four classifiers. In *Multimedia Computing and Systems (ICMCS), 2012 International Conference*, pages 184–187, May 2012.

6 Vinay S. Gupta, and A. Mehra. Gender specific emotion recognition through speech signals. In *Signal Processing and Integrated Networks (SPIN), 2014 International Conference*, pages 727–733, February 2014.

7 I. Bisio, A. Delfino, F. Lavagetto, M. Marchese, and A. Sciarrone. Gender-driven emotion recognition through speech signals for ambient intelligence applications. *IEEE Transactions on Emerging Topics in Computing*, **1** (2):244–257, December 2013.

8 M. Gomathy, K. Meena, and K. R. Subramaniam. Classification of speech signal based on gender: a hybrid approach using neuro-fuzzy systems. *Int. Journal of Speech Technology*, **14** (4):377–391, December 2011.

9 Jose Vina and Ana Lloret. Why women have more Alzheimer's disease than men: gender and mitochondrial toxicity of amyloid-beta peptide. *Journal of Alzheimers Disease*, **20**:527–533, 2010.

10 C. Almqvist, M. Worm, B. Leynaert, and for the working group of GA2LEN WP 2.5 Gender. Impact of gender on asthma in childhood and adolescence: a GA2LEN review. *Allergy*, **63** (1):47–57, 2008.

11 Andrea R. Ennis, Peter McLeod, Margo C. Watt, Mary Ann Campbell, and Nicole Adams-Quackenbush. The role of gender in mental health court admission and completion. *Canadian Journal of Criminology and Criminal Justice*, **58** (1):1–30, 2016.

12 Sarah Bennett, David P. Farrington, and L. Rowell Huesmann. Explaining gender differences in crime and violence: The importance of social cognitive skills. *Aggression and Violent Behavior*, **10** (3):263–288, 2005.

13 Rachael E. Collins. The effect of gender on violent and nonviolent recidivism: A meta-analysis. *Journal of Criminal Justice*, **38** (4):675–684, 2010.

14 John R. Beech and Isla C. Mackintosh. Do differences in sex hormones affect handwriting style? Evidence from digit ratio and sex role identity as determinants of the sex of handwriting. *Personality and Individual Differences*, **39**:459–468, July 2005.

15 M. Genna and A. Accardo. Gender and age influence in handwriting performance in children and adolescents. In *5th European Conference of the International Federation for Medical and Biological Engineering: 14–18 September 2011, Budapest, Hungary*, pages 141–144. Springer, Berlin, 2012.

16 James Hartley. Sex differences in handwriting: a comment on Spear. *British Educational Research Journal*, **17** (2):141–145, 1991.

17 W.N. Hayes. Identifying sex from handwriting. *Perceptual and Motor Skills*, **83** (3 Pt 1):791–800, Dec 1996.

18 Marietta Papadatou-Pastou and Anna Safar. Handedness prevalence in the deaf: Meta-analyses. *Neuroscience & Biobehavioral Reviews*, **60**:98–114, 2016.

19 Quanlei Yu, Qiuying Zhang, Shenghua Jin, Jianwen Chen, Yingjie Han, and Huimi Cao. The relationship between implicit and explicit self-esteem: the moderating effect of handedness. *Personality and Individual Differences*, **89**:1–5, 2016.

20 S. Knecht, B. Dräger, M. Deppe, L. Bobe, H. Lohmann, A. Flöel, E.-B. Ringelstein, and H. Henningsen. Handedness and hemispheric language dominance in healthy humans. *Brain*, **123** (12):2512–2518, 2000.

21 Abdulaziz Al-Musa Alkahtani. The influence of right or left handedness on the ability to simulate handwritten signatures and some elements of signatures: a study of Arabic writers. *Science & Justice*, **53** (2):159–165, 2013.

22 I.C. Friesen, R.A. Dixon, and D. Kurzman. Handwriting performance in younger and older adults: age, familiarity, and practice effects. *Psychology and Aging*, **8** (5):360–370, September 1993.

23 N.A. Lannin, N. van Drempt, and A. McCluskey. A review of factors that influence adult handwriting performance. *Australian Occupational Therapy Journal*, **58** (5):321–328, October 2011.

24 A. McCluskey and D.K. Burger. Australian norms for handwriting speed in healthy adults aged 60–99 years. *Australian Occupational Therapy Journal*, **58** (5):355–363, October 2011.

25 Sara Rosenblum, Batya Engel-Yeger, and Yael Fogel. Age-related changes in executive control and their relationships with activity performance in handwriting. *Human Movement Science*, **32** (2):363–376, 2013.

26 S.H. Cha and S.N. Srihari. A priori algorithm for sub-category classification analysis of handwriting. In *International Conference on Document Analysis and Recognition*, pages 1022–1025, Seattle, WA, September 2001.

27 Edson J.R. Justino, Flávio Bortolozzi, and Robert Sabourin. A comparison of SVM and HMM classifiers in the off-line signature verification. *Pattern Recognition Letters*, **26** (9):1377–1385, July 2005.

28 E. Frias-Martinez, A. Sanchez, and J. Velez. Support vector machines versus multi-layer perceptrons for efficient off-line signature recognition. *Engineering Applications of Artificial Intelligence*, **19** (6):693–704, 2006.

29 Parveen Kumar, Nitin Sharma, and Arun Rana. Handwritten character recognition using different kernel based SVM classifier and MLP neural network (a comparison). *International Journal of Computer Applications*, **53** (11):25–31, September 2012.

30 Chayaporn Kaensar. A comparative study on handwriting digit recognition classifier using neural network, support vector machine and k-nearest neighbor. In *The 9th International Conference on Computing and Information Technology (IC2IT2013):*

9th–10th May 2013 King Mongkut's University of Technology North Bangkok, pages 155–163. Springer, Berlin, 2013.

31 Christopher J.C. Burges. A tutorial on support vector machines for pattern recognition. *Data Mining and Knowledge Discovery*, **2** (2):121–167, 1998.

32 Nesrine Bouadjenek, Hassiba Nemmour, and Youcef Chibani. Local descriptors to improve off-line handwriting-based gender prediction. In *6th International Conference of Soft Computing and Pattern Recognition (SoCPaR), 2014*, pages 43–47, Tunisia, August 2014.

33 Nesrine Bouadjenek, Hassiba Nemmour, and Youcef Chibani. Age, gender and handedness prediction from handwriting using gradient features. In *Document Analysis and Recognition (ICDAR), 2015 13th International Conference*, pages 1116–1120, August 2015.

34 Nesrine Bouadjenek, Hassiba Nemmour, and Youcef Chibani. Histogram of oriented gradients for writer's gender, handedness and age prediction. In *Innovations in Intelligent SysTems and Applications (INISTA), 2015 International Symposium*, pages 220–224, September 2015.

35 Hassiba Nemmour and Youcef Chibani. Multiple support vector machines for land cover change detection: an application for mapping urban extensions. *{ISPRS} Journal of Photogrammetry and Remote Sensing*, **61** (2):125–133, 2006.

36 A. Gattal, Y. Chibani, B. Hadjadji, H. Nemmour, I. Siddiqi, and C. Djeddi. Segmentation-verification based on fuzzy integral for connected handwritten digit recognition. In *Image Processing Theory, Tools and Applications (IPTA), 2015 International Conference*, pages 588–591, November 2015.

37 Gabriela E. Martinez, Patricia Melin, Olivia D. Mendoza, and Oscar Castillo. Face recognition with a Sobel edge detector and the Choquet integral as integration method in a modular neural networks. In *Design of Intelligent Systems Based on Fuzzy Logic, Neural Networks and Nature-Inspired Optimization*, pages 59–70. Springer, Cham, 2015.

38 Jung Soh. Computational method for document object locator combination. *Image and Vision Computing, Proceedings from the 15th International Conference on Vision Interface*, **22** (12):1015–1029, 2004.

39 K. R. Bandi and S. N. Srihari. Writer demographic classification using bagging and boosting. In *Proceedings of International Graphonomics Society Conference*, pages 133–137, Salerno, Italy, 2005.

40 M. Liwicki, A. Schlapbach, P. Loretan, and H. Bunke. Automatic detection of gender and handedness from on-line handwriting. In *Conference of the International Graphonomics Society*, pages 179–183, Melbourne, Australia, 2007.

41 Marcus Liwicki, Andreas Schlapbach, and Horst Bunke. Automatic gender detection using on-line and off-line information. *Pattern Analysis Application*, **14**:87–92, February 2011.

42 Nesrine Bouadjenek, Hassiba Nemmour, and Youcef Chibani. Robust soft biometrics prediction from off-line handwriting analysis. *Applied Soft Computing*, **46**:980–990, 2016.

43 S. Al-Maadeed, F. Ferjani, S. Elloumi, and A. Hassaine. Automatic handedness detection from off-line handwriting. In *GCC Conference and Exhibition (GCC), 2013 7th IEEE*, pages 119–124, Doha, Qatar, November 2013.

44 S. Al Maadeed, W. Ayouby, A Hassaine, and J.M. Aljaam. Quwi: an Arabic and English handwriting data set for offline writer identification. In *International Conference on Frontiers in Handwriting Recognition (ICFHR)*, pages 746–751, Bari, Italy, September 2012.

45 S. Al-Maadeed and A Hassaine. Automatic prediction of age, gender, and nationality in offline handwriting. *EURASIP Journal on Image and Video Processing*, 2014.

46 Somaya Al-Maadeed, Fethi Ferjani, Samir Elloumi, and Ali Jaoua. A novel approach for handedness detection from off-line handwriting using fuzzy conceptual reduction. *EURASIP Journal on Image and Video Processing*, **2016** (1):1–14, 2016.

47 Imran Siddiqi, Chawki Djeddi, Ahsen Raza, and Labiba Souici-Meslati. Automatic analysis of handwriting for gender classification. *Pattern Analysis and Applications*, pages 1–13, 2014.

48 A.S. Ibrahim, A.E. Youssef, and A.L. Abbott. Global vs. local features for gender identification using Arabic and English handwriting. In *Signal Processing and Information Technology (ISSPIT), 2014 IEEE International Symposium*, pages 000155–000160, Dec 2014.

49 Timo Ojala, Matti Pietikäinen, and David Harwood. A comparative study of texture measures with classification based on featured distributions. *Pattern Recognition*, **29** (1):51–59, 1996.

50 Bo Yang and Songcan Chen. A comparative study on local binary pattern (LBP) based face recognition: {LBP} histogram versus {LBP} image. *Neurocomputing*, **120**:365–379, 2013.

51 D. Bertolini, L.S. Oliveira, E. Justino, and R. Sabourin. Reducing forgeries in writer-independent off-line signature verification through ensemble of classifiers. *Pattern Recognition*, **43** (1):387–396, 2010.

52 Yilmaz Kaya, Omer Faruk Ertugrul, and Ramazan Tekin. Two novel local binary pattern descriptors for texture analysis. *Applied Soft Computing*, **34**:728–735, 2015.

53 J.F. Vargas, M.A. Ferrer, C.M. Travieso, and J.B. Alonso. Off-line signature verification based on grey level information using texture features. *Pattern Recognition*, **44** (2):375–385, 2011.

54 Matti Pietikäinen, Abdenour Hadid, Guoying Zhao, and Timo Ahonen. *Computer Vision Using Local Binary Patterns*. Springer-Verlag, London, 2011.

55 N. Dalal and B. Triggs. Histograms of oriented gradients for human detection. In *Computer Vision and Pattern Recognition, 2005. CVPR 2005. IEEE Computer Society Conference*, vol. 1, pages 886–893, June 2005.

56 O. Deniz, G. Bueno, J. Salido, and F. De la Torre. Face recognition using histograms of oriented gradients. *Pattern Recognition Letters*, **32** (12):1598–1603, 2011.

57 M.B. Yilmaz, B. Yanikoglu, C. Tirkaz, and A. Kholmatov. Offline signature verification using classifier combination of HOG and LBP features. In *Biometrics (IJCB), 2011 International Joint Conference*, pages 1–7, October 2011.

58 Ning Jiang, Jiu Xu, Wenxin Yu, and S. Goto. Gradient local binary patterns for human detection. In *IEEE International Symposium on Circuits and Systems (ISCAS), 2013*, pages 978–981, Beijing, China, May 2013.

59 Vladimir N. Vapnik. *The Nature of Statistical Learning Theory*. Springer-Verlag, New York, 1995.

60 Tsau Young Lin, Ying Xie, Anita Wasilewska, and Churn-Jung Liau. *Data Mining: Foundations and Practice*, vol. 118. Springer, Berlin, 2008.

61 Sabri A. Mahmoud, Irfan Ahmad, Mohammad Alshayeb, Wasfi G. Al-Khatib, Mohammad Tanvir Parvez, Gernot A. Fink, Volker Margner, and Haikal El Abed. Khatt: Arabic offline handwritten text database. In *Proceedings of the International Conference on Frontiers in Handwriting Recognition*, pages 449–454, Seattle, WA, 2012. IEEE Computer Society.

62 Sabri A. Mahmoud, Irfan Ahmad, Wasfi G. Al-Khatib, Mohammad Alshayeb, Mohammad Tanvir Parvez, Volker Märgner, and Gernot A. Fink. KHATT: an open Arabic offline handwritten text database. *Pattern Recognition*, **47** (3):1096–1112, 2014.

63 Sarah Hamid and Kate Miriam Loewenthal. Inferring gender from handwriting in Urdu and English. *The Journal of Social Psychology*, **136** (6):778–782, 1996.

6

Brain-Inspired Machine Intelligence for Image Analysis: Convolutional Neural Networks

Siddharth Srivastava and Brejesh Lall

Department of Electrical Engineering, Indian Institute of Technology Delhi, India

6.1 Introduction

The human brain is a sophisticated and wonderful biological marvel, and more than a third of the brain's processing power is dedicated to visual processing. The evolution of humans and their triumph on earth can be attributed to the ability of the visual system to coordinate with the brain and perform complicated pattern analysis, within a fraction of a second. Provided the humongous amount of objects and visual stimuli that humans are able to identify and discern, there is no surprise that the functioning of our visual system and its subsequent processing by the brain have always intrigued everyone. Scientists and philosophers have been pondering over understanding the basic nature and functioning of our visual system for centuries. But it wasn't till the 1940s [1] that any practical system attempted to replicate the working of the human brain or, to be more precise, the biological neurons. Since then, artificial intelligence (AI) practitioners have performed numerous experiments [2–6] to understand this mysterious part of the human anatomy.

Deep learning [7, 8] is the modern buzz word when it comes to algorithms mimicking the functioning of the brain. Deep learning techniques have been swiftly outperforming traditional machine learning and AI algorithms, and by a large margin [9–11]. Therefore, the study of deep learning techniques utilizing images as inputs has become interesting. From being labeled as useless by the research community to being the most successful techniques of all time, deep learning and neural networks have progressed through decades of success and failures. To this end, in the next few paragraphs, the reader is introduced to the evolution of neural networks and deep learning. We believe that in order to appreciate the complexity and strength of the contemporary research in deep learning involving convolutional neural networks (CNNs), the reader should be familiar with fundamentals of CNNs. Therefore, detailed discussion on state-of-the-art work is deferred till the end of this chapter.

In 1943, McCulloch proposed a simple neuron model [1], showing that it can perform basic logic operations (AND/OR/NOT). The work caught the immediate attention of AI researchers since performing logical operations on computers was a significant

Hybrid Intelligence for Image Analysis and Understanding, First Edition.
Edited by Siddhartha Bhattacharyya, Indrajit Pan, Anirban Mukherjee, and Paramartha Dutta.
© 2017 John Wiley & Sons Ltd. Published 2017 by John Wiley & Sons Ltd.
Companion Website: www.wiley.com/go/bhattacharyya/hybridintelligence

achievement back then and it was the ultimate objective of AI. But this model lacked a model for learning. In 1949, Hebb [12] put forth an idea that learning in the brain happens due to formation and changes of synapses among neurons. Using this idea, Rosenblatt in 1958 devised the perceptron [13]. It was a simple mathematical model allowing learning similar to that of biological neurons. Its efficacy was demonstrated on a simple shape classification problem. But more importantly, it was the first model that showed that computers can learn functions based on provided input and expected output data. Although perceptrons yield single output, adding layers to such a model for higher level classification was a straightforward extension as per Rosenblatt and other AI researchers who saw potential in neural networks. But in a book published in 1969 [14] by Prof. Minsky of MIT, through rigorous analysis it was shown that perceptrons have significant limitations such as inability to learn a XOR function. This work is believed to have influenced the downfall of neural networks, which led to a freeze in any further research in this field during that time. In 1986, it was shown that by using back-propagation [15], the multilayer perceptrons can be efficiently trained. This was the time when it was realized that Minsky's work actually meant that for learning complicated functions, multilayer perceptrons were needed instead of a single-layer perceptron. In 1989, Hornik *et al.* [16] showed that multilayer perceptrons can be used to model any mathematical functions. This provided a major thrust to the neural network, and in the same year LeCun applied back-propagation to recognize handwritten digits [17]. It was an important work because, till then, it was believed that only humans have the capability to effectively distinguish among handwritten digits.

Following the ideas of multilayer neural networks, reinforcement learning-based approaches were proposed for playing games such as backgammon [18]. But a few years later, it was shown that such networks fail miserably in learning to play chess [19], and the primary reason behind it was that the network spent most of its time computing parameters. Therefore, nonavailability of computational resources or for that matter efficient techniques that were suitable for hardware available during that era made the adoption of neural networks even more difficult. In addition to this, lack of any efficient method to perform back-propagation on large networks added to the agony of neural networks. During this period, a supervised learning technique, the support vector machines (SVMs) [20], was getting much attention. In fact Lecun, who is now considered among the pioneers in deep learning research, showed that SVMs worked better than neural networks for hand-digit recognition [21]. It was now widely accepted that neural networks are dead and any research effort toward them is a waste of resources. This can be corroborated by the fact that during this period, the funding for research projects related to neural networks dropped significantly and consequently the research in this area was toward a dead-end.

Despite such a negative environment around neural networks, Prof. Hinton was able to secure funding from the Canadian Institute for Advanced Research (CIFAR) for research in this area. And after several years of effort, he along with his team proposed *A fast learning algorithm for deep belief nets* [22], which formally paved the path for deep learning. The main idea was that if the weights in the neural networks are initialized in a specific way instead of randomly, they can be trained very efficiently. The technique primarily involved training each layer sequentially in an unsupervised way. It achieved a state-of-the-art result on the MNIST data set [23]. Following this work, another deep learning maestro of the modern world, Prof. Yoshua Bengio, and his team showed that deep learning algorithms perform better on complex tasks than other learning paradigms [24]. They in fact justified why unsupervised pre-training works and that it learns useful representations from data. Even though deep learning was becoming popular, it was still plagued by limitations of computational resources in research laboratories across the globe. During this time, Andrew Ng of Stanford University collaborated with Google and utilized their enormous computing resources to learn, in an unsupervised way, the object category labels. By this time, both industry and research laboratories had realized the power of deep learning. In 2010, problems behind subpar working of back-propagation was analyzed [25]. It was identified that choice of nonlinear activation and initialization of weights significantly impact the performance of a deep network.

After this discovery, sincere research efforts were put into improvising the deep networks. It resulted in unprecedented progress in AI research and especially computer vision, where CNNs were reborn. Since then, CNNs have demonstrated remarkable capability in tasks related to image analysis [26–29]. Convolutional networks, in general, are motivated from biological processes [30]. This is corroborated by the fact that CNNs are hierarchical networks that abstractly resemble the working of the human mind. The authors in [31] evaluate CNNs for their resemblance with the human brain using functional magnetic resonance imaging (fMRI). They analyze the contribution of each layer of a CNN against the observed brain activity for visual recognition tasks. The study found that the visual processing and layers of CNNs simulate similar hierarchical computational responses in various parts of the brain. An interesting observation from the study is that CNNs are able to explain the intrinsic functioning of the brain. CNNs in their existing form only correspond to the functional flow of information in the visual cortex and the brain. Despite such great advancements, it should be noted here that in practice, CNNs are far from attaining human-level intelligence. It has been shown that traditional CNN architectures can be fooled [32] into recognizing objects in images appearing nonsensical to the human brain. But it is important to note that it is the potential of CNNs that has led the research community to continuously improvise upon such failures and come up with stronger architectures. The struggle to attain human brain-level performance continues, with CNNs being the strongest in achieving it.

6.2 Convolutional Neural Networks

The popularity of CNNs for image-related tasks, as we know today, can be attributed to the work on large-scale image classification by Krizhevsky *et al.* [27] in 2012. They achieved more than 10% reduction in error rates on a data set with 1.3 million images belonging to a thousand classes. Although CNNs have existed for a long time, their practical use has become possible only now when appropriate computational power is available. Table 6.1 compares CNNs and the human brain at a coarse level to bring out the differences in their functioning. We expect that it will also allow the reader to appreciate the fact that simulating brain (especially for image analysis) is an extremely difficult task, and while CNNs may lack in many aspects, they are still the best available today.

To explain the working of CNNs, we start with a discussion on the building blocks of CNNs.

Table 6.1 Conceptual differences between CNN and the brain/visual system

CNN	Human brain/visual system	Difference
Architecture of CNN is inspired from biological processes.	They are the biological organs.	Man vs. Nature
CNNs process images in hierarchical order.	Study shows that brain processes visual stimulus in hierarchical order.	The flow of information could be similar, but the channels through which this information is passed are different.
CNNs work very well if trained with a huge amount of input data.	Can learn with as little as one image.	Human brain has evolved over thousands of years.
Has millions of parameters or neuronal connections	Studies suggest that human brains have billions or trillions of neuronal connections.	The computational complexity and efficiency of the human brain are huge when compared to CNNs.
Generally uses gradient descent	Difficult to determine if it is a single algorithm	Implicit learning vs. explicit learning
Usually trained with a single modality (images)	Can multitask with multiple sensory inputs (audio, visual, touch, etc.)	Unlike a CNN, the brain has various sensory inputs working in coherence.

6.2.1 Building Blocks

The basic architecture of a neural network is an **artificial neuron**. Figure 6.1 depicts a set of inputs $(x_1, x_2, \ldots, x_n)$ to a neuron N, which essentially weighs the input with weights (say, w_i), and sums them. This value may then pass through an activation function (explained later in this chapter) to produce the desired output. The neuron hence yields a single numeric output. This output may also be produced by a set of cascaded layers of neurons, as shown in Figure 6.2. The intermediate layers between the input and output layers of the network are also called **hidden layers**.

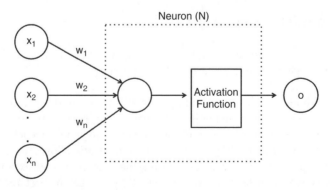

Figure 6.1 A neuron.

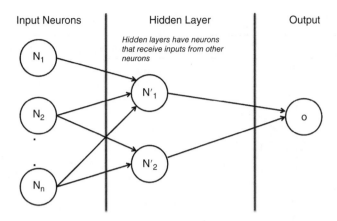

Figure 6.2 Hidden layers.

Mathematically, a neuron can be represented as follows:

$$f(x_i, w_i) = \phi\left(\sum_{i=1}^{n}(w_i.x_i)\right)$$

(6.1)

where x_i is the input, w_i is the weight corresponding to n inputs, and ϕ is the activation function. An **activation function** is meant to bound the output of a neuron. The applications of activation functions are not limited to neural networks. For example, with SVMs, they are used to transform the input feature space to reflect a decision boundary. But what makes them interesting in the context of biologically motivated neural networks is their capability to abstractly represent the rate of change in the electrical membrane potential or, more specifically, the action potential of a cell. Numerous activation functions have been proposed in literature; Table 6.2 shows a few such activation functions and their corresponding plots. An activation function is usually applied at the hidden layers of a neural network and/or at the output layer.

Activation at hidden layers: The *rectified linear units (ReLUs)* [33] have become very popular in the last few years and are now the recommended activation function to be used at the hidden layers of a neural network. Mathematically, it is expressed as in equation (6.2):

$$\phi(x) = max(0, x)$$

(6.2)

The reason for preferring ReLUs over other activation functions are: first, it is a nonsaturating function. It can be observed from Table 6.2 that while other functions saturate to a value of -1, 0, or 1 (vanishing gradient), ReLU doesn't. This results in faster convergence of gradient descent as compared to other activation functions, especially sigmoid/tanh. Second, the computation is simpler and faster since a ReLU only requires thresholding at zero. It also induces sparsity in the hidden units. This becomes significant when we consider that CNNs have a huge number of matrix computations. Other similar activation functions are *parametric ReLU* [34] and *Maxout* [35]. The ReLU function [equation (6.2)] has the problem that the gradients become 0 (due to thresholding) if the learning rate is set too high. The parametric ReLU instead has a

Table 6.2 Activation functions

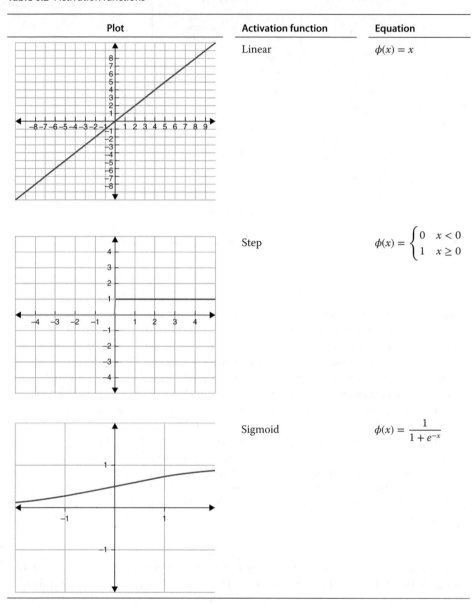

Plot	Activation function	Equation
	Linear	$\phi(x) = x$
	Step	$\phi(x) = \begin{cases} 0 & x < 0 \\ 1 & x \geq 0 \end{cases}$
	Sigmoid	$\phi(x) = \dfrac{1}{1 + e^{-x}}$

small slope (Table 6.2) when the value is less than 0. Maxout function, on the other hand, is a generalized form of the ReLU and parametric ReLU. Though it combines the advantages of both ReLU and parametric ReLU, the number of parameters per neuron are doubled, leading to increased complexity of the network.

Activation at output layers: The de facto choice for the activation function at the output layer in a CNN is a *softmax* function. It computes the probability of the input belonging to each of the possible output classes. The mathematical representation of a

Table 6.2 (Continued)

Plot	Activation function	Equation
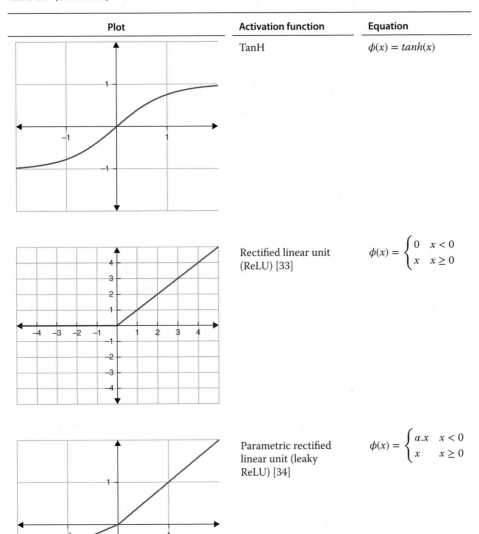	TanH	$\phi(x) = tanh(x)$
	Rectified linear unit (ReLU) [33]	$\phi(x) = \begin{cases} 0 & x < 0 \\ x & x \geq 0 \end{cases}$
	Parametric rectified linear unit (leaky ReLU) [34]	$\phi(x) = \begin{cases} \alpha.x & x < 0 \\ x & x \geq 0 \end{cases}$

softmax function is given as:

$$\phi_i = \frac{e^{z_i}}{\sum e^{z_j}} \tag{6.3}$$

where z_i is the value of the ith output neuron, while z_j represents all the output neurons in the output layers.

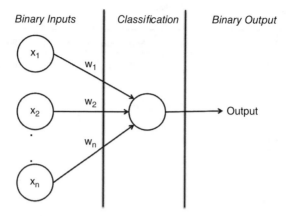

Binary Inputs   Classification   Binary Output

Figure 6.3 A perceptron.

6.2.1.1 Perceptron

While the neuron is the basic unit in a neural network, we still need a model that can learn how to classify the given inputs to a desired set of outputs. One such model is a *perceptron* [13]. More precisely, a perceptron (Figure 6.3) is a supervised learning model for building a linear binary classifier. It takes binary inputs and produces a binary output.

Mathematically, a perceptron is represented as:

$$output = \begin{cases} 1 & \sum_{i=1}^{n} w_i.x_i > threshold \\ 0 & \sum_{i=1}^{n} w_i.x_i \leq threshold \end{cases} \tag{6.4}$$

A careful comparison of Equations (6.1) and (6.4) shows that the output of a perceptron is essentially a threshold applied over the output of a neuron. An alternative way to look at it is that a perceptron weighs up an input only if it is important.

Let us understand the perceptron with the help of an example. Suppose you want to buy a book for studying CNNs. You choose to decide whether you would buy a particular book based on the following factors (other factors are not considered relevant for this discussion):

- Are the authors trustworthy?
- Are the reviews on e-commerce sites positive?
- Does the book mention it is meant for beginners?
- Can you afford it?

Each of the above factors can be represented as a binary variable. For instance, if x_1 symbolizes *Are the authors trustworthy?*, then we would have $x_1 = 1$ if you find the authors trustworthy (*Yes*), and $x_1 = 0$ if otherwise (*No*). Similarly we choose x_2, x_3, and x_4 for the next three factors, respectively. The assignment of variables to various factors is shown in Table 6.3.

Now, if you are a novice in this field and want to understand the basics of CNNs (i.e., *Does the book mention it is meant for beginners ?*), you may be willing to overlook other factors. You would model it by assigning a higher weight to x_3 (i.e., a higher value of w_3). Suppose that you assign $w_1 = 2$, $w_2 = 3$, $w_3 = 10$, and $w_4 = 2$, and you choose

Table 6.3 Factors and assigned variables for the perceptron model

Sample no.	Factor	Input variable	Weight variable
1.	Are the authors trustworthy?	x_1	w_1
2.	Are the reviews on e-commerce sites positive?	x_2	w_2
3.	Does the book mention it is meant for beginners?	x_3	w_3
4.	Can you afford it?	x_4	w_4

the threshold as 8. This infers that output of the perceptron [equation (6.4)] can never be greater than 8 if the book is not meant for beginners (Factor 3). Alternatively, you can say that you will buy the book if and only if it satisfies the third factor. One may get alternative models by suitably changing the weights and threshold.

As perceptrons are the fundamental model behind large neural networks, let us simplify the notation and provide some biological perspective. The simplified notation is given by equation (6.5):

$$output = \begin{cases} 1 & w.x + b > 0 \\ 0 & w.x + b \leq 0 \end{cases} \qquad (6.5)$$

where w and x are the vectors representing the weights (w_i) and inputs (x_i) from equation (6.4); and b is the *bias*, which can be understood as the negative of the threshold being added to the left-hand side of equation (6.4). It can also be interpreted as the ease with which a perceptron can be fired (similar to biological neurons).

The above discussion shows you how a perceptron can weigh various factors to reach a decision. Though it is still far from the decision-making capability of the human brain, still it is an important element in computationally realizing this behavior at a fundamental level. We must also understand that when the perceptron was proposed in the 1950s, the scientific community had limited understanding on the actual functioning of our brain. Therefore, artificial neural networks attempt to replicate the flow of processes and information in the human brain as a black box rather than actually performing similar computations.

6.2.2 Learning

We have discussed the perceptron model. A question that arises is, how does a perceptron learn? Prior to explaining how learning is performed, you should understand that while a perceptron is a good model for conceptual understanding, practically we use functions such as sigmoid, ReLU, maxout, and so on for computing weights at the hidden layers. The reason is that a perceptron outputs only 0 or 1, which makes the tuning of the network a little rigid when a large number of parameters are involved. Moreover, by simple algebraic assumptions, any activation function can be converted to a perceptron having binary outputs.

In Figure 6.2, a multilayer network is shown. Such a network is called a **feedforward network**, as it doesn't contain any feedback into the network. Although feedback to the same network may sound nonintuitive at first, such networks exist and are called **recurrent neural networks**. Theoretically, recurrent neural networks have a closer resemblance to the functioning of the human brain, but they have limited practical application, mostly due to lack of efficient learning algorithms when compared to feedforward networks. Therefore, in this text, we focus on the feedforward networks.

Any type of learning requires a defined **cost function**. In general, a cost function can be represented as:

$$C(w, b) = \min_N f(\|prediction - actual\|) \qquad (6.6)$$

where C defines the cost function over weights w, and biases b; N is the total number of training inputs; f is the loss function expression, such as root mean square (RMS); *prediction* is the predicted output value using the desired activation [in the case of a perceptron, it is given by equation (6.5)]; and *actual* is the output obtained from the network. The objective of the learning algorithm is to learn weights and biases, such that the cost C is minimized.

6.2.2.1 Gradient Descent

Gradient descent is a popular algorithm to compute cost with the constraints discussed previously. It works by slowly moving to the minima of the cost function. For this, it computes the gradient at each weight and minimizes it. Pictorially, it can be represented as in Figure 6.4. Technically, a gradient is the rate of change of a property, that is, the gradient of a function $f(x, y)$ would be a vector given by its partial derivatives [equation (6.7)]:

$$\nabla f = \left[\frac{\partial f}{\partial x}, \frac{\partial f}{\partial y} \right] \qquad (6.7)$$

In general, the algorithm begins with a random guess for a weight w, and then gradually moves toward the minima by moving in the opposite direction of the gradient. Having provided the intuition behind the algorithm, we will not delve into further details of gradient descent, as in practice we use a faster algorithm called *back-propagation* for learning the optimum parameter values for the network, which is described next.

6.2.2.2 Back-Propagation

Back-propagation is the process of propagating errors back to a network. It was the discovery of fast methods for computing the gradient of loss function of a network [15] that led to applications of neural networks to practical tasks.

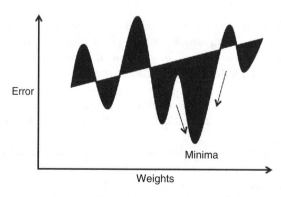

Figure 6.4 Gradient of a single weight.

Back-propagation works in two stages:

- *Propagation*
  - *Forward propagation*: It performs a forward pass to generate output activations based on the provided input.
  - *Backward propagation*: The difference in actual and predicted value (loss) is computed for every neuron in the neural network (hidden, output) based on the output activations.
- *Parameter update*
  - *Gradient computation*: Obtain gradient of the weight from the difference values obtained in the previous step.
  - *Weight update*: Update the weight by subtracting from it a fraction of the gradient value.

The rest of the section describes various steps in detail regarding the sequence in which they appear during a practical implementation. Therefore, Algorithm 6.1 shows pseudo-code for a back-propagation algorithm with one hidden layer to clarify upon the implementation perspective; it sets the context for the upcoming discussion.

Algorithm 6.1   Back-propagation for a network with one hidden layer

---

1. **procedure** Back-propagation
2. **loop**:
3.   **for each training example**
4.   prediction = predicted value // forward pass
5.   actual = actual value
6.   compute error = f(||prediction - actual||)
7.   compute partial derivatives // backward pass
8.   update weights //parameter update
9. **until** stopping criteria //For ex. error rate or number of iterations

---

Propagation: Back-propagation is based on recursively applying the *chain rule* to the gradients to update the parameter values. Let us understand it with the help of an example. Suppose there is a function given as:

$$f(a, b, c) = (a + b) * c$$

The above function may be rewritten as:

$$f(x, c) = x * c \text{ where } x = (a + b)$$

The gradients (or the partial derivatives) are computed as:

$$\frac{\partial f}{\partial x} = c \text{ and } \frac{\partial f}{\partial c} = x. \quad \text{(forward pass)}$$

Since we want the derivatives with reference to a, b, and c, the chain rule comes into play. It says that the required derivatives may be computed as:

$$\frac{\partial f}{\partial c} = x, \frac{\partial f}{\partial a} = \frac{\partial f}{\partial x}\frac{\partial x}{\partial a}, \frac{\partial f}{\partial b} = \frac{\partial f}{\partial x}\frac{\partial x}{\partial b}. \quad \text{(Backward pass)}$$

It can be observed that these equations simply require multiplication of gradient values. (In this particular example, $\frac{\partial x}{\partial a} = 1$ and $\frac{\partial x}{\partial b} = 1$.) To further enhance the understanding, let us work out a numerical example:

Back-Propagation Example

In the discussed example, let $a = -4, b = 5,$ and $c = -2$.

Forward pass
$x = a + b$ (gives x = 1)
$f = x * c$ (gives f = − 2)

Backward pass
$\frac{\partial f}{\partial c} = x$ (gradient on c = 1 as x = 1 from above)

$\frac{\partial f}{\partial x} = c$ (gradient of x = −2 as c = −2 from above)

Back-propagation through x
$\frac{\partial f}{\partial a} = 1.0 * \frac{\partial f}{\partial x} = -2$

$\frac{\partial f}{\partial b} = 1.0 * \frac{\partial f}{\partial x} = -2$

Parameter update: When the weights are adjusted in a network, the error of the network should decrease. There are different ways to update the weights in the network. The weights can be updated using online mode, batch mode, or stochastic gradient descent.

- *Online mode*: In online mode, the weights are updated for each input element. The number of updates is equal to the number of input elements.
- *Batch mode*: The gradients are summed for each input element up to the limit of the specified batch size. The weights are updated once a batch has been processed. The number of updates therefore is $\frac{InputSize}{batchSize}$.
- *Stochastic gradient descent (SGD)*: The third and the most popular (also recommended) way of updating weights is SGD. It can work in both online and offline mode. While training in online mode, the SGD selects an input element at random, computes the gradient, and updates the weight. This continues until the desired error is attained. Similarly, in batch mode, the SGD randomly selects B input elements where B is the prespecified batch size. The process is similar to the batch mode discussed earlier, with the only difference being that each batch has randomly chosen elements.

Mathematically, weights are updated as:

$$\nabla w_i = \alpha \nabla w_{i-1} + (-\eta)\frac{\partial C}{\partial w_i} \qquad (6.8)$$

where, ∇w_i is the weight in i^{th} iteration, C is the error or the cost as given by equation (6.6), and, α is an optional scaling factor. α gives the percentage of the gradient on weight from a previous iteration that should be applied in the current iteration. It helps in algorithms to avoid local minima. η is the *learning rate*. It impacts the speed of the learning process. If it is set too high, then the network may never converge because it may simply be hopping across the minima, while if it is too low, it may take too much time to converge.

Now that we have discussed the basics of neural networks and learning paradigms, we now move toward discussing the convolution operation and its application on images.

6.2.3   Convolution

Convolution is a mathematical operation that is very popular in the domain of signal processing for mixing signals with a specific rule. For discrete signals (we would consider discrete convolutions as images are discrete) f and g, convolution over a set of integers Z is defined as:

$$(f * g)[n] = \sum_{k=-\infty}^{\infty} f[k]g[n-k] = \sum_{k=-\infty}^{\infty} f[n-k]g[k] \tag{6.9}$$

Equation (6.9) shows that convolution of two functions can be computed by sliding one function and performing computation over the overlapping values. A similar concept, when applied in two dimensions, can be used for performing convolution with images.

Applying convolution to images: Convolution to images is applied in two dimensions (i.e., along the width and height). Here the convolution is performed between a region in the image and an equally sized convolution kernel. A *convolution kernel* is a floating point matrix that is applied over the image. For example, if the size of the image is 256 × 256 and the chosen kernel has a size of 3 × 3, then the convolution is performed on equivalent-sized patches from the input images (i.e., 3 × 3). The overlapping values are then multiplied and summed to obtain the resultant value for one pixel (the center pixel of the local patch). The kernel is moved across the image, hence producing a 2D response of the convolution kernel to the image. In terms of deep learning, this 2D output is called the *feature map*. The above process can be mathematically expressed as:

$$response = I * K = \sum_{c=-\frac{n-1}{2}}^{\frac{n+1}{2}} \left(\sum_{r=-\frac{m-1}{2}}^{\frac{m+1}{2}} I(a+r, b+c)K(r,c) \right) \tag{6.10}$$

where I is the input image, K is the *mxn* convolution kernel, (a, b) is the image pixel for which convolution is being computed, and m and n are chosen as odd so that the kernel is symmetric about the center. Figure 6.5 pictorially shows how a convolution kernel is applied over a patch in the image. In the shown example, the convolution kernel is placed at the center of the image (the pixel with a dark black border). The response at this particular pixel is computed by multiplication of corresponding elements and their summation. When this process is repeated for all the pixels, a 2D feature map is formed. It should be noted that padding may need to be applied for computing the response at the border pixels.

The definition of the convolution kernel allows us to perform many useful operations on images. Figure 6.6 shows a few such kernels.

Use of convolution in machine learning: Different convolution kernels can be used to extract specific features from images such as edges, blobs, and so on (Figure 6.6). Such features are generally used in machine learning algorithms to perform a variety of higher level tasks, such as object detection, recognition, classification, and so on. Now, the

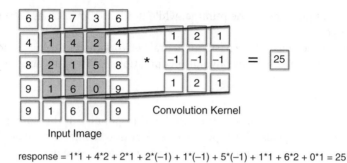

response = 1*1 + 4*2 + 2*1 + 2*(−1) + 1*(−1) + 5*(−1) + 1*1 + 6*2 + 0*1 = 25

Figure 6.5 Example of applying convolution to images.

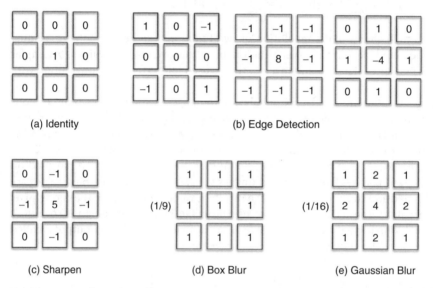

Figure 6.6 Convolution kernels.

performance of such tasks is directly dependent upon the quality of input features. Let us take the edge detection convolution kernel as an example to clarify the importance of convolution in machine learning. As can be seen from Figure 6.6b, that for finding edges at different orientations, one needs to adjust the weights in the kernel. Figure 6.7 shows an example of two edge detection techniques. The Sobel operator (Figure 6.8) is a simple matrix operation where the weights in the matrix govern the detection of edges at a particular orientation, while the Canny edge detector is a multistep edge detection algorithm. Discussion of the Canny edge detector is beyond the scope of this chapter. It has been shown here to emphasize that to obtain relevant features, one needs to identify specific techniques that are suitable to the problem domain. This is also known as feature engineering. Since convolution operations can be used to extract low-level features from images, and since features are crucial for any machine learning algorithm, they become decisive in obtaining state-of-the-art results in almost any problem scenario. In the next section, we will see that CNNs try to solve exactly this problem. They attempt to learn weights of convolution kernels to detect various types of low-level features.

(a) Original Image

(b) Sobel Edge Detector   (c) Canny Edge Detector

Figure 6.7 Edge detection in images.

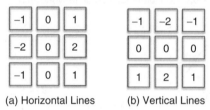

(a) Horizontal Lines   (b) Vertical Lines

Figure 6.8 Sobel operator.

6.2.4 Convolutional Neural Networks: The Architecture

CNNs have multilayer architectures similar to those of the neural networks discussed in the previous section. The CNN architecture, in general, comprises the following layers:

- Convolution (Conv) layer
- Pooling layer
- Fully connected (FC) layer

A general architecture diagram for a CNN is shown in Figure 6.9.

The core working of CNNs is similar to that of neural networks, where the input is given to neurons, certain operations are performed on them, and weights and biases are learned based on a loss function. The only additional assumption is that CNNs explicitly take only raw image pixels as inputs. This assumption allows several modifications in the architecture, resulting in a more efficient network with a far smaller number of parameters as compared to a traditional neural network.

Arrangement of neurons: The input to a CNN are raw images. Correspondingly, the neurons are arranged as a 3D volume I i.e., width, height, and depth). More precisely, for a $N \times M$-sized image with R, G, and B channels (#channels = depth = 3), the neurons at the input layer are arranged in a volume with dimensions $N \times M \times 3$. Additionally, unlike regular neural networks, the layers of CNNs are not necessarily fully connected. This is sometimes also called **sparse connectivity** and is shown in Figure 6.10.

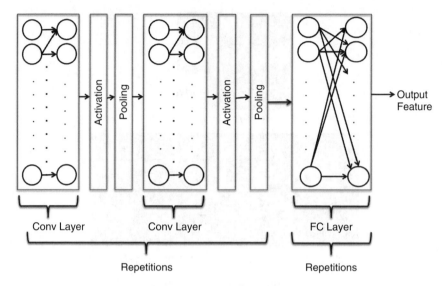

Figure 6.9 Convolutional neural network: architecture.

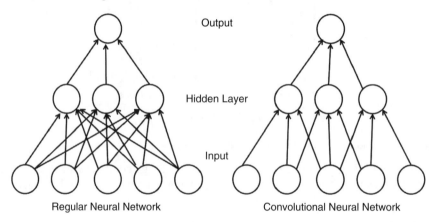

Figure 6.10 Connections in a regular neural network versus a convolutional neural network.

Each layer has a specific purpose and has associated hyperparameters that must be specified for a CNN to work. We now describe the working of each layer.

6.2.4.1 Convolution Layer

The convolution layer is the most important part of a CNN. We begin by providing an intuitive explanation of the working of convolution layers, followed by more intricate details.

Intuitive explanation: The main purpose of the convolution layer is to detect features from images (lines, edges, etc.). The convolution layer consists of a set of filters that can be learned in order to detect these features. The filters are small in terms of width and height, but extend to the complete depth of the input image or, more precisely, the input volume as described in this chapter. Now **convolution** between the input volume and filter is performed by sliding the filter across the width and height of the input volume while computing the dot product on the overlapping values at a location. The dot products over a spatial region result in a 2D activation map containing the filter response at every location. Intuitively, it can be understood that these activation maps are in fact the responses to the features in an image such as edges on the first layer. Now considering that there are multiple convolution layers in a network, at higher layers the activation may correspond to more complex features such as object shapes or patterns. An interesting question to consider here is what must be the output of the convolution layer, since the subsequent layers are generally convolution layers again. As mentioned, the input to the first convolution layer is a 3D volume (RGB image). Therefore, to stack multiple convolution layers, the output has to be a volume as well. Hence, the output volume is formed by stacking along the depth dimension, the 2D activation maps of each of the filters in a convolution layer.

Hyperparameters: There are many parameters that are considered while designing the convolution layers; these parameters are termed as hyperparameters since they decide the size of the output volume.

- *Depth*: A filter is a 2D matrix whose elements are learned by a CNN. The filters are responsible for detection of features, such as edges, lines, blobs, and so on. The depth here is related to the number of filters that would be used. The higher the number of filters, the more complex features a CNN can learn.
- *Size of filter*: The size of the filter in a CNN essentially allows us to convey how local the low-level features are in an image. For example, a filter of size 1×1 essentially means that features are local on the pixel level, where they do not have any relation with the neighboring pixels. On the other side, a filter of the size of the images would mean that there would be connections among every input pixel to every neuron in the convolution layer (i.e., it becomes a fully connected layer). In practice, the size of the filter is kept small (3×3 or 5×5).
- *Stride*: Stride is the number of pixels a filter is shifted while performing the convolution. The stride is chosen such that if the filter begins operating at the top-left pixel, it should be able to reach the bottom right of the image. The stride is at least 1 (else the filter would not move). The common values for stride are 1 or 2. Larger strides are uncommon in the case of images, but the reader should not deduce that they are not useful. Recursive neural networks have been used with larger strides to form tree-like structures and have obtained impressive results for natural language processing applications [36].
- *Zero-padding*: It controls the number of rows/columns of zeros that must be added to the image border. With zero-padding, the convolution can be applied to the border pixels of an image and hence can be used to control the size of the output volume.

Input and output volumes: As discussed in this chapter, the input to the convolution layer is a volume of size $N \times M \times 3$ for an $N \times M$ RGB image. Each filter in the convolution layer sweeps across the input with the given *stride* and *padding*. Here, stride is the steps in which the filter moves across the input, while padding is the number of zero border pixels in the area where the filter is operating. Technically, this is viewed as a convolution of the filter with the input. The output for each filter is stacked together in the dimension of the depth and is said to be the output volume of the convolution layer.

It is important to mention again that both the input and output to the convolution layer are volumes. If the input to the layer are image pixels, then the input volume is $N \times M \times 3$ for an RGB image. The number of weights for an input volume with depth D would be:

$$\#Weights = F_S \times F - S \times D \tag{6.11}$$

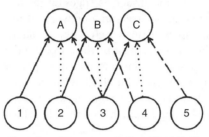

Strong lines (from neurons 1, 2, and 3), dotted lines (from neurons 2, 3, and 4) and broken lines (from neurons 3, 4, 5) show weight sharing among respective connections.

Figure 6.11 Example of parameter sharing among neurons.

For example, an input image and filter of sizes $10 \times 10 \times 37$ (7×7), would result in $7 \times 7 \times 3 = 147$ number of weights. This implies that the filters are connected locally to regions in an image but are fully connected in the input depth dimension.

The size of the output volume ($O \times O$) is governed by the size of the input volume ($I \times I$), the filter size or the receptive field size (F_S), padding (P), and stride (S), and is given as:

$$O = \frac{I - F_S + 2P}{S} + 1 \tag{6.12}$$

Now, if the convolution layer has a depth D, the output volume would have a dimension of $O \times O \times D$. This output volume either is given as is to another convolution layer as input or is down-sampled by passing it through a pooling layer.

Parameter sharing: Parameter sharing helps in controlling the number of parameters. It allows for achieving computational and memory-efficient implementation of CNNs. It works with the assumption that if a patch has been used for computation at one spatial location, it would also be useful for computation at another location. An example of parameter (weight) sharing is shown in Figure 6.11.

Example: To clarify upon the concepts discussed here, we now discuss a numerical example demonstrating the computations being performed in a convolution layer. Let us consider an architecture with the following parameters:

Size of Input Image (I) = 227×227 Size of Filter or Receptive Field Size (F_S) = 13

Stride (S) = 2

Padding (P) = 0

Depth of Convolution Layer (D) = 96

Using equation (6.12), the size of the output volume can be computed as:

$$O = \frac{227 - 13 + 0}{2} + 1 = 108$$

which results in a output volume of size:

Size of Output Volume = $108 \times 108 \times 96$

This means that each of $108 \times 108 \times 96$ neurons in this volume is connected to the same $13 \times 13 \times 3$ region in the input volume. We can also observe that the number of neurons in the output volume is $108 \times 108 \times 96 = 1,119,744$. Each neuron has $13 \times 13 \times 3 = 507$ weights and a bias. This results in $1,119,744 \times 507 = 567,710,208$ parameters in the first convolution layer itself. A billion parameters in the first layer (or any layer for, that matter) is clearly an exorbitantly high number. This is where *parameter sharing*, as discussed earlier, becomes useful. If we consider that all the neurons in a depth level ($D = 96$ in this example) have the same weight and bias, it would result in $96 \times 13 \times 13 \times 3 = 48,672$ weights or 48,768 parameters with 96 biases, which dramatically reduces the computational complexity and memory footprint of the operations being performed on this layer.

6.2.4.2 Pooling Layer

This layer is responsible for progressively down-sampling the input volume from the convolution layer. In other words, it condenses the feature map produced by the convolution layer. It is an optional layer and is put between successive convolution layers. It serves the following purposes:

- Reduces the number of parameters
- Avoids overfitting.

Hyperparameters: It has two hyperparameters:

- Spatial extent (F): It means that patches of $F \times F$ would be scaled down to one pixel.
- Stride (S): This is the number of pixels by which the pooling operation moves across the input.

The commonly used values for the hyperparameters are $F = 2$ and $S = 2$. It is important to understand here that if pooling is performed with very large spatial extents, information from a larger region gets diluted. Therefore, pooling with large receptive fields (spatial extents) is not recommended and in practice has a negative impact on the results.

Input and output volumes: The output of the convolution layer is provided as input to the pooling layer. Since this layer doesn't consist of any weight, training has no effect on it. If an input volume with dimensions $N \times M \times D$ is given as input to the pooling layer, it reduces it to a volume of size $N' \times M' \times D$ with the following mapping functions:

$$N' = \frac{N - F}{S} + 1$$
$$M' = \frac{M - F}{S} + 1$$

Types of pooling: The most common form of pooling layer is *max pooling*. With max pooling and the commonly used setting of $F = 2$, $S = 2$, each 2×2 grid would be replaced with the maximum value in that grid. This also results in 75% loss of pixel information (i.e., a 4×4 grid would become 2×2). *Average pooling* works similarly to max pooling, but instead of replacing the resulting grid with the maximum value, average pooling replaces it with the average of the region. The operation is shown with the help of an example in Figure 6.12. This process is repeated for each of the D feature maps of size $N \times M$ in the input volume.

Though max pooling is the most commonly used, there are other forms as well, such as L2-norm pooling, fractional max pooling, 3×3 pooling region [27], stochastic pooling [37], or just removing the pooling layer altogether [38].

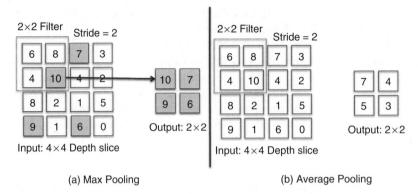

(a) Max Pooling   (b) Average Pooling

Figure 6.12 Example of pooling.

6.2.4.3 Dense or Fully Connected Layer

The final layer in a CNN is generally a dense layer. It is a fully connected layer from the output volume of the convolutional/pooling layer to neurons in this layer. The CNN architecture can contain multiple dense layers, with the last layer performing the task of classification. As discussed in this chapter, the final layer uses softmax activation while other layers may use any activation function, with ReLU being the recommended activation function.

The reason that fully connected layers are used toward the end is twofold. First, the convolution layer is better at exploiting the spatial structure in the input image. Second, fully connected layers require a huge number of parameters, as was explained while discussing the convolution layer. Therefore, if fully connected layers are used at the beginning, not only would it be computationally inefficient but also millions of redundant parameters would be needed.

6.2.5 Considerations in Implementation of CNNs

- *Hardware considerations*:
  - *GPU*: Almost all the modern CNN architecture is based on GPUs. In fact, the progress of CNNs was halted for many years due to unavailability of high-performance computing machines. Most of the recent research reports result on Nvidia's GTX TitanX or GTX 980. These GPUs support efficient convolution operations and hence are the recommended choice for executing experiments involving CNNs.
  - *CPU*: Most of the code of CNNs is executed on GPUs, but the CPU still plays an important role as it is the hardware component that initiates calls to GPU routines, reads and writes variables, executes instructions, and creates mini-batches, among many other low-level tasks. Most of the deep learning libraries utilize a single CPU core (if the GPU is also used). Therefore, it is recommended to select CPUs such that there is one CPU core per GPU. Also, note that it is better to select a CPU that can allocate two threads per GPU since many asynchronous function calls rely on a second CPU thread to relay information.
  - *RAM*: As a rule of thumb, a system supposed to be used for executing CNNs should have at least as much RAM as that of the GPU. Additional RAM is always useful, since the larger the RAM, the larger the chunk of data set (or mini-batches)

that can fit into it. Given that a GPU operates on mini-batches and switches them very frequently, having a number of mini-batches readily available in RAM would help in saving many CPU clock cycles, which may get wasted in disk read/write operations, although it must be noted here that such efficiency is implementation specific.

- *Hard disk drive (HDD)*: The rate at which the GPU processes mini-batches decides efficient read/write operations from HDD are needed. Usually, multiple mini-batches are needed to be read asynchronously. As the GPU speeds are increasing, it is recommended to use solid-state drives (SSDs) to keep pace with it.
- *Computational considerations*: CNNs require a very huge amount of memory, which presents many bottlenecks in their implementation. Therefore, the following factors need to be taken care of while implementing CNNs:
  - *Size of intermediate volumes*: Most of the activations in a CNN are concentrated toward the first few layers. An efficient CNN implementation should only store current activations at any layer. This kind of optimization is most suited at test time.
  - *Parameter size*: Since the data structure storing parameters involve storing network parameters, the gradients during back-propagation, weights, and a few architecture-specific values, it is recommended that while allocating memory for the parameter vector, it should be at least thrice the size of the network parameters.
  - *Split the image data*: Since CNNs operate on very large data sets, it is advisable to split the image data (e.g., into mini-batches).
- *Avoiding overfitting*: Overfitting in a neural network can be reduced by regularization. Popular regularization techniques are L1 regularization, L2 regularization, and dropout [39]. These are discussed here:
  - *L1 regularization*: In this form of regularization, for each weight w, a term $\lambda|w|$ is added to the objective function where λ represents the strength of regularization. L1 normalization causes the weight vectors to become sparse, and hence they become invariant to noisy inputs.
  - *L2 regularization*: In this from, for every weight w, a term $\frac{1}{2}\lambda * w^2$ is added to the objective function. It has the property that it diffuses the weight vectors. This allows the network to utilize a majority of inputs instead of only a few of them.
  - *Dropout*: Dropout means switching off a few neurons while training the network. The idea is that it will force the remaining network to optimize without the dropped neurons. Dropout can be implemented as a layer. Such a layer acts as a regular fully connected layer with periodic dropout of neurons while training. It is important to note that dropout does not permanently remove any neuron, hence the size of the network remains the same after the training. In fact, once the training is over, all the neurons are used by the network to compute the result.

6.2.6 CNN in Action

In this section, we introduce the readers to practical examples using CNNs. To get you started with CNNs, we discuss various tools that can be used for developing CNNs as well as work through some coding examples.

6.2.7 Tools for Convolutional Neural Networks

- *Caffe*: Caffe [40] is a deep learning framework developed by Berkley Vision and Learning Center. The primary programming language is C++, but it has wrappers for many languages. It is one of the most popular tools for implementing CNN-based algorithms, primarily owing to its highly efficient implementation.
- *Torch*: This is an open-source machine learning library [41]. It is implemented in C and uses LuaJIT scripting language for programming.
- *MatConvNet*: MatConvNet [42] is a recent MATLAB Toolbox specifically targeted toward computer vision problems. It is based on MATLAB and has many pretrained models.
- *MATLAB 2016a*: MATLAB 2016a introduces a deep learning toolbox where many of the CNN-related operations are supported.

6.2.8 CNN Coding Examples

We now introduce the reader to MatConvNet and show an example of how to visualize the output of various layers in CNN architecture.

6.2.8.1 MatConvNet

MatConvNet has a simple installation process and easy-to-follow documentation. Here, we introduce the reader to how to set up MatConvNet and a few other basic details, but we recommend an enthusiastic reader to look into the official documentation [43] for more insights. The following code snippet allows you to set up MatConvNet on your system. The commands have to be executed from within your MATLAB Workspace. We have tested these commands on MATLAB 2014b. The example installs MatConvNet and shows you a simple example of how to classify an input image with a pretrained model. We have added comments to the relevant sections of the code to explain their working.

```
% install MatConvNet
untar('<path to matconvnet tar>') ;
cd <path to extracted directory>

% compile MatConvNet
run matlab/vl_compilenn

% download a pre-trained CNN from
% http://www.vlfeat.org/matconvnet/models/
urlwrite(...
  '<path to mat file from
  http://www.vlfeat.org/matconvnet/models/', ...
  'model.mat') ;

% setup MatConvNet
run  matlab/vl_setupnn

% load the pre-trained CNN
net = load('model.mat') ;
```

```
% Here we use a default image
% load and preprocess an image (Example from MatConvNet
   website)
im = imread('peppers.png') ;
im_ = single(im) ; % note: 0-255 range

% resize the image before providing it to CNN
im_ = imresize(im_, net.meta.normalization.imageSize(1:2)) ;

%Normalize an image to reduce noise
im_ = im_ - net.meta.normalization.averageImage ;

% execute CNN
res = vl_simplenn(net, im_) ;

% show the classification result

%extract scores
scores = squeeze(gather(res(end).x)) ;

%find the highest score from the softmax layer
[bestScore, best] = max(scores) ;

%display the image
figure(1) ; clf ; imagesc(im) ;
title(sprintf('%s (%d), score %.3f',...
net.meta.classes.description{best}, best, bestScore));
```

6.2.8.2 Visualizing a CNN

The best way to understand a CNN is to visualize the outputs of its various layers. The following code example loads images for you and extracts the image outputs from various CNN layers.

```
clear all;close all;clc;
run ./matconvnet-1.0-beta17/matlab/vl_setupnn

net = load('imagenet-vgg-m.mat') ;

% read image
I=imread('lena.jpg');
imshow(I);
feat = [];

im_ = single(I) ;
im_ = imresize(im_, net.meta.normalization.imageSize(1:2)) ;
im_ = im_ - net.meta.normalization.averageImage ;
figure(5) ; clf ; imagesc(im_)
```

```
res = vl_simplenn(net, im_) ;

% find feature vector
% first layer res(5), for 2nd layer  res(9)
% for 3rd layer res(11), for 4th layer  res(13) and so on
% because each convolution layer has four sublayers
featureVector = res(5).x;

featureVector = featureVector(:);
feature = [feature; featureVector'];
fprintf('extract %d image\n\n', i);
feat_norm = feature;

%% for 2d image matrix  for all output image
ft=featureVector;

%% for first layer image dimension is 54*54 and the depth
    is 96
ty=54*54;
d = zeros(54,54,96);

for i=1:96
        n=i-1
    a= ft((ty*n+1):(n+1)*ty);
    b=vec2mat(a,54);
    d(:,:,i) = b';
end

% for displaying image
figure;
n=0 % 1,2,3,4,5,6,7,8,9
%display first nine images
% the reader may change this to visualize other images
for i= 1:9
   subplot(3,3,i);
   J=(i+n*9)
   imshow(d(:,:,J));
end
```

The above code uses VGGNet [44] Architecture implementation of CNN, which earned second place in the 2014 ImageNet Large Scale Visual Recognition Challenge (ILSVRC) [45]. It showed that the depth of a network is an important factor for the performance of a CNN. Their network consisted of 16 convolutional and fully connected layers. Surprisingly, the architecture is very simple, performing 3×3 convolution operations (Stride = 1, Padding = 1) and 2×2 pooling operations (max pooling; Stride = 2) throughout the network. Due to its simplicity and ability to give strong features, we have chosen it for demonstration in this example. MatConvNet and Caffe both provide pretrained models for this architecture.

Here, we show and discuss the output of VGGNet at various layers for an example image with the help of the above code. Due to space considerations, we will show only

Figure 6.13 Input image.

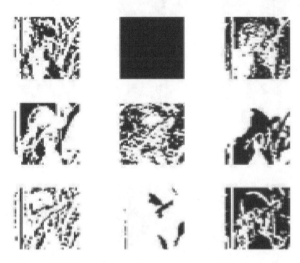

Figure 6.14 Output from the first sublayer of the first layer.

nine output images from each layer. The reader is encouraged to execute the code to visualize all the output images. Figure 6.13 is our input image to the code. The first layer (i.e., the convolutional layer of the network) has four sublayers: convolution (conv1), ReLU (relu1), normalization (norm1), and pooling (pool1). Figure 6.14 shows the output of the first sublayer (i.e. the conv1 layer), while Figure 6.15 shows the output after the completion of the first layer. As can be observed, the first layer extracts the edges from the image. This is a crucial part in any image analysis or, more specifically, visual recognition task. The research for identification and localization of objects spans decades [46–49]. Figures 6.16–6.18, and 6.19 show the output up to the fifth layer of the network. As you can observe, the identified elements in the images go on becoming

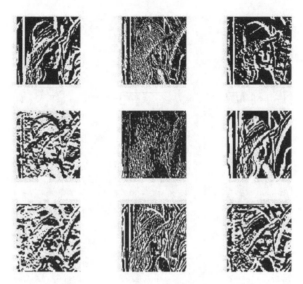

Figure 6.15 Output from the first layer.

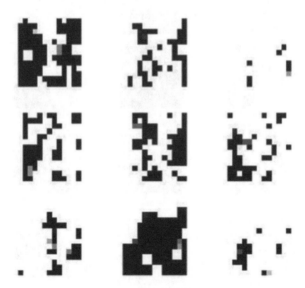

Figure 6.16 Output from the second layer.

coarser, with the fifth layer showing rectangular boxes over the regions of interest. This behavior can be compared with basic image segmentation and object localization tasks where a rectangular window is moved over an image as a mask to identify regions of interest. In fact, this is how our visual system works as well [3, 50]. This demonstrates that CNNs also learn filters, which agrees with the fundamental works in the field of machine learning and artificial intelligence, with a twist that they learn it without specifying what filters to use (as if humans have bestowed CNNs with powers to learn by themselves, but as with everything, it is still far from being ideal!).

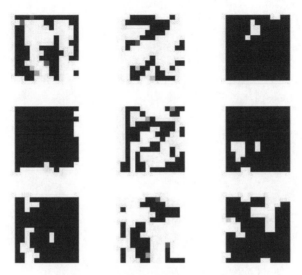

Figure 6.17 Output from the third layer.

Figure 6.18 Output from the fourth layer.

6.2.8.3 Image Category Classification Using Deep Learning

For this coding example, we use the framework provided by MATLAB 2016a and MatConvNet.

```
% Download the compressed data set from the following
   location
url = 'http://www.vision.caltech.edu/Image_Datasets/
   Caltech101/
101_ObjectCategories.tar.gz';
% Store the output in a temporary folder
% define output folder
outputFolder = fullfile(tempdir, 'caltech101');

% download only once
```

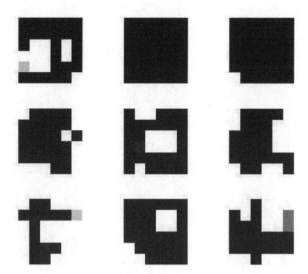

Figure 6.19 Output from the fifth layer.

```
if ~exist(outputFolder, 'dir')
    disp('Downloading 126MB Caltech101 data set...');
    untar(url, outputFolder);
end

rootFolder = fullfile(outputFolder, '101_ObjectCategories');
categories = {'airplanes', 'ferry', 'laptop'};

imds = imageDatastore(fullfile(rootFolder, categories),
'LabelSource', 'foldernames');

tbl = countEachLabel(imds);

% determine the smallest amount of images in a category
minSetCount = min(tbl{:,2});

% Use splitEachLabel method to trim the set.
imds = splitEachLabel(imds, minSetCount, 'randomize');

% Notice that each set now has exactly the same number of
    images.
countEachLabel(imds);

% Find the first instance of an image for each category
airplanes = find(imds.Labels == 'airplanes', 1);
ferry = find(imds.Labels == 'ferry', 1);
laptop = find(imds.Labels == 'laptop', 1);
```

```
figure
subplot(1,3,1);
imshow(imds.Files{airplanes})
subplot(1,3,2);
imshow(imds.Files{ferry})
subplot(1,3,3);
imshow(imds.Files{laptop});

%Download Pre-trained Convolutional Neural Network (CNN)
% Location of pre-trained "AlexNet"
cnnURL = 'http://www.vlfeat.org/matconvnet/models/beta16/
imagenet-caffe-alex.mat';
% Store CNN model in a temporary folder
cnnMatFile = fullfile(tempdir, 'imagenet-caffe-alex.mat');

if ~exist(cnnMatFile, 'file') % download only once
    disp('Downloading pre-trained CNN model...');
    websave(cnnMatFile, cnnURL);
end

% Load MatConvNet network into a SeriesNetwork
convnet = helperImportMatConvNet(cnnMatFile);

% Set the ImageDatastore ReadFcn
imds.ReadFcn = @(filename)readAndPreprocessImage(filename);

[trainingSet, testSet] = splitEachLabel(imds, 0.3,
    'randomize');

% Get the network weights for the second convolutional layer
w1 = convnet.Layers(2).Weights;

% Scale and resize the weights for visualization
w1 = mat2gray(w1);
w1 = imresize(w1,5);

% Display a montage of network weights.
% There are 96 individual sets of weights in the first
    layer.
figure
montage(w1)
title('First convolutional layer weights')

featureLayer = 'fc7';
trainingFeatures = activations(convnet, trainingSet,
    featureLayer, ...
```

```
    'MiniBatchSize', 32, 'OutputAs', 'columns');

% Train a multi-class SVM
% Get training labels from the trainingSet
trainingLabels = trainingSet.Labels;

% Train multiclass SVM classifier using a fast linear
    solver,
%and set 'ObservationsIn' to 'columns' to match the
    arrangement used for
% training features.
classifier = fitcecoc(trainingFeatures, trainingLabels, ...
    'Learners', 'Linear', 'Coding', 'onevsall',
    'ObservationsIn',
    'columns');

% Classifier Evaluation
% Extract test features using the CNN
testFeatures = activations(convnet, testSet, featureLayer,
'MiniBatchSize',32);

% Pass CNN image features to trained classifier
predictedLabels = predict(classifier, testFeatures);

% Get the known labels
testLabels = testSet.Labels;

% Tabulate the results using a confusion matrix.
confMat = confusionmat(testLabels, predictedLabels);

% Convert confusion matrix into percentage form
confMat = bsxfun(@rdivide,confMat,sum(confMat,2))

% Display the mean accuracy
mean(diag(confMat))

%% Test the classifier
newImage = fullfile(rootFolder, 'airplanes', 'image_0690
    .jpg');

% Pre-process the images as required for the CNN
img = readAndPreprocessImage(newImage);

% Extract image features using the CNN
imageFeatures = activations(convnet, img, featureLayer);

% Make a prediction using the classifier
```

```
label = predict(classifier, imageFeatures)

function Iout = readAndPreprocessImage(filename)

        I = imread(filename);
        if ismatrix(I)
            I = cat(3,I,I,I);
        end

        % Resize the image as required for the CNN.
        Iout = imresize(I, [227 227]);

    end
```

6.3   Toward Understanding the Brain, CNNs, and Images

We have already discussed neural networks and CNNs, and have related their working to abstract biological functioning of the human brain and visual system. Therefore, we can now move toward mentioning a few works that give more insight into CNNs, their similarity with the functioning of the human brain, and their applications. We believe that a reader wanting to go deeper into working with CNNs and brain-inspired algorithms should definitely go into the details of these works, which would provide a pathway for the reader to explore further.

6.3.1   Applications

Nearly all the problem areas in computer vision have witnessed the application of CNNs. In fact, much of the success of CNNs can be attributed to the industry, which contributed significantly by providing high-performance computing infrastructure and open-source implementations of various deep learning architectures. It would be not only unfair but also difficult to differentiate industrial research from academic research, as most of the recent research work has happened with a swift collaboration among them. Therefore, we discuss the state-of-the-art work categorized by application areas where CNNs have had a dramatic impact.

- *Object classification*: The hysterical popularity of CNNs in recent times can be attributed to the work on ImageNet classification in 2012 [27]. Despite being a recent paper and provided that, in the last few years, CNNs have already seen better and bigger architectures [51–53] (discussed in Section 6.3.2), this work stands out as classic.
- *Object memorability*: Object memorability is a concept that aims at finding entities in an image that are worth remembering. What makes it interesting is that while our brain is exceptionally good at selective extraction of information from visual stimulus, a recent research paper entitled "What makes an object memorable?" [54] by a group at the Massachusetts Institute of Technology (MIT) has shown that CNNs can be used to some extent to perform such kind of selection. It extends upon previous works on image memorability [55] and photograph memorability [56].

- *Object segmentation*: The authors in [57] obtained CNN-based region proposals from images and subsequently used them for object segmentation. This is among the first work to use CNNs for region proposals. The authors in [58] proposed a fully convolutional network for semantic segmentation, which is obtained by using a skip architecture augmented by a shallow layer trained with appearance models. Following this, the authors in [59] propose an end-to-end architecture for obtaining object boundaries. The architecture is based upon the VGGNet [44].

  Additionally, we would like to remind the reader that CNNs are still far from achieving human-level intelligence. Hence, we also discuss a few attempts by the research community in trying to find limiting cases of CNNs or exploring a different learning paradigm altogether.

- *Fooling CNNs*: With the help of evolutionary algorithms, the authors in [32] demonstrated that CNNs can be made to incorrectly recognize images having no meaning to a human. In fact, the images had no pattern at all. However, despite this failure of CNNs, there are two key takeaways. First, a CNN extracts more low- and middle-level features as compared to features such as boundary, shape, and so on, which are considered high-level features. Second, the primary reason why it was fooled was because it learned patterns irrespective of what an image actually contains.

- *Hierarchical and temporal machine*: There is another school of thought that says that if humans can learn from single-input instances, why can't a machine do the same? Such techniques are based on the observation that the brain works with both temporal and spatial patterns rather than only spatial ones (in the case of visual stimulus). For example, everything that the brain sees is a temporal pattern or simply a short video sequence, while traditional learning architectures including CNNs are based on static images. Theis class of techniques claiming to actually function like the brain is based on hierarchical and temporal models. These techniques haven't yet evolved to a stage where they may work with images, but they should be a useful read as we still do not understand the brain completely. The PhD thesis of Dileep George [60] from Stanford University provides in-depth analysis of this paradigm along with comparison to contemporary algorithms.

6.3.2 Case Studies

There is a CNN architecture behind the success of algorithms in each of the application areas discussed in this chapter. Therefore, in this section, we give a brief overview of the evolution of CNN architectures and differences among them. The architectures are based on the fundamental organization discussed in Section 6.2.4. For detailed explanation, we recommend readers to refer to the corresponding publications.

- *LeNet* [61]: This architecture was developed in the 1990s for digit recognition in zip codes and the like. This was the first framework that formalized the use of CNNs in practical applications with the use of convolutional, max pooling, and dense layers. This architecture was similar to the fundamental architecture discussed in this chapter.

- *AlexNet* [27]: The architecture was proposed in 2012 and can be actually considered as the baseline for the modern CNN's popularity in computer vision tasks. AlexNet won the ILSVRC in 2012 by significantly outperforming the contemporary techniques. It acheived a top-five error rate of 16%, as compared to 26% achieved by the runner-up.

The primary difference was that this network had more layers, had a larger number of parameters, and utilized the power of GPU computation. In fact, the authors had written an extremely efficient implementation of convolution on the GPU. The architecture ran on two 3 GB GPUs with clever offloading of computations among them to gain significant performance optimizations. Interestingly, they showed that ignoring certain computations (e.g., those among layers not residing on the same GPU) allowed them to achieve better error rates.

- *ZFNet* [62]: This architecture won the ILSVRC in 2013. It tweaked various hyper-parameters, specifically reducing the stride and size of the filters in the first few layers. It also increased the size of the intermediate convolution layers, which resulted in significant performance gain.
- *GoogleNet* [53]: GoogleNet won the ILSVRC in 2014. It proposed an *inception module* that significantly reduced the number of parameters. As discussed in Section 6.2.5, the network parameters present a huge computational bottleneck. In fact, if we consider that the progress of CNNs was at halt for decades due to these bottlenecks, one can better appreciate the contribution of GoogleNet. It reduced the number of parameters to 4 million as compared to AlexNet's 60 million while consisting of 22 layers.
- *VGGNet* [44]: This architecture secured second place in the ILSVRC in 2014. It showed that the depth of a network critically impacts the performance of a network. Throughout the network, convolution is performed by 3×3 filters and 2×2 pooling operations, but it is computationally very expensive and requires almost 2.5 times the number of parameters of AlexNet.
- *ResNet* [63]: ResNet won the ILSVRC in 2015. In order to avoid the vanishing gradient problem, it skips connection while also utilizing batch normalization very heavily. At the time of writing this text, ResNets and its variations are the state-of-the-art CNN architectures, and the reader is recommended to use ResNets for solving practical problems in computer vision.

6.4 Conclusion

In this chapter, we introduced convolutional neural networks as an algorithmic way of understanding how the brain might work. We also explained concepts to allow readers to appreciate that CNNs may not reflect the exact way that brains work but do provide an intuition for development of advanced theories in this already complicated area of understanding the human brain, the visual system, and their behavior. Lastly, we presented coding examples and pointers to future reading materials so that interested readers can expand upon the knowledge acquired from this chapter.

References

1 McCulloch, W.S. and Pitts, W. (1943) A logical calculus of the ideas immanent in nervous activity. *The Bulletin of Mathematical Biophysics*, **5** (4), 115–133.
2 Thorpe, S., Fize, D., Marlot, C. *et al.* (1996) Speed of processing in the human visual system. *Nature*, **381** (6582), 520–522.

3 Blakemore, C.T. and Campbell, F. (1969) On the existence of neurones in the human visual system selectively sensitive to the orientation and size of retinal images. *The Journal of Physiology*, **203** (1), 237.

4 Field, D.J., Hayes, A., and Hess, R.F. (1993) Contour integration by the human visual system: evidence for a local "association field." *Vision Research*, **33** (2), 173–193.

5 Ungerleider, L.G., Courtney, S.M., and Haxby, J.V. (1998) A neural system for human visual working memory. *Proceedings of the National Academy of Sciences*, **95** (3), 883–890.

6 Isik, L., Meyers, E.M., Leibo, J.Z., and Poggio, T. (2014) The dynamics of invariant object recognition in the human visual system. *Journal of Neurophysiology*, **111** (1), 91–102.

7 Bengio, Y. (2009) Learning deep architectures for ai. *Foundations and Trends® in Machine Learning*, **2** (1), 1–127.

8 Schmidhuber, J. (2015) Deep learning in neural networks: an overview. *Neural Networks*, **61**, 85–117.

9 Lee, H., Pham, P., Largman, Y., and Ng, A.Y. (2009) Unsupervised feature learning for audio classification using convolutional deep belief networks, in *Advances in Neural Information Processing Systems*, pp. 1096–1104.

10 Glorot, X., Bordes, A., and Bengio, Y. (2011) Domain adaptation for large-scale sentiment classification: a deep learning approach, in *Proceedings of the 28th International Conference on Machine Learning (ICML-11)*, pp. 513–520.

11 Arel, I., Rose, D.C., and Karnowski, T.P. (2010) Deep machine learning: a new frontier in artificial intelligence research [research frontier]. *Computational Intelligence Magazine, IEEE*, **5** (4), 13–18.

12 Hebb, D.O. (1949) The organization of behavior: a neuropsychological theory. *New York*, **4**.

13 Rosenblatt, F. (1958) The perceptron: a probabilistic model for information storage and organization in the brain. *Psychological Review*, **65** (6), 386.

14 Minsky, M. and Papert, S. (1969) *Perceptrons*. MIT Press, Cambridge, MA.

15 Rumelhart, D.E., Hinton, G.E., and Williams, R.J. (1988) Learning representations by back-propagating errors. *Cognitive Modeling*, **5** (3), 1.

16 Hornik, K., Stinchcombe, M., and White, H. (1989) Multilayer feedforward networks are universal approximators. *Neural Networks*, **2** (5), 359–366.

17 LeCun, Y., Boser, B., Denker, J.S., Henderson, D., Howard, R.E., Hubbard, W., and Jackel, L.D. (1989) Backpropagation applied to handwritten zip code recognition. *Neural Computation*, **1** (4), 541–551.

18 Tesauro, G. (1995) TD-Gammon: a self-teaching backgammon program, in *Applications of Neural Networks*, Springer, pp. 267–285.

19 Thrun, S. (1995) Learning to play the game of chess. *Advances in Neural Information Processing Systems*, **7**.

20 Cortes, C. and Vapnik, V. (1995) Support-vector networks. *Machine Learning*, **20** (3), 273–297.

21 LeCun, Y., Jackel, L., Bottou, L., Brunot, A., Cortes, C., Denker, J., Drucker, H., Guyon, I., Muller, U., Sackinger, E. *et al.* (1995) Comparison of learning algorithms for handwritten digit recognition, in *International Conference on Artificial Neural Networks*, Vol. 60, pp. 53–60.

22 Hinton, G.E., Osindero, S., and Teh, Y.W. (2006) A fast learning algorithm for deep belief nets. *Neural Computation*, **18** (7), 1527–1554.

23 LeCun, Y., Cortes, C., and Burges, C.J. (1998) The MNIST database of handwritten digits. Available from http://yann.lecun.com/exdb/mnist/

24 Bengio, Y., Lamblin, P., Popovici, D., Larochelle, H. *et al.* (2007) Greedy layer-wise training of deep networks. *Advances in Neural Information Processing Systems*, **19**, 153.

25 Glorot, X. and Bengio, Y. (2010) Understanding the difficulty of training deep feed-forward neural networks, in *Aistats*, vol. 9, pp. 249–256.

26 Razavian, A., Azizpour, H., Sullivan, J., and Carlsson, S. (2014) CNN features off-the-shelf: an astounding baseline for recognition, in *Proceedings of the IEEE Conference on Computer Vision and Pattern Recognition Workshops*, pp. 806–813.

27 Krizhevsky, A., Sutskever, I., and Hinton, G.E. (2012) Imagenet classification with deep convolutional neural networks, in *Advances in Neural Information Processing Systems*, pp. 1097–1105.

28 LeCun, Y. and Bengio, Y. (1995) Convolutional networks for images, speech, and time series. *The Handbook of Brain Theory and Neural Networks*, **3361** (10), 1995.

29 Dosovitskiy, A., Tobias Springenberg, J., and Brox, T. (2015) Learning to generate chairs with convolutional neural networks, in *Proceedings of the IEEE Conference on Computer Vision and Pattern Recognition*, pp. 1538–1546.

30 Matsugu, M., Mori, K., Mitari, Y., and Kaneda, Y. (2003) Subject independent facial expression recognition with robust face detection using a convolutional neural network. *Neural Networks*, **16** (5), 555–559.

31 Ramakrishnan, K., Scholte, S., Lamme, V., Smeulders, A., and Ghebreab, S. (2015) Convolutional neural networks in the brain: an fMRI study. *Journal of Vision*, **15** (12), 371–371.

32 Nguyen, A., Yosinski, J., and Clune, J. (2015) Deep neural networks are easily fooled: high confidence predictions for unrecognizable images, in *Computer Vision and Pattern Recognition (CVPR), 2015 IEEE Conference*, IEEE, pp. 427–436.

33 Nair, V. and Hinton, G.E. (2010) Rectified linear units improve restricted Boltzmann machines, in *Proceedings of the 27th International Conference on Machine Learning (ICML-10)*, pp. 807–814.

34 He, K., Zhang, X., Ren, S., and Sun, J. (2015) Delving deep into rectifiers: surpassing human-level performance on ImageNet classification, in *Proceedings of the IEEE International Conference on Computer Vision*, pp. 1026–1034.

35 Goodfellow, I., Warde-Farley, D., Mirza, M., Courville, A., and Bengio, Y. (2013) Maxout networks, in *Proceedings of the 30th International Conference on Machine Learning*, pp. 1319–1327.

36 Socher, R., Perelygin, A., Wu, J.Y., Chuang, J., Manning, C.D., Ng, A.Y., and Potts, C. (2013) Recursive deep models for semantic compositionality over a sentiment treebank, in *Proceedings of the Conference on Empirical Methods in Natural Language Processing (EMNLP)*, vol. 1631, Citeseer, p. 1642.

37 Zeiler, M.D. and Fergus, R. (2013) Stochastic pooling for regularization of deep convolutional neural networks. *ICLR 2013*.

38 Springenberg, J., Dosovitskiy, A., Brox, T., and Riedmiller, M. Striving for simplicity: the all convolutional net, in *ICLR (workshop track)*.

39 Srivastava, N., Hinton, G.E., Krizhevsky, A., Sutskever, I., and Salakhutdinov, R. (2014) Dropout: a simple way to prevent neural networks from overfitting. *Journal of Machine Learning Research*, **15** (1), 1929–1958.

40 Jia, Y., Shelhamer, E., Donahue, J., Karayev, S., Long, J., Girshick, R., Guadarrama, S., and Darrell, T. (2014) Caffe: convolutional architecture for fast feature embedding. *arXiv preprint arXiv:1408.5093*.

41 Torch. (2016) Available from http://torch.ch/#

42 Vedaldi, A. and Lenc, K. (2015) Matconvnet convolutional neural networks for Matlab, in *Proceeding of the ACM International Conference on Multimedia*.

43 MatConvNet. (2016). Available from http://www.vlfeat.org/matconvnet/.

44 Simonyan, K. and Zisserman, A. (2014) Very deep convolutional networks for large-scale image recognition. *CoRR*, abs/1409.1556.

45 Russakovsky, O., Deng, J., Su, H., Krause, J., Satheesh, S., Ma, S., Huang, Z., Karpathy, A., Khosla, A., Bernstein, M., Berg, A.C., and Fei-Fei, L. (2015) ImageNet Large Scale Visual Recognition Challenge. *International Journal of Computer Vision (IJCV)*, **115** (3), 211–252, 10.1007/s11263-015-0816-y.

46 Lowe, D.G. (1999) Object recognition from local scale-invariant features, in *Computer Vision, 1999. The Proceedings of the Seventh IEEE International Conference*, vol. 2, IEEE, pp. 1150–1157.

47 Warrington, E.K. and Taylor, A.M. (1973) The contribution of the right parietal lobe to object recognition. *Cortex*, **9** (2), 152–164.

48 Lowe, D.G. (1987) Three-dimensional object recognition from single two-dimensional images. *Artificial Intelligence*, **31** (3), 355–395.

49 Srivastava, S., Mukherjee, P., and Lall, B. (2015) Characterizing objects with sika features for multiclass classification. *Applied Soft Computing*.

50 Field, G.D., Gauthier, J.L., Sher, A., Greschner, M., Machado, T.A., Jepson, L.H., Shlens, J., Gunning, D.E., Mathieson, K., Dabrowski, W. *et al.* (2010) Functional connectivity in the retina at the resolution of photoreceptors. *Nature*, **467** (7316), 673–677.

51 Simonyan, K. and Zisserman, A. (2014) Very deep convolutional networks for large-scale image recognition. *arXiv preprint arXiv:1409.1556*.

52 Ovtcharov, K., Ruwase, O., Kim, J.Y., Fowers, J., Strauss, K., and Chung, E.S. (2015) Accelerating deep convolutional neural networks using specialized hardware. Available from http://research.microsoft.com/apps/pubs/default.aspx?id=240715.

53 Szegedy, C., Liu, W., Jia, Y., Sermanet, P., Reed, S., Anguelov, D., Erhan, D., Vanhoucke, V., and Rabinovich, A. (2015) Going deeper with convolutions, in *Proceedings of the IEEE Conference on Computer Vision and Pattern Recognition*, pp. 1–9.

54 Dubey, R., Peterson, J., Khosla, A., Yang, M.H., and Ghanem, B. (2015) What makes an object memorable?, in *Proceedings of the IEEE International Conference on Computer Vision*, pp. 1089–1097.

55 Isola, P., Xiao, J., Torralba, A., and Oliva, A. (2011) What makes an image memorable?, in *Computer Vision and Pattern Recognition (CVPR), 2011 IEEE Conference*, IEEE, pp. 145–152.

56 Isola, P., Xiao, J., Parikh, D., Torralba, A., and Oliva, A. (2014) What makes a photograph memorable? *Pattern Analysis and Machine Intelligence, IEEE Transactions*, **36** (7), 1469–1482.

57 Girshick, R., Donahue, J., Darrell, T., and Malik, J. (2014) Rich feature hierarchies for accurate object detection and semantic segmentation, in *Proceedings of the IEEE Conference on Computer Vision and Pattern Recognition*, pp. 580–587.

58 Long, J., Shelhamer, E., and Darrell, T. (2015) Fully convolutional networks for semantic segmentation, in *Proceedings of the IEEE Conference on Computer Vision and Pattern Recognition*, pp. 3431–3440.

59 Ren, S., He, K., Girshick, R., and Sun, J. (2015) Faster r-CNN: towards real-time object detection with region proposal networks, in *Advances in Neural Information Processing Systems*, pp. 91–99.

60 George, D. (2008) *How the brain might work: a hierarchical and temporal model for learning and recognition*, PhD thesis, Stanford University.

61 LeCun, Y., Bottou, L., Bengio, Y., and Haffner, P. (1998) Gradient-based learning applied to document recognition. *Proceedings of the IEEE*, **86** (11), 2278–2324.

62 Zeiler, M.D. and Fergus, R. (2014) Visualizing and understanding convolutional networks, in *European Conference on Computer Vision*, Springer, pp. 818–833.

63 He, K., Zhang, X., Ren, S., and Sun, J. (2015) Deep residual learning for image recognition, in *Proceedings of the IEEE Conference on Computer Vision and Pattern Recognition*.

7

Human Behavioral Analysis Using Evolutionary Algorithms and Deep Learning

Earnest Paul Ijjina and Chalavadi Krishna Mohan

Visual Learning and Intelligence Group (VIGIL), Department of Computer Science and Engineering, Indian Institute of Technology Hyderabad (IITH), Hyderabad, Telangana, India

7.1 Introduction

Human behavior analysis refers to the use of machine learning techniques and computer vision to recognize and classify human behavior. One may classify human behavior into gesture, event, action, and activity based on the duration for which the subjects motion is analyzed. Depending on the region of interest, this can be further classified into facial expression, hand-gesture, or upper/lower-body action. In the last decade, recognizing actions in videos has gained a lot of interest in the computer vision research community due to its applications in ambient assisted living, health monitoring, video analytics, sports analysis, robotics, and automatic video surveillance. Various surveys are proposed in the literature, analyzing and categorizing existing approaches to human action recognition based on the characteristic considered for recognition. For instance, [1] covers various approaches for recognizing actions of a single subject, [2] discusses approaches for multiview human action recognition, and [3] presents approaches that categorize full-body motion into spatial and temporal structures. A survey of various video data sets proposed over the years for human action recognition is available in [4]. These data sets differ in objective of creation like detection of realistic interactions, actions, and multiview actions. The traditional approaches to action recognition consider features extracted from input observations to provide the necessary discriminative information. The most commonly used features in the literature are bag of visual words [5], histograms-oriented gradient [6], histograms of optical flow [6], motion boundary histograms [7], action bank representation [8], and dense trajectories [9]. Due to the task-specific design of these features, these features may not be efficient for other visual recognition tasks that lead to data-driven approaches like deep learning, and that can automatically learn a hierarchy of discriminative features from input representation [10].

Deep learning is an emerging machine learning technique that has seen rapid evolution in the last decade. The availability of general-purpose graphic processing units (GPGPUs) has aided this growth by enabling the parallel implementation of these deep learning models. The implementation of deep learning models by neural

Hybrid Intelligence for Image Analysis and Understanding, First Edition.
Edited by Siddhartha Bhattacharyya, Indrajit Pan, Anirban Mukherjee, and Paramartha Dutta.
© 2017 John Wiley & Sons Ltd. Published 2017 by John Wiley & Sons Ltd.
Companion Website: www.wiley.com/go/bhattacharyya/hybridintelligence

networks makes them referred to as deep neural network (DNN) models. Even though neural network models were proposed in the early 1940s, due to the lack of effective training algorithms, only shallow neural networks were used till the early 2000s. Some of the seminal work by pioneers in deep learning like Yann LeCun, Yoshua Bengio, and Geoffrey Hinton has set the groundwork on which the field of deep learning is evolving. A convolutional neural network (CNN) proposed by Yann LeCun *et al.* in [11, 12] uses weight sharing to reduce the number of free parameters to be learned during training. The greedy layer-wise training of deep belief networks [13, 14] is some of the most extensively used work by the deep learning community. The deep learning models are trained on fairly raw data, to learn a hierarchy of features for achieving a recognition task. This ability to extract discriminative features from raw data led to their usage on various modalities of data like text, audio, and video, to name a few. Some of the well-known applications of convolutional neural networks (CNN) is the MNIST digit recognition [15], ILSVRC challenge for object recognition [16–18], face detection [19], ACM Facial Expression recognition in the Wild (AFEW) challenge [20], and action recognition [21, 22]. Among the deep learning models, CNN is the most widely used approach for visual recognition that consists of alternating layers of convolution and subsampling.

In contrast to the conventional techniques that rely on (hand-crated) features for discriminative information, deep learning models extract a hierarchy of discriminative features from input data. Baccouche *et al.* [23] utilized the spatiotemporal evolution of features generated by a 3D CNN for a sequence of frames with a recurrent neural network to recognize human actions in the KTH data set. Shuiwang Ji *et al.* [21] proposed a 3D CNN model using gray, optical-flow, and gradient information along x and y directions, features for recognizing human actions in surveillance videos. In [24], Keze Wang *et al.* extended CNNs by incorporating a structure to manipulate the activation of neurons. During recognition, the variation in temporal composition of activities is handled by the partial activation of neural network configuration. The videos are decomposed into temporal segments of subactivities by a spatiotemporal CNN to recognize human actions in [25]. The radius-margin regularization is used in the learning algorithm employed to iteratively optimize the human action recognition model for RGBD videos. A recurrent neural network is used for modeling the temporal evolution of state dynamics in [26] for action recognition. A two-stream CNN to recognize human actions from still frames and motion between frames is proposed in [27]. The softmax scores across the streams are combined to determine the action label.

Even though deep learning models are effective for various classification tasks [28], effective training deep architecture [29] is still a challenge. In a fully connected DNN trained using a gradient descent algorithm, the gradient decreases as the error is back-propagated from the output to the input layer, thereby causing the vanishing gradient problem [30]. As a result, the weights in the beginning layers won't be optimized. This is addressed in a CNN by reducing the number of distinct weights at each layer and the number of neurons from which a neuron computes its output (i.e., local connectivity). A CNN trained using back-propagation algorithms (BPAs) may not be optimal as it may get stuck in a local optima, and its performance after training depends on its initial weights. Therefore, the optimization of a CNN classifier is now translated to finding optimum initial weights of the neural network. There are several other optimization and regularization techniques like: (1) early stopping [31] may avoid overfitting the data, (2) co-adaptation can be avoided using dropout, (3) rectified linear units (ReLUs), and (4) efficient weight initialization through pre-training [32] proposed in the literature. Evolutionary techniques were also used in the last few years to optimize DNN. David *et al.* [33] used evolutionary algorithm (EA)-assisted back-propagation to optimize the weights of a sparse autoencoder. Fedorovici *et al.* [34] used evolutionary optimization techniques like the gravitational search algorithm [35] and particle swarm optimization [36] for optimizing CNN weights.

This chapter proposes a hybrid deep learning approach to determine the weights of CNN classifiers by exploiting the local and global search capabilities of gradient descent and EAs for human action recognition in videos. The novelty of this work lies in: (1) using EAs for exploring different basins, (2) utilizing the effectiveness of the gradient descent algorithm in finding the local optimum for a given basin, and (3) combining complementary information across classifiers generated by the evolutionary framework. In this chapter, Section 7.2 describes the evolutionary deep learning approach to action recognition. The experimental evaluation of this work is given in Section 7.3, followed by conclusions and future work in Section 7.4.

7.2 Human Action Recognition Using Evolutionary Algorithms and Deep Learning

In this section, we present a DNN model optimized using EA for human action recognition in videos. The block diagram of the proposed action recognition framework is shown in Figure 7.1. As shown in the figure, the EA (with evolutionary strategy) is used for initializing the weights of the CNN classifier that recognizes actions from action bank representation of videos. As the convolution kernels impact the effectiveness of a CNN classifier, we use the value of kernels and the seed value given to the random

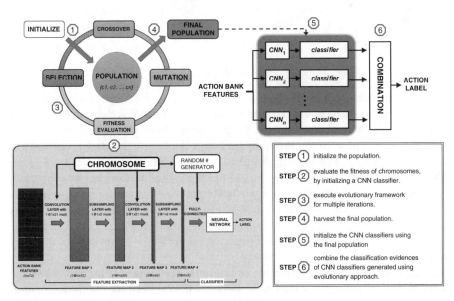

Figure 7.1 Pictorial representation of various steps in the proposed classification system. Best viewed in color.

number generator of fully connected neural networks classifier as the chromosome. As multiple (n) solutions are generated by evolutionary algorithms, the evidence across these classifiers is combined using fusion rules to determine the class label.

An introduction to EA for search optimization is presented in Section 7.2.1. Next, we discuss action bank features in Section 7.2.2, followed by the DNN architecture used to recognize human actions in Section 7.2.3. As the effectiveness of DNN models trained using BPAs is dependent on weight initialization, we aim to find an optimum weight initialization using EAs. We conclude with results, analysis, and future work of the proposed action recognition framework. The following subsection introduces EAs and elaborates on how the fitness of the population improves over EA iterations.

7.2.1 Evolutionary Algorithms for Search Optimization

EAs are based on Charles Darwin's principle of *survival of the fittest* [37]. EA model an optimization problem by capturing the most significant attributes of the system as a chromosome. For instance, the kernels of a CNN that determine the effectiveness of a CNN classifier can by used as the chromosome to optimize the CNN classifier. These chromosomes representing various candidate solutions need to be evaluated using a cost function that represents their fitness as it is used in crossover and mutation operations. A crossover operation constructs new chromosomes from existing ones, and mutation disrupts the genes in the existing chromosomes. When using EA to optimize the weights of a CNN classifier, mutation will be more effective in keeping the solution from getting stuck in a local optima. Hence, the (with evolutionary strategy, i.e., mutation probability is greater than crossover probability) is used in this work to optimize the weights of a CNN classifier. As the weights of a practical neural network are real numbers, using only EA to learn the weights would take a greater number of iterations. To overcome this challenge, we use BPA to optimize the weights initialized by EAs, before computing its fitness. As a result, an evolutionary DNN model leveraging the fast convergence of BPAs and search capabilities of EAs is produced. The next section introduces action recognition using action bank features.

7.2.2 Action Bank Representation for Action Recognition

Action bank representation for human action recognition was proposed by Sadanand *et al.* in [8]. Similar to a bag of visual words (BoVW) approach, action bank representation consists of a set of video templates for the actions to be recognized. To compute action bank features of a video, it is compared with the template videos to generate a 1×73 vector corresponding to each template, which captures the similarity of video with the template. If an action bank with m templates is used in this computation, then an action bank representation of size $m \times 73$ gets generated. As different observations of the same action may have similar motion, action bank representation of observations may contain similar local patterns. From the action bank features of observations the KTH data set in Figure 7.2, it can be observed that observations of the same action may have similar local patterns.

As the location and extent of similarity of action bank representation depend on the similarity of motion between the video and the templates, a visual recognition algorithm is needed to learn the patterns associated with each class. As a CNN classifier can learn the necessary discriminative local features from the input data, we use a CNN classifier

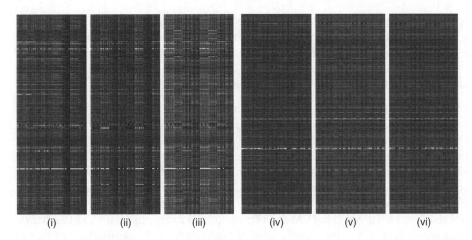

| (i) | (ii) | (iii) | (iv) | (v) | (vi) |

Figure 7.2 Action bank features of observations in KTH: (i–iii) are videos of boxing action and (iv–vi) are videos of running.

to recognize actions from action bank representation. The next section elaborates the architecture of CNN classifier considered to recognize human actions.

7.2.3 Deep Convolutional Neural Network for Human Action Recognition

The architecture of the CNN classifier considered for recognizing human actions from action bank representation of videos is shown in Figure 7.3. The first 72 features of action bank representation are used to represent videos as a $m \times 72$ image. Here, m represents the number of templates in the action bank used to generate action bank representation. As each row in this image represents the similarity of the video with one template, there is no correlation between pixels along the vertical axis. Hence, we use 1×21 kernels in convolution layers and 1×2 kernels in sub sampling layer. Refer to [38] for an overview to the working of the CNN classifier. The (Convnet) features generated by the deep CNN are used by a classifier for action recognition.

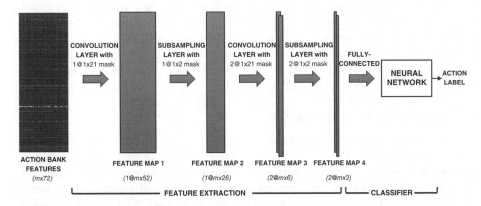

Figure 7.3 CNN classifier for recognizing actions in videos.

As the convolution kernels in a CNN are the feature detectors, optimizing these matrices is essential for the effectiveness of the CNN classifier. The next subsection describes the proposed hybrid training approach for optimizing CNN classifiers.

7.2.4 CNN Classifier Optimized Using Evolutionary Algorithms

As the effectiveness of a CNN classifier trained using a gradient descent algorithm is dependent on its initial weights, optimal initialization of DNN is essential for designing an effective CNN classifier. Even though the DNN model trained using a BPA converges quickly, the solution might get trapped in a local minima. Hence, an evolutionary search algorithm is employed to identify the optimum weight initialization that maximizes performance. To avoid overfitting using BPAs and to reduce the number of iterations to find the optimum weights of the DNN using EAs, we use a hybrid approach. We train the CNN classifier initialized using EAs for a small number of epochs (p) using BPAs before evaluating its fitness. As a result, we exploit the fast convergence capability of BPAs and the search capability of EAs. The proposed system aims to find the global optimum solution by using EA to identify various basins in the error surface and BPAs to find the local optimum quickly in a given basin. The next section describes the experimental study on the KTH and UCF50 data sets.

7.3 Experimental Study

The details of the experimental setup used to evaluate the proposed approach on the KTH [39] and UCF50 [40] data sets are discussed in this section. Precomputed action bank features for these data sets, available at [41], are used in this work for comparability of results with existing approaches. The proposed system is implemented in MATLAB using [42] to implement the CNN classifier shown in Figure 7.3 and the native optimization functionality for designing EAs with evolutionary strategy. The range of weights of CNN kernels is set to vary between -100 and 100. Similarly, the seed value of the random number generator is set to vary between 0 to 5000. The population size (n) of 20 is used in EA, and the algorithm is run for five iterations with a mutation probability of 0.8. The optimum value for these parameters is determined empirically. The outputs of the CNN classifiers is binary coded, and an input is assigned the class label corresponding to the output with the highest value. The fusion of evidence across classifiers applies the fusion rule on the corresponding indexes of outputs of CNN classifier, to obtain the binary coded outputs of the fusion model. The outputs are interpreted similar to a classifier with binary coded outputs to determine the class label. The next subsections describe the experimental setup used for the UCF50 and KTH data sets.

7.3.1 Evaluation on the UCF50 Data Set

The UCF50 data set is a collection of realistic videos corresponding to 50 human actions gathered from Youtube. The large number of classes and unconstrained realistic nature of videos make this a challenging data set. We consider the pre-computed action bank features available at [41] computed with an action bank of 205 (m) templates, which also ensures comparability of results with existing approaches. The data set is evaluated using fivefold cross-validation (CV), where the entire data is split into five sets, and four sets are used for training and one set as test data. This process is repeated five times changing the set considered for testing in each execution. An n-fold CV splits the data into n parts and uses one part for testing and the remaining parts for training, performed n times and changing the part considered for testing each time. We use the notation *Set-i* for representing an execution using the ith set for testing and the remaining sets as training data. The overall performance across the n executions is considered as the accquracy for n-fold cross-validation. This is the most widely used evaluation strategy as it ensured that each observation is used for testing. The CNN classifier initialized by the evolutionary algorithm is trained by the gradient descent algorithm (BPA) in batch mode for 50 (p) epochs before evaluating its fitness. The parameters are optimized empirically. In this work, we aim to reduce the number of iterations EA needs for convergence, by expediting the local search in a given basin using the steepest-descent algorithm.

For five-fold CV on UCF50, the first four sets are trained using a batch size of 10 and the fifth set is trained with a batch size of 9. The variation in fitness value of the population over EA iterations, for UCF50, is given in Table 7.1. The steady decrease in fitness of the population over iterations indicate the convergence of EA and thereby the proper selection of parameters. This completes step 4 of Figure 7.1, which results in the generation of 20 (n) candidate weights for CNN classifiers for each fold. The performance of these candidate solutions {$c01$, $c02$,..., $c20$} is given in Table 7.2. It can be observed that the average performance of candidate solutions across the splits is less than 97%.

As a single number is considered for optimizing the weights of a neural network (NN) classifier in Figure 7.1, its performance may not be optimum. Experiments are conducted by replacing the NN classifier with an ELM classifier [43] for all the candidate solutions. The performance of all the candidate solutions generated for all the splits using the ELM classifier is given in Table 7.3. It can be observed that average performance of candidate solutions across all the splits is more than 98.8%, which is better than the one using the NN classifier given in Table 7.2, which could be due to better generalization ability of ELM.

As discussed in step 6 of Figure 7.1, the classification evidence generated by the n candidate solutions is combined using various fusion rules. The performance of the proposed hybrid approach on the UCF50 dataset using different fusion functions and ELM classifier is shown in Table 7.4. It can be observed that the accuracy across all fusion rules is above 99% and the best performance of 99.9% is achieved with *Max* fusion-rule, which could be due to the low deviation in accuracy across classifiers used in fusion.

Table 7.1 The variation in fitness value of EA populations over iterations for UCF50

Iteration	Set-1		Set-2		Set-3		Set-4		Set-5	
	best	mean	best	mean	best	mean	best	mean	best	mean
1	17.5	48.6	26.5	45.3	2.8	24.2	21.3	50.3	14.4	41.4
2	13.6	35.9	5.3	28.6	2.4	5.7	5.4	15.9	6.1	19.4
3	5.2	17.4	3.9	10.0	2.4	3.5	4.9	10.9	5.9	7.9
4	5.1	9.1	3.8	8.7	2.4	3.2	4.6	6.8	5.6	8.1
5	3.0	5.6	3.8	8.4	2.4	3.2	4.4	6.3	5.6	7.4

Table 7.2 Accuracy (in %) of candidate solutions generated for UCF50 using neural network classifier

Candidate solution	Classification accuracy				
	Set-1	*Set-2*	*Set-3*	*Set-4*	*Set-5*
c01	95.91	96.14	97.13	95.06	94.36
c02	94.80	90.98	96.98	91.79	90.93
c03	94.50	94.62	97.13	93.31	91.62
c04	91.30	90.61	97.21	93.84	92.91
c05	95.91	95.23	97.13	92.24	92.61
c06	96.95	94.32	97.36	94.45	93.06
c07	96.80	94.55	97.06	90.04	92.00
c08	93.09	93.03	95.09	92.17	92.30
c09	91.38	90.53	97.43	94.30	92.84
c10	93.31	91.82	96.83	94.37	92.38
c11	91.75	90.38	97.36	93.69	92.30
c12	96.58	92.50	96.68	92.47	92.00
c13	91.82	83.71	97.21	91.86	93.45
c14	95.84	94.85	89.36	94.30	92.23
c15	90.71	91.52	96.98	93.00	92.53
c16	96.13	94.92	96.91	92.47	94.28
c17	95.84	91.74	96.53	94.22	93.22
c18	94.50	86.44	96.45	94.30	92.23
c19	92.57	92.80	96.75	92.93	92.23
c20	94.20	78.94	94.64	91.79	92.23
Mean	**94.19 ± 2.06**	**91.48 ± 4.22**	**96.41 ± 1.81**	**93.13 ± 1.27**	**92.58 ± 0.82**

The confusion matrix for UCF50 corresponding to 99.90% accuracy is shown in Table 7.5. The true class labels are given on the vertical axis, and the predicted class labels on the horizontal axis. Thus, all nondiagonal elements in this matrix represent the misclassified observations. It can be observed that only 6 out of 6617 observations got misclassified by the proposed approach. The performance of the existing and proposed approaches for UCF50 is given in Table 7.6. The high performance for the proposed approach indicates the effectiveness in identifying the optimum weights of the CNN. It can be observed that the proposed approach using an evolutionary algorithm for weight initialization has better performance when compared to random weight initialization used by Earnest Ijjina *et al.* in [44]. The next section describes the experimental study conducted on the KTH data set.

7.3.2 Evaluation on the KTH Video Data Set

The KTH video data set, proposed in 2004 by Christian Schuldt *et al.* in [39], consists of six actions (*running, walking, jogging, hand clapping, hand waving,* and *boxing*) performed in an outdoor environment by 25 subjects. The action bank representation is generated with an action bank of 202 templates. The CNN classifiers initialized by EA are

Table 7.3 Performance (in %) of candidate solutions generated for UCF50 using ELM classifier

Candidate	Classification accuracy				
solution	*Set-1*	*Set-2*	*Set-3*	*Set-4*	*Set-5*
c01	100.00	99.92	100.00	99.85	99.39
c02	100.00	98.94	100.00	98.56	98.93
c03	100.00	99.77	100.00	99.47	98.17
c04	100.00	98.71	100.00	99.16	98.48
c05	100.00	99.70	100.00	99.24	99.54
c06	99.70	99.85	100.00	99.70	98.48
c07	99.93	99.85	100.00	97.49	99.31
c08	100.00	99.62	99.85	99.47	98.70
c09	100.00	98.48	100.00	99.70	98.86
c10	100.00	99.92	100.00	99.77	98.78
c11	100.00	98.64	100.00	99.39	98.78
c12	99.78	99.55	100.00	99.70	99.01
c13	100.00	96.74	100.00	98.71	99.16
c14	100.00	99.55	98.87	99.70	98.48
c15	100.00	98.79	99.92	99.85	99.01
c16	99.93	99.70	100.00	99.70	99.54
c17	100.00	99.62	100.00	99.70	99.09
c18	100.00	97.20	99.92	99.70	98.86
c19	100.00	99.70	100.00	99.85	99.01
c20	100.00	93.03	99.70	98.33	99.54
Mean	**99.97 ± 0.08**	**98.86 ± 1.63**	**99.91 ± 0.26**	**99.35 ± 0.62**	**98.96 ± 0.39**

Table 7.4 Accuracy (in # misclassified observations) using fusion on UCF50

Data	No.	Fusion-rule					*Majority*
fold	observations	*Min*	*Max*	*Avg*	*Prod*	*Median*	voting
Set-1	1345	0	0	0	0	0	0
Set-2	1320	3	0	0	0	1	2
Set-3	1325	0	0	0	0	0	0
Set-4	1315	3	3	3	4	3	3
Set-5	1312	9	3	8	8	11	11
Total	6617	15	6	11	12	15	16
Accuracy (in %) =		99.77	**99.90**	99.83	99.81	99.77	99.75

Table 7.5 Confusion matrix for UCF50

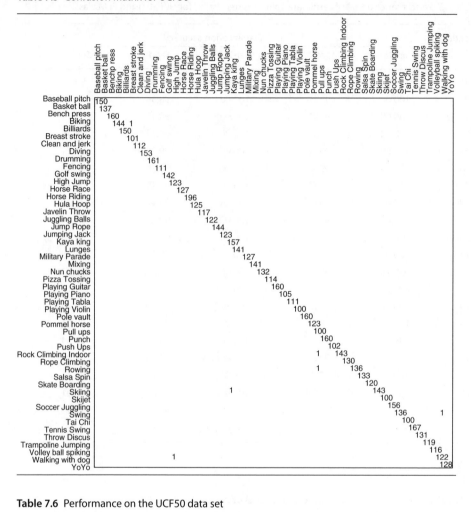

Table 7.6 Performance on the UCF50 data set

Approach	Accuracy %
Sreemanananth *et al.* [8]	57.9
Kliper-Gross *et al.* [45]	68.5
Feng *et al.* [46]	71.7
Wang *et al.* [47]	71.7
Wang *et al.* [9]	75.7
Zhou *et al.* [48]	80.2
Ijjina *et al.* [49]	94.0
Ballas *et al.* [50]	94.1
Ijjina *et al.* [44]	99.68
Proposed approach	**99.90**

trained using a gradient-descent algorithm in batch mode for 15 (p) epochs before evaluating their fitness. The parameters are optimized empirically. Similar to the previous data set, the steps in Figure 7.1 are also followed for this data set. The average accuracy of the population using NN and ELM classifiers is shown in Table 7.7. The evidences of all candidate solutions generated by the proposed approach with ELM classifier combined using an *Avg*-fusion rule is 96.76%, as given in this table. The accuracy of CNN classifier trained using BPA is very low (16.77%), which could be due to the possibility of the NN solution getting stuck in a local optima when trained using BPA. The performance of various approaches for this data set is shown in Table 7.8. It can be observed that the accuracy of the proposed approach is comparable with existing state-of-the-art approaches. The relatively low performance of the proposed approach for the KTH data set (when compared to UCF50) could be due to the use of a fixed training set with a smaller number of training observations. As the performance of neural network models depends on the number of observations used to train them, the proposed approach with neural network implementation has low accuracy for the KTH data set, when compared to (99.9% for) the UCF50 data set. The next section presents the analysis of results generated by this approach.

Table 7.7 Performance (in %) of CNN features with NN and ELM classifiers generated by the proposed approach [with back-propagation algorithm (BPA) and evolutionary algorithms (EA)] on the KTH data set

Training approach	Accuracy
CNN classifier with **only EA** (i.e., initialized using EA)	84.07 ± 7.31
CNN classifier **without EA** (i.e., trained using BPA)	16.77
CNN features with **NN classifier** (generated using **both EA and BPA**)	94.12 ± 1.77
CNN features with **ELM classifier** (generated using **both EA and BPA**)	96.22 ± 0.50
Fusion of CNN features with **ELM classifier** (generated using **both EA and BPA**)	96.76

Table 7.8 Performance (in %) on the KTH data set

Approach	Accuracy
Kliper-Gross *et al.* [51]	83.3
Ji *et al.* [21]	90.2
Ryoo *et al.* [52]	91.1
Laptev *et al.* [53]	91.8
Le Quoc *et al.* [54]	93.9
Iosifidis *et al.* [55]	93.52
Amer *et al.* [56]	96.8
Proposed approach	**96.76**

7.3.3 Analysis and Discussion

In this section, we analyze the various components used in the proposed system to identify their significance. The performance of the CNN classifier is given in Figure 7.3 when trained using gradient-descent algorithms (BPAs and evolutionary algorithms (EA), and the hybrid approach is shown in Table 7.9. It can be observed that the proposed hybrid training approach achieves better performance than the rest of the alternatives for the UCF50 data set. Similar observations can be made from Table 7.7 for the KTH data set. These tables report the average performance across classifiers generated using EAs. The experiments corresponding to using only EAs were conducted for five iterations with a population size of 200. Thus, by training a CNN initialized using EAs with BPAs, optimal weights were found in less iterations using a small population. The average accuracy of the candidate solutions generated by the proposed approach using NN and ELM classifiers is given in Table 7.10. The ELM classifier is found to perform better than the NN classifier. Similar observations can be made from Table 7.7 for the KTH data set. From Table 7.4, the proposed approach has an accuracy of 99.9% due to misclassification of 6 out of 6617 observations.

The major issues with training a deep neural network model using a gradient-descent algorithm are: (1) overfitting, and (2) the solution getting stuck in a local optima. Neural networks trained for a large number of epochs will overfit to the training data, thereby producing small errors for training data and large errors for test data, as shown by the red dotted line in Figure 7.4. To visualize the effectiveness of EAs in initializing the CNN and the impact of training using BPAs before fitness evaluation, we plot the accuracy of CNN classifier when it is initialized and after training with a BPA for p epochs. In this 2D plot, each CNN classifier is represented by a circle with its (x, y) coordinates representing the classification error for training and test data, respectively. We also color these circles to indicate the EA iteration in which it is explored.

Table 7.9 Accuracy (in %) of CNN classifiers using BPAs, EAs and the hybrid approach for UCF50

Training	Set-1	Set-2	Set-3	Set-4	Set-5	Average
CNN classifier with **only EA** (i.e., initialization using EA)	18.06	20.15	25.43	13.15	27.28	20.81
CNN classifier **without EA** (i.e., training using BPA)	86.02	87.50	18.26	80.30	23.39	59.19
CNN classifier **with EA** (with EA and BPA)	94.19	91.48	96.41	93.12	92.58	93.55

Table 7.10 Accuracy (in %) of the proposed approach using NN and ELM classifiers for UCF50

Classification	Set-1	Set-2	Set-3	Set-4	Set-5	Average
Proposed approach using **NN classifier**	94.19	91.48	96.41	93.12	92.58	93.55
Proposed approach using **ELM classifier**	99.96	98.86	99.91	99.34	98.95	99.40

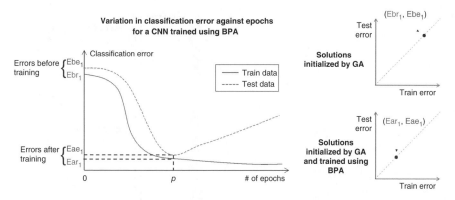

Figure 7.4 Analysis of classification errors of solutions generated by the proposed hybrid training approach.

In Figure 7.5 part i, we plot the classification error of solutions generated by EAs for Set-1 of the UCF50 data set before they are trained using BPAs. Figure 7.5 part ii depicts the solutions in Figure 7.5 part i after they are trained with gradient-descent algorithms for 50 (p) epochs. Each circle in the plots represents a CNN classifier generated by an EA. As shown in the color scale of these plots, blue is used for solutions generated in the first iteration and yellow for solutions generated in the final iteration. The same methodology is used to visualize the solutions of *Set-2* to *Set-5* given in Fig 7.6. Some of the observations from the subfigures of Fig 7.5 and Fig 7.6, are: (1) the proposed hybrid training approach improves the accuracy of the action recognition framework significantly, (2) the location of solutions on the 45° line suggests the similarity of patterns in both training and test data (which could be the result of using k-fold CV), and (3) the concentration of yellow circles at the bottom-left corner in the subfigures with BPA-trained solutions indicates the effectiveness of training. Similarly, the candidate solutions explored by the proposed approach for the KTH data set are shown in Figure 7.7. From the plots, it can be observed that the solutions have fewer training errors when compared to test errors, which could be due to the fixed training set and smaller number of training observations, when compared to n-fold CV of the UCF50 data set. This could be the reason behind the low accuracy of the proposed approach for the KTH data set when compared to the UCF50 data set with 99.9%. The pictorial representation of chromosomes generated by the proposed approach for the KTH and UCF50 data sets is shown in Figure 7.8. From this visualization, it can be observed that the candidate solutions generated by the proposed hybrid training approach are near identical, which could be the reason behind the low deviation in performance across solutions in the population generated for these data sets, as given in Table 7.2 and Table 7.7. The next section discusses the computational aspects of this work.

7.3.4 Experimental Setup and Parameter Optimization

The proposed approach is implemented in MATLAB 2016a by using the in-built global optimization toolbox for defining the evolutionary strategy and extending the CNN implementation of [42] for rectangular masks. The double vector representation of chromosomes is used with range constraints, tournament selection strategy, intermediate

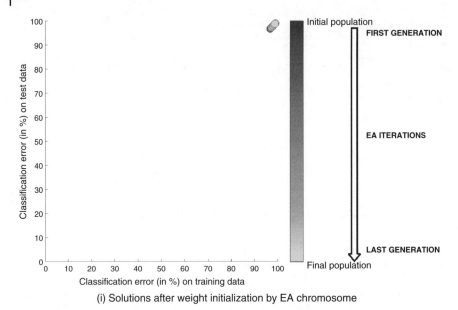

(i) Solutions after weight initialization by EA chromosome

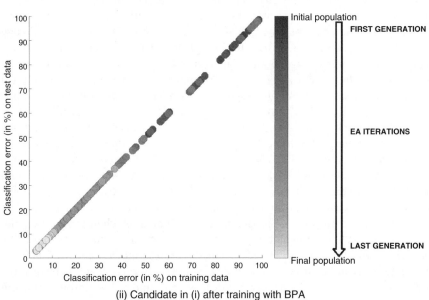

(ii) Candidate in (i) after training with BPA

Figure 7.5 Visualization of EA population generated by the proposed approach for *Set-1* of UCF50: (i) after initialization by EA and (ii) after training the solutions with BPA for *p* epochs.

crossover operation, and a mutation probability of 0.8 for the experimental study in this work. We empirically determine the seed value used for initializing the EA population and the optimum number of epochs (p) the CNN classifier should be trained for convergence, without overfitting the training data (as shown in Figure 7.4). The number of epochs (p) should be optimized to avoid undertraining or overfitting of CNN classifier,

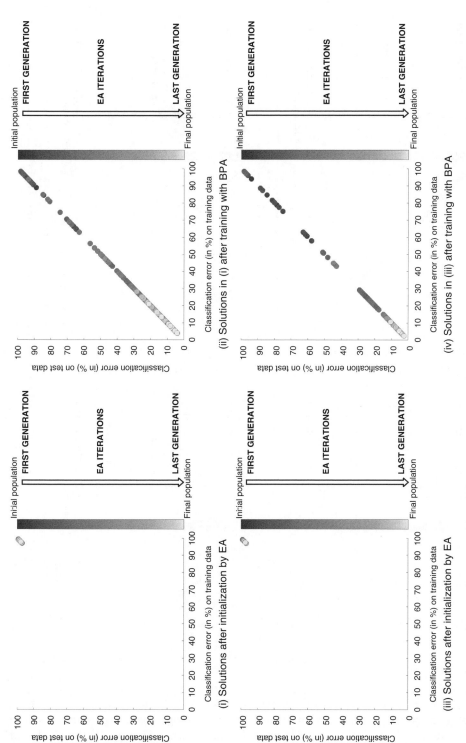

Figure 7.6 Visualization of EA population generated by the proposed approach for *Set-2* to *Set-5* of UCF50. The sub-figures (i), (ii) correspond to *Set-2*; (iii), (iv) are for *Set-3*; (v), (vi) correspond to *Set-4* and (vii), (viii) are for *Set-5*.

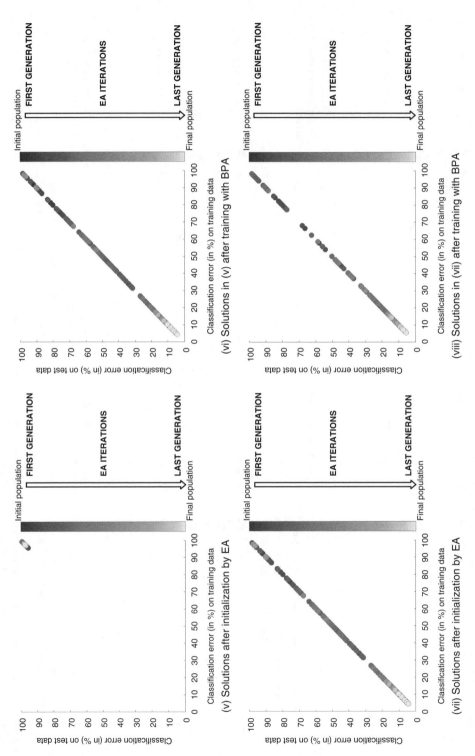

Figure 7.6 (*Continued*)

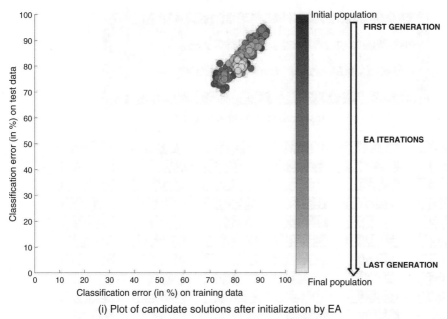

(i) Plot of candidate solutions after initialization by EA

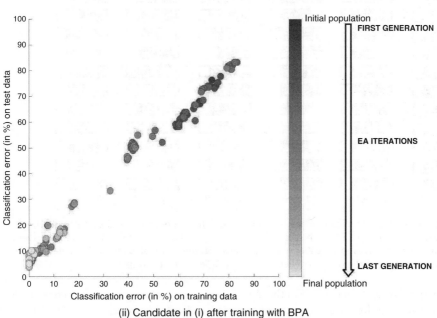

(ii) Candidate in (i) after training with BPA

Figure 7.7 Visualization of EA population generated for KTH: (i) visualization after initialization by EA and (ii) visualization after training with BPA.

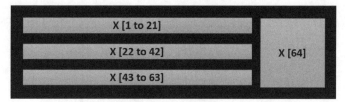

(a) Representation of EA chromosome **X**

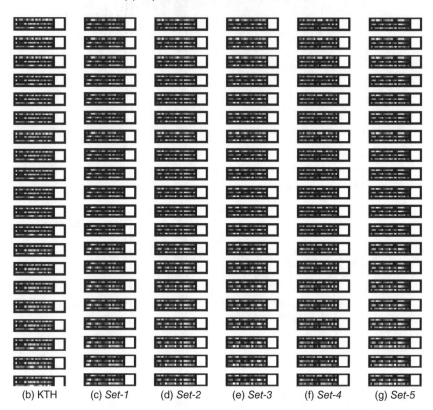

(b) KTH   (c) *Set-1*   (d) *Set-2*   (e) *Set-3*   (f) *Set-4*   (g) *Set-5*

Figure 7.8 Visualization of chromosomes (candidate solutions) generated by the proposed hybrid training approach for UCF50: (a) The pictorial representation of convolution masks and seed values corresponding to a chromosome **X** of size 64; (b) chromosomes generated for the KTH data set; and (c–g) correspond to chromosomes generated for UCF50 *Set-1* to *Set-5*, respectively.

which impacts the fitness value of the chromosome and in turn the convergence of EAs. The next section discusses the computational aspects of this work.

7.3.5   Computational Complexity

In this chapter, we use EAs and CNN classifier to design a human action recognition system. The fitness evaluation of candidate solutions in EA can be parallelized, and efficient GPU-based implementation of CNN classifiers like cuDNN can be used to reduce the training time. As the CNN is trained using a gradient-descent algorithm for p epochs (i.e., 50 epochs for UCF50), the CNN classifier gets trained in a short time. Several other multi-GPU-based implementations of CNN like NVIDIA's DIGITS, Theano, Torch, and Berkeley's Caffe are also available. The CNN classifiers are today used for many real-time applications like image recognition and speech processing in mobile devices. These systems are trained on high-end GPU platforms and deployed on low-end mobile hardware, as training is more computationally expensive than testing. Hence, this approach can be used to find the optimum weights of CNNs, which can be deployed for real-time use.

7.4 Conclusions and Future Work

In this work, we proposed a deep neural network model optimized using evolutionary algorithms for recognizing human actions in videos. As the performance of a DNN model (trained using a back-propagation algorithm) depends on its initial weights, we aim to optimize the dnn architecture by using an evolutionary algorithm to search for the optimum initial weights. As a result, the DNN is initialized to avoid getting stuck in a local minimum and overfitting to training data. The fusion across models generated using evolutionary algorithms overcomes the constraints of individual models, by combining complementary information across classifiers. The experimental study on KTH and UCF50 data sets suggests the effectiveness of the proposed approach (accuracy of 99.9% for UCF50). The future work will extend this approach to other spatiotemporal features like 3DHOG features [57].

References

1 Aggarwal, J. and Ryoo, M. (2011) Human activity analysis: a review. *ACM Computing Surveys*, **43** (3), 1–43.

2 Holte, M.B., Tran, C., Trivedi, M.M., and Moeslund, T.B. (2011) Human action recognition using multiple views: a comparative perspective on recent developments, in *Joint ACM Workshop on Human Gesture and Behavior Understanding*, ACM, New York, J-HGBU'11, pp. 47–52.

3 Weinland, D., Ronfard, R., and Boyer, E. (2011) A survey of vision-based methods for action representation, segmentation and recognition. *CVIU*, **115** (2), 224–241.

4 Chaquet, J.M., Carmona, E.J., and Fernández-Caballero, A. (2013) A survey of video datasets for human action and activity recognition. *CVIU*, **117** (6), 633–659.

5 Foggia, P., Percannella, G., Saggese, A., and Vento, M. (2013) Recognizing human actions by a bag of visual words, in *SMC*, pp. 2910–2915.

6 Laptev, I., Marszalek, M., Schmid, C., and Rozenfeld, B. (2008) Learning realistic human actions from movies, in *CVPR*, pp. 1–8.

7 Dalal, N., Triggs, B., and Schmid, C. (2006) Human detection using oriented histograms of flow and appearance, in *ECCV - vol. pt. 2*, Springer-Verlag, Berlin, Heidelberg, ECCV 06, pp. 428–441.

8 Sadanand, S. and Corso, J.J. (2012) Action bank: A high-level representation of activity in video, in *CVPR*, pp. 1234–1241.

9 Wang, H., Klaser, A., Schmid, C., and Liu, C.L. (2011) Action recognition by dense trajectories, in *CVPR*, pp. 3169–3176.

10 Bengio, Y., Courville, A., and Vincent, P. (2013) Representation learning: a review and new perspectives. *PAMI*, **35** (8), 1798–1828.

11 LeCun, Y., Kavukcuoglu, K., and Farabet, C. (2010) Convolutional networks and applications in vision, in *ISCAS*, pp. 253–256.

12 Lecun, Y., Bottou, L., Bengio, Y., and Haffner, P. (1998) Gradient-based learning applied to document recognition. *Proc. of the IEEE*, **86** (11), 2278–2324.

13 Bengio, Y., Lamblin, P., Popovici, D., Larochelle, H., Montréal, U.D., and Québec, M. (2007) Greedy layer-wise training of deep networks, in *NIPS*, MIT Press.

14 Hinton, G.E., Osindero, S., and Teh, Y.W. (2006) A fast learning algorithm for deep belief nets. *Neural Computation*, **18** (7), 1527–1554.

15 Lecun, Y. and Cortes, C. (2009) The MNIST database of handwritten digits. Available from http://yann.lecun.com/exdb/mnist/.

16 Krizhevsky, A., Sutskever, I., and Hinton, G.E. (2012) Imagenet classification with deep convolutional neural networks, in *NIPS*, pp. 1097–1105.

17 Girshick, R.B., Donahue, J., Darrell, T., and Malik, J. (2013) Rich feature hierarchies for accurate object detection and semantic segmentation. *CoRR*, **abs/1311.2524**.

18 Russakovsky, O., Deng, J., Su, H., Krause, J., Satheesh, S., Ma, S., Huang, Z., Karpathy, A., Khosla, A., Bernstein, M.S., Berg, A.C., and Fei-Fei, L. (2014) Imagenet large scale visual recognition challenge. *CoRR*, **abs/1409.0575**.

19 Lawrence, S., Giles, C., Tsoi, A.C., and Back, A. (1997) Face recognition: a convolutional neural-network approach. *IEEE Transactions on Neural Networks*, **8** (1), 98–113.

20 Matsugu, M., Mori, K., Mitari, Y., and Kaneda, Y. (2003) Subject independent facial expression recognition with robust face detection using a convolutional neural network. *Neural Networks*, **16** (5-6), 555–559.

21 Ji, S., Xu, W., Yang, M., and Yu, K. (2013) 3d convolutional neural networks for human action recognition. *PAMI*, **35** (1), 221–231.

22 Tran, D., Bourdev, L.D., Fergus, R., Torresani, L., and Paluri, M. (2014) C3D: generic features for video analysis. *CoRR*, **abs/1412.0767**.

23 Baccouche, M., Mamalet, F., Wolf, C., Garcia, C., and Baskurt, A. (2011) Sequential deep learning for human action recognition, in *ICHBU*, Springer-Verlag, pp. 29–39.

24 Wang, K., Wang, X., Lin, L., Wang, M., and Zuo, W. (2014) 3D human activity recognition with reconfigurable convolutional neural networks, in *ACMMM*, ACM, New York, NY, USA, MM'14, pp. 97–106.

25 Lin, L., Wang, K., Zuo, W., Wang, M., Luo, J., and Zhang, L. (2015) A deep structured model with radius-margin bound for 3D human activity recognition. *IJCV*, pp. 1–18.

26 Veeriah, V., Zhuang, N., and Qi, G. (2015) Differential recurrent neural networks for action recognition. *CoRR*, **abs/1504.06678**.

27 Simonyan, K. and Zisserman, A. (2014) Two-stream convolutional networks for action recognition in videos. *CoRR*, **abs/1406.2199**.

28 Bengio, Y. and Delalleau, O. (2011) On the expressive power of deep architectures, in *ALT*, Springer-Verlag, Berlin, pp. 18–36.

29 Bengio, Y. (2009) Learning deep architectures for AI. *Foundation and Trends in Machine Learning*, **2** (1), 1–127.

30 Bengio, Y., Simard, P., and Frasconi, P. (1994) Learning long-term dependencies with gradient descent is difficult. *IEEE Transactions on Neural Networks*, **5** (2), 157–166.

31 Prechelt, L. (1997) Early stopping - but when? In *Neural networks: tricks of the Trade, vol. 1524 of LNCS, ch. 2*, Springer-Verlag, pp. 55–69.

32 Erhan, D., Bengio, Y., Courville, A., Manzagol, P.A., Vincent, P., and Bengio, S. (2010) Why does unsupervised pre-training help deep learning? *JMLR*, **11**, 625–660.

33 David, O.E. and Greental, I. (2014) Genetic algorithms for evolving deep neural networks, in *GECCO*, ACM, New York, NY, USA, pp. 1451–1452.

34 Fedorovici, L.O., Precup, R.E., Dragan, F., and Purcaru, C. (2013) Evolutionary optimization-based training of convolutional neural networks for ocr applications, in *ICSTCC*, pp. 207–212.

35 Rashedi, E., Nezamabadi-pour, H., and Saryazdi, S. (2009) GSA: a gravitational search algorithm. *Information Sciences*, **179** (13), 2232–2248.

36 Kennedy, J. and Eberhart, R.C. (1995) Particle swarm optimization, in *Int. Conf. on Neural Networks, vol 4*, pp. 1942–1948.

37 Bascom, J. (1871) Darwin's theory of the origin of species. *American Theological Review*, **3**, 349–379.

38 Wikipedia. (2016), Convolutional neural network. Available from https://en .wikipedia.org/wiki/Convolutional_neural_network.

39 Schuldt, C., Laptev, I., and Caputo, B. (2004) Recognizing human actions: a local SVM approach, in *ICPR*, vol 3, pp. 32–36.

40 Reddy, K.K. and Shah, M. (2012) Recognizing 50 human action categories of web videos. *Machine Vision and Applications*, **24** (5), 971–981.

41 Corso, J.J. (2004), Action bank™: a high-level representation of activity in video. Available from http://web.eecs.umich.edu/jjcorso/r/actionbank/.

42 Palm, R.B. (2012) Prediction as a candidate for learning deep hierarchical models of data, Master's thesis, Technical University of Denmark, Asmussens Alle, Denmark.

43 Huang, G.B., Zhou, H., Ding, X., and Zhang, R. (2012) Extreme learning machine for regression and multiclass classification. *SMC, Part B: Cybernetics*, **42** (2), 513–529.

44 Ijjina, E.P. and Mohan, C.K. (2016) Hybrid deep neural network model for human action recognition. *Applied Soft Computing*, **46**, 936–952.

45 Kliper-Gross, O., Gurovich, Y., Hassner, T., and Wolf, L. (2012) Motion interchange patterns for action recognition in unconstrained videos, in *ECCV - vol. pt. 6*, Springer-Verlag, Berlin, Heidelberg, ECCV'12, pp. 256–269.

46 Shi, F., Petriu, E., and Laganiere, R. (2013) Sampling strategies for real-time action recognition, in *CVPR*, pp. 2595–2602.

47 Wang, L., Qiao, Y., and Tang, X. (2013) Motionlets: Mid-level 3D parts for human motion recognition, in *CVPR*, pp. 2674–2681.

48 Zhou, Q., Wang, G., Jia, K., and Zhao, Q. (2013) Learning to share latent tasks for action recognition, in *ICCV*, pp. 2264–2271.

49 Ijjina, E.P. and Mohan, C. (2014) Human action recognition based on recognition of linear patterns in action bank features using convolutional neural networks, in *ICMLA*, pp. 178–182.

50 Ballas, N., Yang, Y., Lan, Z.Z., Delezoide, B., Preteux, F., and Hauptmann, A. (2013) Space-time robust representation for action recognition, in *ICCV*.

51 Niebles, J.C., Wang, H., and Fei-Fei, L. (2008) Unsupervised learning of human action categories using spatial-temporal words. *IJCV*, **79** (3), 299–318.

52 Ryoo, M.S. and Aggarwal, J.K. (2009) Spatio-temporal relationship match: video structure comparison for recognition of complex human activities, in *ICCV*, pp. 1593–1600.

53 Laptev, I., Marszalek, M., Schmid, C., and Rozenfeld, B. (2008) Learning realistic human actions from movies, in *CVPR*, pp. 1–8.

54 Le, Q., Zou, W., Yeung, S., and Ng, A. (2011) Learning hierarchical invariant spatio-temporal features for action recognition with independent subspace analysis, in *CVPR*, pp. 3361–3368.

55 Iosifidis, A., Tefas, A., and Pitas, I. (2013) Minimum class variance extreme learning machine for human action recognition. *CSVT*, **23** (11), 1968–1979.

56 Amer, M.R. and Todorovic, S. (2016) Sum product networks for activity recognition. *PAMI*, **38** (4), 800–813.

57 Klaser, A., Marszałek, M., and Schmid, C. (2008) A spatio-temporal descriptor based on 3D-gradients, in *BMVC*, British Machine Vision Association, pp. 275–281.

8

Feature-Based Robust Description and Monocular Detection: An Application to Vehicle tracking

Ramazan Yíldíz and Tankut Acarman

Computer Engineering Department, Galatasaray University, Istanbul, Turkey

8.1 Introduction

This study is focused in particular on the state-of-art feature extraction methods used in computer vision, their comparison in the context of object description and implementation into real-time recognition and tracking of land vehicles. Throughout this chapter, scale-invariant feature transform (SIFT) [1, 2], speeded-up robust features (SURF) extraction [3], biological visual cortex inspired features (the HMAX model) [4], and Haar-like features [5] are elaborated. On one hand, SIFT and SURF extraction are used for local image analysis; both methods extract distinctive local image features. On the other hand, the HMAX model and Haar-like features are global image features; each feature is constituted by thickness and orientation of contours, junctions, intensity changes, geometry of shapes, and objects. The HMAX model uses multiscale Gabor filters applied in different orientations. In [5], Haar-like features are presented, and they are used to describe object models. In this study, motivated by [5], Haar-like features are implemented with Adaboost classifier to detect land vehicles in a road traffic video.

In this chapter, we present a new method by fusing the information provided by global and local features. The first approach is based on local features for image object description and tracking. A SIFT-based model is presented to assure robust description with an improved object tracking performance versus the state-of-the-art SIFT method. The model is tested with the Multi-View Car Dataset of the Computer Vision Laboratory (CVLAB), École Polytechnique Fédérale de Lausanne, France [6], and it is implemented in the context of object tracking in a video.

The second approach uses global features for image object description, detection, and real-time tracking. A video object detection and tracking algorithm is developed and elaborated. Haar-like features are used for detection purposes, and regions of interest (ROIs) are defined. Validation algorithms are developed based on low-level image analysis, enabling the presented detection scheme to be robust and safe. Finally, a novel algorithm leveraging temporal information of detection history is presented to perform tracking. To illustrate the effectiveness of this algorithm, a set of videos recorded under different traffic conditions from a monocular camera mounted on a vehicle traveling on the Transit European Motorway (TEM) and Europe 5 (E5) freeway,

Hybrid Intelligence for Image Analysis and Understanding, First Edition.
Edited by Siddhartha Bhattacharyya, Indrajit Pan, Anirban Mukherjee, and Paramartha Dutta.
© 2017 John Wiley & Sons Ltd. Published 2017 by John Wiley & Sons Ltd.
Companion Website: www.wiley.com/go/bhattacharyya/hybridintelligence

and the well-known vehicle detection data set LISA-Q Front FOV [7], are used for evaluation purposes. Results are discussed, and performance metrics are calculated and compared with respect to the state-of-the-art methods.

8.2 Extraction of Local Features by SIFT and SURF

SIFT extracts local particularities [2]. Local particularities are robust to changes in illumination, noise, and viewpoint up to 30°, and hence they are distinctive and can be identified iteratively in different scenes. But SIFT is sensitive and fragile when object texture and geometry deformations suddenly occur, and in the domain of vehicle tracking in abruptly changing traffic scenes, these deformations frequently occur. Haar responses in SURF provide image information that is also invariant to illumination changes. SURF is invariant against noise and image rotation changes, but it is sensitive and inaccurate with viewpoint changes.

Although SIFT is suitable to image scene description tasks, feature extraction on a CPU cannot fulfill the requirements of real-time applications; for instance, feature extraction of an image with a size of 1000×700 pixels requires processing of approximately 4000 key points. This task costs 5 seconds of processing time on a Intel Quad-Cores i5 processor with a 2.70 GHz single core frequency, and 2 MB smart cache, and 2 GB memory. In terms of responsiveness, SURF provides three times faster description of images in comparison to SIFT [8]. Furthermore, implementation of SIFT on GPUs allows performing computer vision tasks on a real-time basis [9, 10].

Feature extraction in the SIFT model generates vectors or so-called descriptors. Resemblance between two vectors is calculated by using Euclidean distance [4]. Calculation of the dot products of two vectors is computationally inexpensive; for instance, computation of arccosines of the two unit vectors' dot products is simply calculated by the angles' ratio between them, which determines their resemblance level. In [2], Lowe's SIFT model computes the ratio of two vectors' angle by computing the angle ratio; and, when it is a small value, the resemblance is concluded to be very close to the resemblance calculated by Euclidean distance. In Figure 8.1, Lowe's matching method is illustrated; descriptors are normalized to a unit vector. Query image SIFT descriptors are denoted by $d_1, \ldots, d_n$, and image SIFT descriptors to be compared are denoted by $d_1, \ldots, d_m$. Dot products of d_i for $i = 1, \ldots, m$ and the set of $d_1, \ldots, d_m$ are calculated. After sorting and ranking the most elevated two resemblances, for instance, r_1 and r_2 are chosen for illustration purposes in Figure 8.1; if their ratio (i.e., $\frac{r_1}{r_2}$) is less than 0.6, then d_i for $i = 1, \ldots, n$ cannot assure a confident correspondence. In other words, it is not distinctively matched because it has another candidate matching descriptor. Otherwise, d_i is distinctively matched. The set of matching vectors yields the resemblance of compared images, and matching with existing objects over sequential frames is repeated.

Generic points extraction is presented in [11]. These generic key points represent the corresponding originals that are called robust SIFT descriptors. Generic points and their corresponding points are matched to obtain the preprocessed frame descriptors. Matching the set of generic points aided robust description (GPRD) descriptors is similar to Lowe's matching, but the resemblance value of 0.75 is chosen for the distance ratio, instead of the resemblance value of 0.6 that is the default ratio chosen to match irrelevant random images over classic SIFT descriptors. These factors are

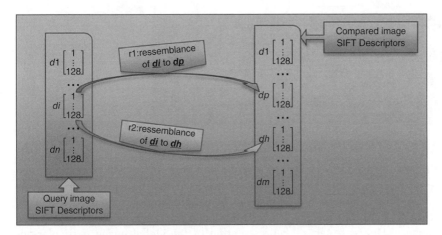

Figure 8.1 Illustration of Lowe's matching method.

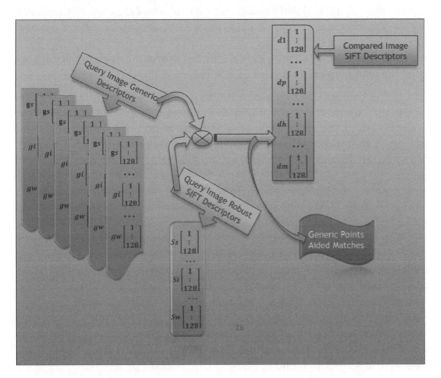

Figure 8.2 Generic points aided robust description (GPRD) matching scheme.

experimentally validated and have improved the false match elimination algorithm results (see, e.g., [11]). In the case of GPRD matching, the query image is pretreated subject to a perspective planar transformation, and transformed images are compared with the query image. This matching scheme is illustrated in Figure 8.2.

Our method is inspired by the extraction stage describing objects for object category detection in [8]. This is an intermediate step for the HMAX model C1 global image;

Figure 8.3 Matching with Lowe's SIFT description.

oriented filters are used to extract the responses of contours in four directions, and two different filter sizes of 11×11 and 13×13 are used. At the end of this step, the significant contour responses are extracted. Then the dilatation operation is applied to these extracted contour images. Dilated contour images become roughly described contour images. Hence, a general representation of the global image particularities such as contours and junctions is generated. By using this representative model, we are able to compare other objects' global features in order to detect the class of the object. By following the same approach, representation of a SIFT descriptor subject to some logical deformations is achieved.

This method is applied on a multiview data set [6]. Some preliminary comparisons are made by using the matching method presented in [11]:

- The object of the car rear is compared with its identical but 40°-rotated version. Since SIFT points are invariant and insensitive subject to a limited object rotation invariance of 30–45°, a matching result after a given rotation of 45° is unreliable for SIFT descriptors. In this comparison, the matching result of the original SIFT implementation when the distance ratio is equal to 0.6 yields 11 correct matches, as plotted in Figure 8.3.
- When the same object is matched by using the GPRD approach, the number of correct matches is significantly higher when the distance ratio is chosen to be equal to 0.6 and it is increased when the ratio is chosen to be 0.7, 0.75, or 0.8. The GPRD matching results are plotted in Figure 8.4 for different ratio values. For the ratio of 0.8, 30 correct matchings and three incorrect matchings are generated. When the ratio is chosen to be lower than or equal to 0.75, matchings are always correct.
- When the identical object to be matched is subject to strong shadows, rotation more than 45°, or cluttered background scenes, SIFT matching is not accurate and these results are not reliable. In these perturbed scenes, SIFT GPRD matching outperforms the classical SIFT. In Figure 8.5, the same object is matched with its perturbed identical object, and visual matching points are plotted for both GPRD and Lowe's SIFT description method.

GPRD assures more robust tracking in comparison with other feature extraction methods.

8.3 Global Features: Real-Time Detection and Vehicle Tracking

A boosted cascade of simple Haar-like rectangular features is used to detect vehicles. This method was originally introduced by Viola and Jones for face detection in [5]. In

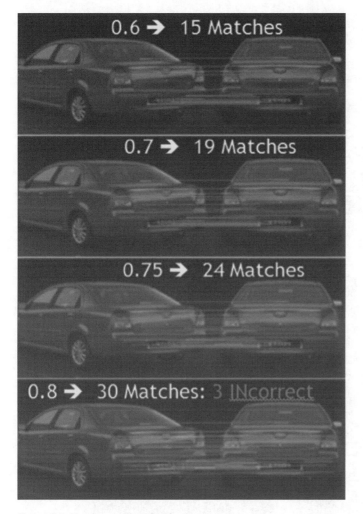

Figure 8.4 Matching with GPRD descriptors.

general terms, Haar-like rectangular features are well suited for object shape detection. These features are sensitive to image global features like edges, bars, vertical and horizontal details, and symmetric structures. The original algorithm used by Viola and Jones allows for rapid object detection that can be suitable for implementation on a real-time basis. The usage of integral images provides fast and efficient feature extraction. Extracted resulting values are effective weak learners, which are classified by Adaboost. Adaboost performs classification based on a weighted majority vote of weak learners. It is a discriminative learning algorithm. A cascade of classifier stages is constructed with Adaboost learning. Scores computed from feature extraction make the decision about rejection at each stage. Candidate objects are eliminated at each stage within the cascaded approach, and remaining candidates at the final stage are taken as positive detections.

Integral images, introduced by [5], are defined like look-up tables in the form of a matrix having the same size as the original image. Each element of the matrix contains

Figure 8.5 Visual comparison of GPRD and Lowe's SIFT description: The frame on the left side (blue colored) is the result of GPRD, and the frame on the right side (red colored) is the result of the SIFT description.

the sum of all pixels located on the upper-left region of the original image. This provides effective processing by using only four look-ups. Haar-like features are extracted by using box filters, which tend to behave like Haar wavelets of degree one. A Haar-like feature is extracted by summing up the pixel intensities over two adjacent rectangles and then subtracting the two sums. Basically, it is the difference between pixel intensity sums over two rectangles; total pixel intensity change in the adjacent region presents a global feature that is used by a weak learner like Adaboost. This difference is then used to categorize subsections of an image. For example, let's consider an image database of human faces; a common observation is that among all faces, the region of the eyes is darker in comparison to the region of the cheeks. Therefore, a common Haar feature for face detection is a set of two adjacent rectangles that lie above the eye and the cheek region. The position of these rectangles is defined in relation to a detection window that acts like a bounding box to the target object (that is simply the face in this example). In the detection phase of [5], a window of the target size is moved over the input image, and for each subsection of the image the Haar-like feature is calculated. This difference is then compared to a learned threshold that separates nonobjects from objects. Because such a Haar-like feature is only a weak learner, its detection quality is slightly better than random guessing, and a large number of Haar-like features is required to describe an object with a sufficient level of accuracy. In [5] again, the Haar-like features are organized, such as in a cascaded classifier, to form a strong learner or classifier.

But still, detection using Haar-like features with Adaboost is not robust with respect to challenging scene conditions; this detection scheme may produce a significant amount of false positives besides true positives. For laboratory testing purposes, we used Haar-like features in the context of face recognition. Subject to object pose changes, false positives occur frequently. Further processing is required in order to achieve a reliable recognition prototype. For instance, through sequences plotted in Figure 8.6, the big circle (colored blue) is the ROI, which is detected by using Haar-like features. Toward vehicle recognition and tracking, we develop further validation processes to achieve a reliable recognition system.

Figure 8.6 Assessment of Haar-like features in the context of face recognition.

8.4 Vehicle Detection and Validation

Low-level image analysis is applied to enhance detection. Namely, texture-based X-symmetry analysis and edge-based prominent horizontal line search are integrated in the overall detection and tracking solution given in Figure 8.7. In addition, detection history is used for validation and tracking purposes while creating a short-term memory dedicated to enhancement in tracking the sudden disappearance of vehicles due to occlusions by other surrounding vehicles and their possible reappearance. In the block diagram plotted in Figure 8.7, a video frame is subjected to a smoothing operation and then subsampled. A subsampled color image is then subjected to Canny edge detection and Haar detection with Adaboost. Haar detection provides candidate object locations that are labeled as object bounding boxes (OBBs).

8.4.1 X-Analysis

Colored image data inside the bounding box of a detected candidate object (in our case, it is a land vehicle) is analyzed, and based on the image data, symmetry is searched around the vertical axis. To accomplish this task, two adjacent sliding windows are used. A sliding window is scaled laterally by a minimum of one-fifth to the maximum of half-width of an OBB window with the same height. While the window is sliding from left to right at each iteration, the difference between the sum of the image data corresponding to the left and right window is calculated. The minimum image data difference

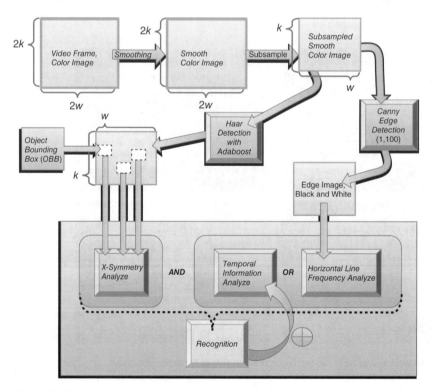

Figure 8.7 Recognition and tracking scheme.

over all iterations is considered as the candidate symmetry axis. If the difference value corresponding to the candidate symmetry axis is not smaller than a threshold value, then symmetry detection is rejected. The threshold value is chosen to be 15. Color image data is used to detect the symmetry axis. We used three colored channels and subsampled smooth images. Since the three channels and a single channel yield to the same decision result, a single channel is considered at each iteration to reduce computational costs.

Alternatively, to detect the symmetry axis, we analyzed edge images, which are single-channel images containing only white- or black-colored contours of the original image. Edge images are extracted with the Canny edge detection algorithm. Before applying the Canny algorithm, to reduce noise and increase performance, colored images are blurred and subsampled as presented in Figure 8.7. Again, the sliding window approach is applied, and at each iteration, the ratio between the count of white pixels and the whole pixel's count corresponding to the sliding window area is compared. Similarly, the left-side window is scanned and the same ratio is extracted. The two ratios are compared at each iteration, and the ratio pair providing maximal similitude is considered to detect symmetry axis. If the two ratios are not sufficiently close to each other (i.e., the ratio is below the threshold), then symmetry detection is rejected.

8.4.2 Horizontal Prominent Line Frequency Analysis

While analyzing low-level, global vehicle image features, one can exploit that the shape of a vehicle produces more horizontal edges than on-road background textures, and these horizontal edges are significantly longer and condensed in parallel inside the vehicle bounding box.

Quantification of prominent horizontal edges is a good indicator when validating a candidate vehicle given within a bounding box. Toward detection of vehicle horizontal lines, edge images are used at the first stage. Images are blurred and then subsampled by two; irrelevant noise data is eliminated. Afterward, the Canny edge detection algorithm is used to extract edge images consisting of only white- and black-colored data, containing only relevant edges marked with white color over a black background; see, for instance, the edge images plotted in Figure 8.8.

The edge image (candidate vehicle bounding box) is scanned by beginning from the bottom to the top of the frame, and horizontal edges are sought. As illustrated in the left side of Figure 8.8, the scanning procedure is started from point A and iteratively goes along the bounding box symmetry axis denoted by S until the top point, B. Scanning resolution is 1 pixel, and at each iteration [i.e., for a given central point (x, y) located on

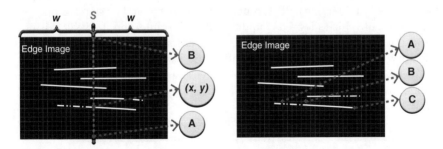

Figure 8.8 Scan for lines' vertical response (right), and discontinuity on lines (left).

the symmetry axis denoted by S], a continuous line search is executed by scanning in the horizontal direction from left to right. Left-side horizontal scanning is started from the point (x, y), terminated at the point $(x - w, y)$; and right-side horizontal scanning is started from (x, y), terminated at $(x + w, y)$. At each iteration of horizontal scanning, the pixel value at (x, y) is checked. If the pixel value of (x, y) is white and both $(x, y - 1)$ and $(x, y + 1)$ are black, then the next iteration is continued with $(x + 1, y)$. Otherwise, the value of the adjacent pixel $(x, y + 1)$ and $(x, y - 1)$ is checked. If $(x, y + 1)$ is white, then the next step is continued with $(x, y + 1)$; otherwise, if $(x, y - 1)$ is white, then the next step is continued with $(x, y - 1)$; and, if x cannot be incremented more than one iteration, then scanning is terminated. This seeking approach provides tolerance for detection of horizontal edges.

Another approach to seek horizontal edges is implemented by setting a gap of two pixels. According to the walking algorithm illustrated on the left side of Figure 8.8, if (x, y) is black, then the adjacent pixel located at $(x, y - 1)$ and $(x, y + 1)$ is checked. If those upper and lower neighbor pixels are also black, iteration continues at $(x + 1, y)$. If a white pixel is not detected, the gap at the edges is increased to two pixels. Gap over edges may be discontinued, as denoted by B on the right side of Figure 8.8. The points denoted by A and C are scan end points. At the end points, the obliquity of the line is checked by comparing the angle between the line and horizontal axis, and if this difference is more than 10°, the detected horizontal edge is rejected. The strength of the line can be considered as a constraint for decision making, and if the detected horizontal line's length is shorter than 10 pixels, then it is rejected.

We consider prominent lines to enhance reliability of validation information. This constraint increases robustness, but it also decreases the detection rate of distant vehicles subject to lower scaling. If a candidate vehicle has an active state, then we decrease 10 pixels in length with a weighted factor determined with the input from previous frames in the detection history of this vehicle. This provides adaptation against scale change when the relative distance between the tracked vehicle and the monocular camera sensor is increased or decreased.

8.4.3 Detection History

Temporal information is useful for enhancing tracking performance. In our implementation, temporal information is used for both tracking and detection. At each frame processing, history is updated, only vehicle objects detected within the last 30 frames are conserved, and others are removed from the detection history log file.

When a vehicle is detected, it is also searched in the history, and if it has been already detected within the last 30 frames, then its position is updated and its status is marked as active. The state of vehicles logged is updated at each frame, and if a vehicle is not redetected within the last 15 frames, then it is marked as inactive. But it is not discarded from the history until it is not redetected within the last 30 frames. And, all active vehicles in the history are weighted for voting. This voting weight is calculated by using the detection rate over the last 15 frames, which is simply calculated as the detection counts over 15.

Overall, the detection and tracking algorithm in Figure 8.7 can be summarized as follows:

1. *Step 1*: If the candidate vehicle bounding box does not satisfy an X-symmetry axis, then detection is rejected.

Figure 8.9 Screenshots of the recognition and tracking results on the Istanbul TEM highway.

2. *Step 2*: Vehicle edges may produce horizontal lines. At least four horizontal lines with a width of 10 pixels should be inside the bounding box.
3. *Step 3*: Detection history is used as an indicator for tracking and detection purposes. The candidate vehicle's bounding box center must have 28 close matchings over the last 30 frames in which detection has occurred. The distance between this center and the center of the detected vehicle captured from history must be less than two times the radius of the bounding box. This redundancy attenuates false positives.
4. If either Step 2 or Step 3 is not verified, then detection is rejected.

Two captured samples illustrate the detection performance as the application summary of the presented method in Figure 8.9.

8.5 Experimental Study

This section is divided in three parts. In the first part, local features are evaluated, and GPRD-based description and tracking results are compared to the SIFT and SURF results. The second part is oriented toward global features, and the results of local and global features–based hybrid feature tracking solutions are presented in the final part.

8.5.1 Local Features Assessment

To evaluate local feature–based description methods, a multiview vehicle rear data set is used [6]. This data set includes images subject to shadows, rotation, different objects in the background, and cluttered scenes. In Figure 8.10, 30 snapshots are created by rotating the original vehicle view with a resolution of 3° between −45° and 45°. Local features about a vehicle's rear view are compared during 20 sequences, and at least three correct matchings are decided to be the true detection. SIFT GPRD outperforms the SIFT and SURF methods. In Figure 8.11, the responses of false matching are compared, and SIFT GPRD generates fewer false matchings.

A set of features in the ROI of the vehicle is detected and tracked by implementing SIFT GPRD to a road traffic video recorded on the TEM. Tracking results are plotted in Figure 8.12, and inside the ROIs, matching and unmatched points are plotted with green and red dots, respectively.

8.5.2 Global Features Assessment

To evaluate the performance of recognition and tracking by using global features, the following metrics are used: the true positive rate (TPR), false detection rate (FDR), true

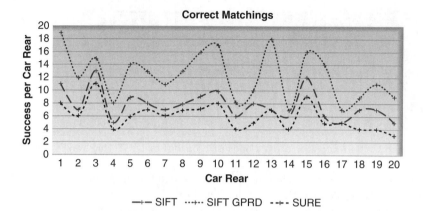

Figure 8.10 The responses of true matchings for SIFT, SURF, and SIFT GPRD–based detection.

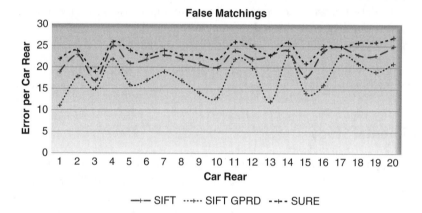

Figure 8.11 The responses of false matchings.

positive per frame (TPF), and false positive per frame (FPF).

$$TPR = \frac{t_p}{N} \tag{8.1}$$

$$FDR = \frac{f_p}{t_p + f_p} \tag{8.2}$$

$$TPF = \frac{t_p|_F}{N|_F} \tag{8.3}$$

$$FDR = \frac{f_p|_F}{N|_F} \tag{8.4}$$

TPR is a measure of recall and localization. In contrast, FDR is a measure of precision and localization. TPF is a measure of robustness; FPF is a measure of robustness, localization, and scalability. For instance, consider a road scene where N vehicles are traveling. True positives (t_p) refer to the number of vehicles in this scene that are correctly recognized and tracked by using global features, while false positives (f_p) refer

Figure 8.12 Evaluation of SIFT GPRD in a road traffic video.

the number of vehicles not in this scene that are incorrectly recognized and tracked. Similarly, true positives per frame $(t_p|_F)$ are the average number of vehicles correctly recognized and tracked versus the number of successive frames denoted by $N|_F$, and $(f_p|_F)$ is the average number of vehicles incorrectly recognized and tracked versus the number of successive frames.

To evaluate the local and global features–based hybrid feature tracking solution, video records captured under different traffic conditions from a monocular camera mounted on a vehicle traveling on the TEM and the E5 freeway, and the well-known vehicle detection data set LISA-Q Front FOV, are used. Three videos were recorded during different hours of the day and on different road segments of the TEM and E5 roads. These three video recordings are used for evaluation purposes. The metrics are calculated and given in Table 8.1.

The presented hybrid system tracks other vehicles on the highway with a high rate of true positives per frame, and results are plotted in Figure 8.13. Figure 8.13 and Figure 8.14 present the captured results of the presented recognition and tracking method.

Table 8.1 Performance results

Video dataset	TPR	FDR	TP/frame	FP/frame
TEM/E5 1	98.8%	4.6%	1.7	0.011
TEM/E5 2	93.5%	8.2 %	5.8	0.151
TEM/E5 2	96.4 %	5.5 %	2.2	0.039

Figure 8.13 Screenshots of recognition and tracking system result for TEM highway during rush hour.

Figure 8.14 Screenshots of recognition and tracking system results for LISA-Q Front FOV 1 during rush hour.

Publicly available video recordings are used for comparison purposes. For different conditions of traffic and road structure, the presented system is tested with the videos entitled in [12] as Lisa-Q Front FOV 1 (delivering rush hour traffic conditions), Lisa-Q Front FOV 2 (for highway scenes), and Lisa-Q Front FOV 3 (for urban driving). The results are given in Table 8.2.

Validation stages presented in Section 8.4 improve the matching accuracy of the presented recognition and tracking method, and fewer error-prone results are generated in comparison with the active learning scheme presented by ALVeRT in [12]. By comparing the metric results given in Table 8.2, the true positive rate (TPR) results are higher than ALVeRT for the two videos named Lisa-Q Front FOV 2 (highway traffic) and Lisa-Q Front FOV 3 (urban traffic). When comparing with a passively trained model for all three driving scenarios such as rush hour, highway, and urban traffic, the

Table 8.2 Performance results (video data set belonging to [12])

Video dataset	TPR	FDR	TP/frame	FP/frame
LISA-Q FOV 1: rush hour	83.3%	15.3%	1.6	2.2
LISA-Q FOV 1: highway	96.3%	5.6 %	2.6	0.03
LISA-Q FOV 1: urban	99.6 %	1.6 %	0.98	0.01

false detection rate (FDR) of our recognition and tracking system is significantly lower than the passively trained results presented in [12]. Lower FDR validates the higher level in accuracy for detection and tracking of land vehicles for various traffic scenes. Results of recognition and tracking system during rush hour are plotted in Figure 8.14.

8.5.3 Local versus Global Features Assessment

Performance regarding true positives and false positives per frame is evaluated for tracking by using local and global features applied to the videos recorded on the TEM highway. GPRD is used for tracking of local features as descriptors, and it outperforms the SIFT descriptors at the matching stage. The true positive per frame metrics for the case of global and local features are evaluated on the three different videos, and the results are plotted in Figure 8.15. In Figure 8.16, false positive per frame results are given, and object detection is not evaluated by using local features.

Considering the usage of computational resources, real-time processing of local features with four frames per second requires a custom CUDA-enabled NVIDIA GPU, whereas global features are more suitable for real-time processing, and 15 frames per second processing performance is achievable on a Intel Core i5 CPU.

8.6 Conclusions

In this section, state-of-the-art image feature extraction methodologies and their implementations are presented. Instead of local feature extraction methodologies sensitive to local image data changes that happen frequently for vehicle surfaces moving on a

Figure 8.15 Comparison of true positive per frame metrics for global and local features.

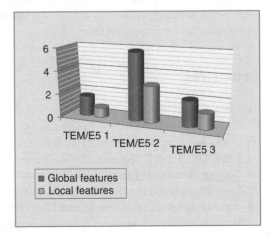

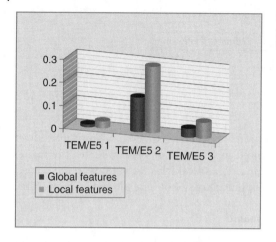

Figure 8.16 Comparison of false positive per frame metrics for global and local features.

road scene, global feature extraction methodologies are used in the context of object description. Haar-like features are adopted for global feature extraction and used for describing object models. A set of vehicle rear-view image data is provided for training purposes, and obtained indexation is used as a preliminary object detection step in our implementation with the Adaboost classifier. This preliminary detection provides ROIs involving global features resembling the features of the trained object model. But this preliminary detection is not accurate and generates a significant amount of false positives. A validation algorithm is developed to enhance accuracy without increasing the computational cost of the method. We examined the low-level image characteristics of the vehicle rear-view images as the objects in our study. Vehicle images present symmetric features, particularly on the rear side. Color channels are used to seek a texture-based symmetry axis. In addition to symmetry search, contour images are extracted from detected ROIs. Vehicle textures generate horizontal lines remarkably significantly in the road scene. Preliminary detection is validated depending on symmetric features and horizontal line frequency. A tracking algorithm is implemented using temporal detection history in which previous vehicle detections are classified as either active or inactive. As a voting mechanism, the temporal detection history is used for preliminary validation as well. An efficient hybrid method is presented toward active safety of intelligent vehicles.

References

1 Lowe, D.G. (1999) Object recognition from local scale-invariant features, in *Computer Vision, 1999. The Proceedings of the Seventh IEEE International Conference*, vol. 2, pp. 1150–1157, doi: 10.1109/ICCV.1999.790410.

2 Lowe, D.G. (2004) Distinctive image features from scale-invariant keypoints. *Int. J. Comput. Vision*, **60** (2), 91–110, doi: 10.1023/B:VISI.0000029664.99615.94. Available from http://dx.doi.org/10.1023/B:VISI.0000029664.99615.94.

3 Bay, H., Ess, A., Tuytelaars, T., and Gool, L.V. (2008) Speeded-up robust features (SURF). *Computer Vision and Image Understanding*, **110** (3),

346–359, doi:http://dx.doi.org/10.1016/j.cviu.2007.09.014. Available from http://www.sciencedirect.com/science/article/pii/S1077314207001555.

4 Serre, T., Wolf, L., and Poggio, T. (2005) Object recognition with features inspired by visual cortex, in *Computer Vision and Pattern Recognition, 2005. CVPR 2005. IEEE Computer Society Conference*, vol. 2, pp. 994–1000, doi: 10.1109/CVPR.2005.254.

5 Viola, P. and Jones, M. (2001) Rapid object detection using a boosted cascade of simple features, in *Computer Vision and Pattern Recognition, 2001. CVPR 2001. Proceedings of the 2001 IEEE Computer Society Conference*, vol. 1, pp. I–511–I–518, doi: 10.1109/CVPR.2001.990517.

6 Ozuysal, M., Lepetit, V., and Fua, P. (2009) Pose estimation for category specific multiview object localization, in *Conference on Computer Vision and Pattern Recognition*, Miami, FL.

7 Trivedi, M.M. (2014) Computer Vision and Robotics Research Laboratory, University of California, San Diego.

8 Moreno, P., Marin-Jimenez, M.J., Bernardino, R., and Blanca, N.P.D.L. (2007) A comparative study of local descriptors for object category recognition: SIFT vs HMAX.

9 Sinha, S.N., Frahm, J.M., Pollefeys, M., and Genc, Y. (2006) GPU-based video feature tracking and matching, *Tech, Rep.*, in Workshop on Edge Computing Using New Commodity Architectures.

10 Ziegler, G., Tevs, A., Theobalt, C., and Seidel, H.P. (2006) GPU point list generation through histogram pyramids, in *Proceedings of VMV*, Aachen, Germany, pp. 137–141.

11 Yíldíz, R. and Acarman, T. (2012) Image feature based video object description and tracking, in *Vehicular Electronics and Safety (ICVES), 2012 IEEE International Conference*, pp. 405–410, doi: 10.1109/ICVES.2012.6294255.

12 Sivaraman, S. and Trivedi, M.M. (2010) A general active-learning framework for on-road vehicle recognition and tracking. *IEEE Transactions on Intelligent Transportation Systems*, **11** (2), 267–276.

9

A GIS Anchored Technique for Social Utility Hotspot Detection

Anirban Chakraborty[1], J.K. Mandal[2], Arnab Patra[1], and Jayatra Majumdar[1]

[1] *Department of Computer Science, Barrackpore Rastraguru Surendranath College, Barrackpore, Kolkata, West Bengal, India*
[2] *Department of Computer Science & Engineering, University of Kalyani, Kalyani, Nadia, West Bengal, India*

9.1   Introduction

Criminals and crime are not new, since such a sense of cruelty is originated right from the epic age of Mahabharata and Ramayana, or even before. The merciless and ruthless professions of crime prove to be very much effective in triggering a puff of fear and revulsion in people's mind. But why does someone choose such heartless professions? One of the main reasons can be poverty. In a country like India, the sentence "The rich get richer, and the poor get poorer" proves to be very true as poverty is successful in climbing at a very snappy pace over time. Because of this, mortals suffering from not being able to fulfill even the basic amenities resort to illegal means that may include snatching property and even taking the lives of innocent people.

The proposed technique makes an attempt for eradicating crime. It aims at predicting the spread of criminal activities and fruitful measures for combating crime. The proposed technique enlightens one of the most common and important measures for controlling criminal activities: uniformly setting up police stations, especially in the crime-sensitive zones. Initially, 21 Anchals under seven police stations of the state of West Bengal in India (i.e., Egra, Barrackpore, Uttarpara, Mogra, Ranaghat, Matigara, and Coochbehar) have been chosen as the case study area, to draw a relationship between population and number of crimes (with associated weighted factors). All of the 32 Anchals under the Egra police station, which is situated at Purba Medinipur in West Bengal, India, have been picked for detailed analysis. All information (including the map of Egra) related to the proposed work is acquired from the Egra police station. The map of Egra is shown in Figure 9.1.

The proposed task targets the crime-keen areas for the years 2011, 2012, and 2013. The areas detected as crime prone (i.e., hotspots) are highlighted by red color on the map, making it understandable for all users. Using the method of the fuzzy *c*-means (FCM) clustering technique, the present methodology also clusters the pointed hotspot zone and suggests the suitable locations for setting up new police stations for eradication of crime.

Hybrid Intelligence for Image Analysis and Understanding, First Edition.
Edited by Siddhartha Bhattacharyya, Indrajit Pan, Anirban Mukherjee, and Paramartha Dutta.
© 2017 John Wiley & Sons Ltd. Published 2017 by John Wiley & Sons Ltd.
Companion Website: www.wiley.com/go/bhattacharyya/hybridintelligence

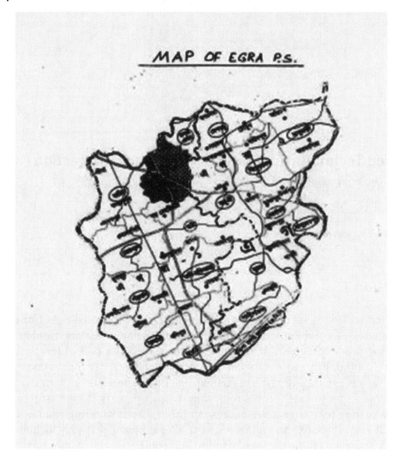

Figure 9.1 Map of Egra police station, West Bengal, India.

Along with its numerous applications in a variety of areas, GIS could be considered as a very powerful tool for creation or analysis of spatial data and, moreover, as smart visualization of the results on the map that has been fed input. A diverse number of industries in various sectors are thus beneficiaries by GIS. Nowadays, there is positive awareness regarding the impact on the society that GIS can have. Identification of hotspots is just one of them. A hotspots have some special statistical significance. On the basis of some statistical data analysis reports, a hotspot is heavily distinguished from the rest of a given study area. Till now, only criminal hotspots, which are areas sensitive with respect to crime, have received concentrated attention. But identification of earthquake hotspots (areas that have a high probability of an earthquake occurring) and identification of accident hotspots (regions having a high probability of road accidents) are also very helpful for the modern society and prevent loss of human lives (a loss that cannot be compensated), properties, and money.

The above-mentioned hotspot-revealing methodology can also be extended for unmasking many other similar hotspot-disclosing jobs, such as:

- Detection of earthquake-prone areas
- Detection of landslide-sensitive areas
- Detection of accident-prone (road accidents) areas.

Table 9.1 Hotspot-detecting tasks, depending on respective factors

Hotspot-detecting tasks with respect to	Influencing factors
Earthquakes	• Level of seismic hazards [3] • Changes in Vp/Vs [4] • Radon emission [4] • Electromagnetic variations [4]
Landslides [5]	• Proximity of earthquake • Encumbering on rock material • Banishment of lateral support • Changes in the water sample of rock or soil frame • Clogged drainage

Note: Vp: velocity of a seismic wave passing through rock; and Vc: velocity of a secondary or shear wave.

However, the factors causing all such additional hotspot detection tasks to be conducted would be unalike. For example, the influencing factors that would prompt the disclosure of earthquake-prone spots and landslide-sensitive areas are simulated in Table 9.1.

A number of existing hotspot detection methods exist, having varying application areas. ISODATA [10, 11] and STAC [9] are used for identification of criminal hotspots. ISODATA uses the software TNT-mips 6.4 for handling maps, and to get efficient results, values of a number of parameters such as lumping parameters, initial cluster means, splitting parameters, and so on need to be specify beforehand. Although STAC doesn't need these types of initial values to specify, it mainly generates circular clusters always; also, this mainframe program has not yet been tested in real life. For detection of earthquake hotspot zones, RADIUS [12] [13] is an existing technique that works with the help of MS Excel, but unfortunately it gives very general approximations and the output is not properly visualized. Multicriteria evaluation (MCE) [15] is a tool that helps in detection of landslide hotspot zones. As the name reveals, it is based on many analytical hierarchy processes (AHPs), making it much too complex.

Section 9.2 of this chapter throws light on the methodology used, Section 9.3 discusses the case study areas adapted for the purpose, Section 9.4 shows the result obtained after application of the technique on the case study area, Section 9.5 deals with analysis of the proposed method, as well as comparison with other existing techniques. Finally, conclusions are drawn in Section 9.6.

9.2 The Technique

The present technique (which is basically a rank-based clustering method) uses the crime data report obtained from field study as its backbone. There is no strict rule for determining the number of case study areas to be chosen, but increasing the number of study areas increases the accuracy of results obtained from regression analysis. It is better to choose the study areas far apart from each other for better results. All these areas should be grouped into three major categories in terms of criminal activities: very sensitive, moderately sensitive, and least sensitive. A numerical rank-based value (which may be a 10-point scale) is tagged thereafter. Unprejudiced answers obtained from police personnel through a method of questionnaire are very effective for the above ranking. To overcome human bias, quite a large number of persons should be interviewed in each study area, the average of which forms the final output. For ranking various criminal activities, the penal code of any country could be a pioneer. A numerical value is also assigned with each type of crime, depicting how heinous is it, driven by the country's penal code. This forms the basis of the regression analysis.

The nth-degree polynomial of the form $y = a_0 x^n + a_1 x^{n-1} + \dots a_n$ is considered for the purpose of regression analysis. To evaluate the unknown variables, data obtained (population and criminal activities reported to the police station for the considered year) from field study is taken into account. Evaluating the unknowns enables to rank any unknown region on a particular scale (say, 10), if only its population and crime reports are known.

For the entire large region intended for clustering on the basis of criminal hotspots, each of its constituent smaller areas should be ranked on a particular scale, using the above procedure. Let the ranking be done on a 10-point scale. Then, not only the regions with rank 10 would be considered as hotspots, but a region more than half of whose neighbors are hotspots also would be considered alarming for the future and is pointed out as a hotspot.

The objective of the present method is not only to figure out the crime hotspots but also to control the criminal activities by suggesting proper locations for police posting (i.e., the construction of new beat houses). The following steps are carried out in order to determine the locations for construction of police beat houses. To achieve this task, the following trails need to be worked through:

The entire crime hotspot zone already demarcated is considered, with two different cases:

- *Case 1:* If the zone contains more than three constituent regions, then for every five regions (obtained by rounding up the result total_number_of_regions / 5) one beat house is proposed, that is, for 4 to 7, one beat house is proposed; for 8 to 12, two beat houses; for 13 to 17, three beat houses; and so on. Let this number of proposed beat houses be denoted by n. Thus, the target is to divide the zone into n clusters, for which FCM clustering could be helpful because it produces compact clusters.

  For the present purpose, hard FCM has been considered where each data sample is assigned to only a particular class. U is a 2D matrix denoting the membership value of any data point to any class. Let this membership assignment of kth data sample in the ith class be denoted by λ_{ij}. d_{ik} is a Euclidean distance measure between the kth data sample and i^{th} cluster, given by $d_{ik} = d(x_k - v_i)$, where v_i is the ith cluster center.

  Thus, after determining the number of clusters $c(2 \leq c < n)$ for partitioning n data points, the membership matrix is initialized, and cluster centers are calculated accordingly. At each iteration, characteristic functions (for all i, k) are updated, using $\lambda_{ik} = 1$ for $d_{ik}^{(r)} = \min\{d_{jk}^{(r)}\}$ for all $j \in c$, or else $\lambda_{ik} = 0$. The iteration stops when $\|U^{(r+1)} - U^{(r)} \leq \xi\|$ (tolerance level).

  Centroid of each cluster is the target location of construction of police beat houses.

- *Case 2:* If the cluster contains less than four constituent regions, the centroid of the entire hotspot zone is the proposed location for a beat house.

  As discussed in Section 9.1, the intended work can also be very successful in unmasking many other hotspot detection tasks (e.g., earthquake-prone areas, landslide-sensitive zones, a-prone areas, etc.). The only variation lies in the influencing factors to take care of. However, it is not always necessary to select the scale of 10 units for accomplishing the proposed work, since any unit of scale (e.g., 5, 8, 15, etc.) can be picked up for accomplishing the intended work.

9.3   Case Study

The already-mentioned technique has successfully been applied for determining criminal hotspots of the Egra police station area, situated at Purba Medinipur in West Bengal, India. The stepwise procedure applied helps in better understanding the technique just discussed. Egra comprises 32 Anchals. The existing police stations are situated at Kosba-1, Vivekananda-1, and Chatri-2, as depicted in Figure 9.2.

Detection of crime hotspots (i.e., the crime-sensitive areas) of Egra is done for the years 2011, 2012, and 2013. In order to accomplish this task, a procedure of tying up rank (an integer value) to each region is performed. The execution of associating rank to each region is rendered, based on the criminal activities of that specific region. The rank assessment is performed on the scale of 10. When the rank of a region is 10, it is regarded as the "highest crime-prone," area and similarly a region that is the least crime prone is tagged with rank 1. To evaluate the rank associated with a particular region, information is gathered from seven distinct police stations throughout the state of West Bengal. The highest, moderate, and least crime-prone areas (as shown in Table 9.2), coming under the wing of each of the respective police stations, are grasped, based on data supplied by a number of police personnel after surveying them through a separate questionnaire, from each of the seven police stations, which serves as the base of the study.

Figure 9.2 Existing police stations of Egra.

Table 9.2 Table depicting the least, moderate, and highest crime-prone regions situated in seven police stations throughout West Bengal State, India

Police stations	Crime-sensitive areas		
	Highest sensitive	Moderate sensitive	Least sensitive
1. Egra	Jumki-1	Chetri-1	Kosba-1
2. Barrackpore	Ardali Bazar	Aamtala	Anandapuri
3. Uttarpara	Kanaipur Colony	Kotrong	Bhadrakali
4. Mogra	Kalitala	Damra	Joypur
5. Ranaghat	Kuparse	Nasra	Payradanga
6. Matigara	Phansideoa	Sivmandir	Babupara
7. Coochbehar	Bamanhat	Pundibari	Sadar

Table 9.3 Classification of crime types with respect to their rankings

Ranking	Types of crime
10	Murder
9	Dacoity
8	Arms Act Rape (376 IPC)
7	Crime against women attempt to murder (498/302 IPC)
6	Kidnapping (363A/366A IPC) Dowry death (498A/304B IPC) Rioting Offence Against Women (OAW)
5	Robbery
4	Crime against women (498A/323IPC) Assemble for preparation of dacoity (399/402 IPC) Crime related to drug (NDPS ACT)
3	Burglary
2	Theft
1	Molestation (354 IPC) Others (such as pickpocketing)

For the present study, 10 different crime types have been considered. Each type is adorned with a rank, the evaluation of which is done on the basis of the Indian Penal Code as shown in Table 9.3, where the higher the rank, the more heinous is the crime.

A weight in a 5-point scale is associated with each item of Table 9.3. Thus, a crime with rank 10 has a weight of 5, and that of rank 1 is 0.5.

The intended work is then buttoned up using the regression analysis strategy [2]. For the regression analysis procedure, ninth-degree polynomial curve fitting has been considered, the form of which is shown in Equation (9.1):

$$y = a + bx + cx^2 + dx^3 + ex^4 + fx^5 + gx^6 + hx^7 + ix^8 + jx^9 \tag{9.1}$$

In order to cook out Equation (9.1), the following simultaneous equations need to be executed:

$$na + b\sum_{k=1}^{n} x_k + c\sum_{k=1}^{n} x_k^2 + d\sum_{k=1}^{n} x_k^3 + e\sum_{k=1}^{n} x_k^4 + f\sum_{k=1}^{n} x_k^5$$
$$+ g\sum_{k=1}^{n} x_k^6 + h\sum_{k=1}^{n} x_k^7 + i\sum_{k=1}^{n} x_k^8 + j\sum_{k=1}^{n} x_k^9 - \sum_{k=1}^{n} y_k = 0 \tag{9.1a}$$

$$a\sum_{k=1}^{n} x_k + b\sum_{k=1}^{n} x_k^2 + c\sum_{k=1}^{n} x_k^3 + d\sum_{k=1}^{n} x_k^4 + e\sum_{k=1}^{n} x_k^5$$
$$+ f\sum_{k=1}^{n} x_k^6 + g\sum_{k=1}^{n} x_k^7 + h\sum_{k=1}^{n} x_k^8 + i\sum_{k=1}^{n} x_k^9 + j\sum_{k=1}^{n} x_k^{10} - \sum_{k=1}^{n} x_k y_k = 0 \tag{9.1b}$$

$$a\sum_{k=1}^{n} x_k^2 + b\sum_{k=1}^{n} x_k^3 + c\sum_{k=1}^{n} x_k^4 + d\sum_{k=1}^{n} x_k^5 + e\sum_{k=1}^{n} x_k^6$$
$$+ f\sum_{k=1}^{n} x_k^7 + g\sum_{k=1}^{n} x_k^8 + h\sum_{k=1}^{n} x_k^9 + i\sum_{k=1}^{n} x_k^{10} + j\sum_{k=1}^{n} x_k^{11} - \sum_{k=1}^{n} x_k^2 y_k = 0 \tag{9.1c}$$

$$a\sum_{k=1}^{n} x_k^3 + b\sum_{k=1}^{n} x_k^4 + c\sum_{k=1}^{n} x_k^5 + d\sum_{k=1}^{n} x_k^6 + e\sum_{k=1}^{n} x_k^7$$
$$+ f\sum_{k=1}^{n} x_k^8 + g\sum_{k=1}^{n} x_k^9 + h\sum_{k=1}^{n} x_k^{10} + i\sum_{k=1}^{n} x_k^{11} + j\sum_{k=1}^{n} x_k^{12} - \sum_{k=1}^{n} x_k^3 y_k = 0 \tag{9.1d}$$

$$a\sum_{k=1}^{n} x_k^4 + b\sum_{k=1}^{n} x_k^5 + c\sum_{k=1}^{n} x_k^6 + d\sum_{k=1}^{n} x_k^7 + e\sum_{k=1}^{n} x_k^8$$
$$+ f\sum_{k=1}^{n} x_k^9 + g\sum_{k=1}^{n} x_k^{10} + h\sum_{k=1}^{n} x_k^{11} + i\sum_{k=1}^{n} x_k^{12} + j\sum_{k=1}^{n} x_k^{13} - \sum_{k=1}^{n} x_k^4 y_k = 0 \tag{9.1e}$$

$$a\sum_{k=1}^{n} x_k^5 + b\sum_{k=1}^{n} x_k^6 + c\sum_{k=1}^{n} x_k^7 + d\sum_{k=1}^{n} x_k^8 + e\sum_{k=1}^{n} x_k^9$$
$$+ f\sum_{k=1}^{n} x_k^{10} + g\sum_{k=1}^{n} x_k^{11} + h\sum_{k=1}^{n} x_k^{12} + i\sum_{k=1}^{n} x_k^{13} + j\sum_{k=1}^{n} x_k^{14} - \sum_{k=1}^{n} x_k^5 y_k = 0 \tag{9.1f}$$

$$a\sum_{k=1}^{n} x_k^6 + b\sum_{k=1}^{n} x_k^7 + c\sum_{k=1}^{n} x_k^8 + d\sum_{k=1}^{n} x_k^9 + e\sum_{k=1}^{n} x_k^{10}$$
$$+ f\sum_{k=1}^{n} x_k^{11} + g\sum_{k=1}^{n} x_k^{12} + h\sum_{k=1}^{n} x_k^{13} + i\sum_{k=1}^{n} x_k^{14} + j\sum_{k=1}^{n} x_k^{15} - \sum_{k=1}^{n} x_k^6 y_k = 0 \tag{9.1g}$$

$$a\sum_{k=1}^{n} x_k^7 + b\sum_{k=1}^{n} x_k^8 + c\sum_{k=1}^{n} x_k^9 + d\sum_{k=1}^{n} x_k^{10} + e\sum_{k=1}^{n} x_k^{11}$$

$$+f \sum_{k=1}^{n} x_k^{12} + g \sum_{k=1}^{n} x_k^{13} + h \sum_{k=1}^{n} x_k^{14} + i \sum_{k=1}^{n} x_k^{15} + j \sum_{k=1}^{n} x_k^{16} - \sum_{k=1}^{n} x_k^7 y_k = 0 \quad (9.1\text{h})$$

$$a \sum_{k=1}^{n} x_k^8 + b \sum_{k=1}^{n} x_k^9 + c \sum_{k=1}^{n} x_k^{10} + d \sum_{k=1}^{n} x_k^{11} + e \sum_{k=1}^{n} x_k^{12}$$

$$+f \sum_{k=1}^{n} x_k^{13} + g \sum_{k=1}^{n} x_k^{14} + h \sum_{k=1}^{n} x_k^{15} + i \sum_{k=1}^{n} x_k^{16} + j \sum_{k=1}^{n} x_k^{17} - \sum_{k=1}^{n} x_k^8 y_k = 0 \quad (9.1\text{i})$$

$$a \sum_{k=1}^{n} x_k^9 + b \sum_{k=1}^{n} x_k^{10} + c \sum_{k=1}^{n} x_k^{11} + d \sum_{k=1}^{n} x_k^{12} + e \sum_{k=1}^{n} x_k^{13}$$

$$+f \sum_{k=1}^{n} x_k^{14} + g \sum_{k=1}^{n} x_k^{15} + h \sum_{k=1}^{n} x_k^{16} + i \sum_{k=1}^{n} x_k^{17} + j \sum_{k=1}^{n} x_k^{18} - \sum_{k=1}^{n} x_k^9 y_k = 0 \quad (9.1\text{j})$$

where n is the number of data points.

The tactics behind extracting the value of x and y in order to carry out the above equations [and thereby executing Equation (9.1)] is on the basis of the data collected from seven police stations, with three regions existing beneath the wrap of each particular police station, classified as the highest, moderate, and least crime-prone regions.

- The region with the "maximum" crime-prone tag has a y value of 10.
- The region with the "moderate" crime-prone tag has a y value of 5.
- The region with the "least" crime-prone tag has a y value of 1.

The x's associated with each of these y's is determined using the following two factors:

- Total population of the particular region
- Total cases received with each classified crime types against the same region.

Finally, a particular x value (with respect to a particular y) is plucked out using Equation (9.2).

$$x = \frac{\text{Total population of a particular region}}{(\sum_{i=1}^{10} C_i \times J) \text{ of the same region}} \quad (9.2)$$

where J = weighted value, as obtained from Table 9.3; and C_i = total number of cases received with respect to the ith ranked crime type. Table 9.4 portrays the required data for computing Equation (9.2).

Now, by dividing numerators by the respective denominators, the x values are obtained from the above table and the corresponding y values are fetched from Table 9.2 (by assigning values high=10, moderate=5, and least=1, as discussed). Tables 9.5, 9.6, and 9.7 give a fleeting look on the final computed x's and y's forming the basis of regression analysis.

Thus, here the process of regression analysis is dealing with 21 data points (i.e., the value of n).

Table 9.4 Table depicting the numerator and denominator values of respective regions calculated with respect to the crime types (as obtained from Table 9.3)

Region name	Value of the numerator of eq(b)	Crime types										Value of the denominator of eq(b)
		Weighted factors										
		CH1	CH2	CH3	CH4	CH5	CH6	CH7	CH8	CH9	CH10	
		5	4.5	4	3.5	3	2.5	2	1.5	1	0.5	
Kosba-I	14,490	0	0	0	0	1	0	0	1	5	1	10
Anandapuri	17,064	0	0	0	0	0	0	1	3	1	9	12
Bhadrakali	15,235	0	0	0	0	0	0	0	5	2	2	10.5
Joypur	12,027	0	0	0	0	0	0	0	0	2	13	8.5
Payradanga	15,829	0	0	0	0	1	0	1	2	3	0	11
Babupara	13,547	0	0	0	1	0	0	1	0	4	0	9.5
Sadar	15,130	0	0	1	0	2	0	0	0	0	1	10.5
Chatri-I	11,792	0	0	0	0	2	1	0	1	3	6	16
Aamtala	18,399	0	2	0	0	3	1	2	0	0	0	24.5
Kotrong	12,176	0	0	1	1	1	0	2	1	0	0	16
Damra	13,072	0	0	0	0	0	5	0	3	0	1	17.5
Nasra	15,603	0	1	1	1	0	1	0	2	7	2	21
Sivmandir	14,781	1	0	2	2	1	1	0	1	0	1	19.5
Pundibari	17,155	0	3	1	1	0	0	1	3	0	0	23.5
Jumki-I	15,334	4	4	14	14	22	10	0	10	0	1	225.5
Ardali Bazar	10,325	3	10	1	1	30	32	2	14	2	21	295
Kanaipur Colony	11,067	0	0	28	28	11	13	14	9	0	0	217
Kalitala	11,560	3	2	0	0	10	3	16	15	32	83	189.5
Kuparse	11,895	2	12	5	5	16	0	26	32	36	39	305
Phansideoa	10,626	12	4	0	0	0	28	14	22	13	18	231
Bamanhat	19,345	6	10	3	3	15	14	40	0	0	7	265

Table 9.5 Table delineating the final values of x's and y's (data for $y=1$)

x	1449	1422	1451	1415	1439	1426	1441
y	1	1	1	1	1	1	1

Table 9.6 Table delineating the final values of x's and y's (data for $y=5$)

x	737	751	761	747	743	758	730	737
y	5	5	5	5	5	5	5	5

Table 9.7 Table delineating the final values of x's and y's (data for $y=10$)

x	68	35	51	61	39	46	73	68
y	10	10	10	10	10	10	10	10

Conclusively, $\sum_{k=1}^{n=21} x_k = 15{,}647$ and $\sum_{k=1}^{n=21} y_k = 112$ (obtained from Tables 9.5, 9.6, and 9.7) and the following values could also be computed:

$$\sum_{k=1}^{n} x_k = 15647 \sum_{k=1}^{n} x_k^2 = 18339438, \sum_{k=1}^{n} x_k^3 = 23602892464,$$

$$\sum_{k=1}^{n} x_k^4 = 3.18663E + 13, \sum_{k=1}^{n} x_k^5 = 4.42386E + 16, \sum_{k=1}^{n} x_k^6 = 6.23811E + 19,$$

$$\sum_{k=1}^{n} x_k^7 = 8.87138E + 22, \sum_{k=1}^{n} x_k^8 = 1.26738E + 26, \sum_{k=1}^{n} x_k^9 = 1.815E + 29,$$

$$\sum_{k=1}^{n} x_k^{10} = 2.60266E + 32, \sum_{k=1}^{n} x_k^{11} = 3.73484E + 35, \sum_{k=1}^{n} x_k^{12} = 5.36175E + 38,$$

$$\sum_{k=1}^{n} x_k^{13} = 7.69931E + 41, \sum_{k=1}^{n} x_k^{14} = 1.10579E + 45, \sum_{k=1}^{n} x_k^{15} = 1.58834E + 48,$$

$$\sum_{k=1}^{n} x_k^{16} = 2.28172E + 51, \sum_{k=1}^{n} x_k^{17} = 3.27809E + 54, \sum_{k=1}^{n} x_k^{18} = 4.70994E + 57,$$

$$\sum_{k=1}^{n} y_k = 112, \sum_{k=1}^{n} x_k y_k = 39933, \sum_{k=1}^{n} x_k^2 y_k = 34151972,$$

$$\sum_{k=1}^{n} x_k^3 y_k = 35284771798, \sum_{k=1}^{n} x_k^4 y_k = 4.05878E + 13, \sum_{k=1}^{n} x_k^5 y_k = 5.07567E + 16,$$

$$\sum_{k=1}^{n} x_k^6 y_k = 6.72536E + 19, \sum_{k=1}^{n} x_k^7 y_k = 9.2357E + 22,$$

$$\sum_{k=1}^{n} x_k^8 y_k = 1.29462E + 26, \sum_{k=1}^{n} x_k^9 y_k = 1.83538E + 29$$

Table 9.8 Values of the unknown variables used in Equation (9.1)

Unknowns	Values
a	10.230838005070808
b	−0.0041658198089687196
c	−3.3045749026962936E-6
d	−1.265402658730446E-9
e	1.639773297320292E-14
f	6.076272376250422E-16
g	4.14913478269917E-19
h	4.201979398479235E-22
i	−9.735005727709414E-26
j	−1.4005417625307426E-28

The next step is to calculate the unknowns [i.e., $a, b, \dots, j$ of Equation (9.1)]. To accomplish this task, the Gaussian elimination method is used. The final values of the unknowns are listed in Table 9.8.

Conclusively, Equation (9.1) now becomes:

$$y = 10.230838005070808 + (-0.0041658198089687196)x$$
$$+ (-3.3045749026962936E - 6)x^2 + (-1.265402658730446E - 9)x^3$$
$$+ (1.639773297320292E - 14)x^4 + (6.076272376250422E - 16)x^5$$
$$+ (4.14913478269917E - 19)x^6 + (4.201979398479235E - 22)x^7$$
$$+ (-9.735005727709414E - 26)x^8 + (-1.4005417625307426E - 28)x^9 \quad (9.3)$$

As already discussed, the Egra police station area (total population: 397,612), under the Purba Medinipur district of West Bengal State, India, comprising 32 Anchals, has been considered as the study area. For year 2013, data regarding the number of crimes held in each of the 32 Anchals was collected. After multiplying with their respective weights and summing up, the denominator of Equation (9.2) is found. Considering the population of the region [numerator of Equation (9.2)], the value of x, corresponding to the region, is cooked out. Finally, putting the value of x into Equation (9.1), the corresponding y (with rounding up to the nearest integer) is obtained; this is nothing but the rank of that region, reflecting how crime prone the area is, as depicted in Table 9.9.

Based on the basis of their ranks, the Anchals are marked, as shown in Figures 9.3 and 9.4.

The present technique also tries to point out those regions that are not crime prone (i.e., rank 10) presently, but have a tendency to become more crime-prone areas in near future. The procedure for achieving this goal is as follows.

For any region with rank > 7, if more than 50% of its neighbor areas are crime prone (i.e., rank 10), then it is also regarded as a crime hotspot for near future (equivalent to rank 10). This is done by checking the adjacency of each region satisfying the following condition:

$$7 < \textbf{rank of a region} \leq 9 \ \dots \ \textbf{Condition(1)}$$

Table 9.9 The ranks of respective Anchals (arranged in descending order of rank)

Anchal	Population	Crime types (from Table 9.3)										Rank
		CH1	CH2	CH3	CH4	CH5	CH6	CH7	CH8	CH9	CH10	acquired
		Weighted factors										
		5	4.5	4	3.5	3	2.5	2	1.5	1	0.5	
Jumki-1	15,334	4	4	8	14	22	10	0	10	0	1	10
Panchrol-2	8618	3	3	4	4	5	5	6	8	10	14	10
Sahara-2	14,065	6	6	7	10	8	11	13	18	26	27	10
Basudebpur-1	7598	4	5	13	12	11	10	9	0	1	4	10
Paniparul-2	8023	10	3	2	5	16	12	0	0	7	0	10
Bathuyari-1	17,559	0	15	10	36	0	0	0	0	0	19	10
Dubda-2	18,500	7	8	13	18	11	19	0	0	0	45	10
Sahara-1	16,477	2	0	0	16	0	0	0	0	17	0	9
Paniparul-1	11,399	0	2	0	0	2	0	0	5	11	29	9
Chatri-2	11,011	0	3	0	2	0	0	0	0	4	19	8
Manjushree-2	9920	0	0	1	0	5	2	0	0	3	0	8
Jumki-2	16,071	3	2	2	2	0	1	0	0	0	0	8
Deshbandhu-2	13,047	0	2	1	3	2	0	0	0	0	7	8
Basudebpur-2	11,919	0	1	2	0	0	3	0	0	0	7	7
Rishi Bankimchandra-2	8969	0	0	4	0	0	0	0	0	1	0	7
Dubda-1	13,065	0	1	0	4	2	0	0	0	1	0	7
Sarboday-2	8852	0	2	0	0	0	0	0	0	5	0	6
Deshbandhu-1	15,298	0	0	0	5	2	0	0	0	1	0	6
Vivekananda-1	11,759	3	0	0	0	1	0	0	0	0	1	6
Panchrol-1	13,035	0	0	0	0	0	8	0	0	1	0	6
Sarboday-1	8351	0	0	0	0	3	0	0	0	2	0	5
Chatri-1	11,792	0	0	0	0	2	1	0	1	3	6	5
Rishi Bankimchandra-1	10,462	0	0	0	0	0	0	5	1	0	5	5
Bathuyari-2	14,232	0	2	2	0	0	0	0	1	0	0	5
Jerthan-2	9760	0	1	1	0	0	0	0	0	3	0	4
Vivekananda-2	12,574	2	0	0	0	0	0	0	1	0	7	4
Jerthan-1	13,056	0	0	2	1	0	0	0	0	2	0	3
Manjushree-1	11,415	0	0	3	0	0	0	0	0	0	0	3

Let A be a region satisfying Condition(1). The following two counting gimmicks need to be performed:

- Counting the number of adjacent regions of A.
- Counting the number of adjacent regions of A having rank=10.

If the latter counting yields a value that is 50% or more than the former one, that is, **Total number of adjacent regions with rank 10 ≥ 50% of Total number of adjacent**

Figure 9.3 Cluster formation after acquiring ranks.

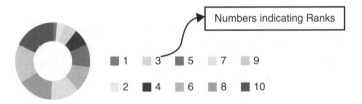

Figure 9.4 Rank indicator.

regions ... Condition (2), then region A will also be treated as a crime hotspot. After considering all such regions, the formation of a hotspot zone is depicted in Figure 9.5. This task is performed to expose the spread of criminal activity in future.

After applying FCM clustering, for deciding suitable locations of police beat houses, the outcome depicted in Figure 9.6 is obtained.

Here the original red hotspot zone, comprising 10 regions, is clustered into two, as shown in Figure 9.7. Finally, the centroid position of each of these clusters, C_1 and C_2, is considered to be the most preferential location for setting up beat houses (as shown in Figure 9.8).

Figure 9.5 Hotspot zone formation after considering Condition(1) and Condition(2).

Figure 9.8 shows the most favorable locations for building up beat houses.

Now, if the location of a predicted beat house is close enough to an existing one, then the predicted location will be of no use, because the existing police station is sufficient enough to control that particular red zone. The Manhattan distance formula is used to conduct this step. If the distance between an existing and a predicted police station is less than 6 km (which has been properly converted to be fitted with the present resolution of the map), that predicted location will no longer exist, as per our technique. Figures 9.9 and 9.10 demonstrate the outcome of this step.

If the technique proposes the construction of a new beat house at the almost same location constantly for three years, then the matter should be seriously considered by the respective authority. For example, in Figure 9.20, a beat house was suggested to the left hotspot in 2011 only, but not in the following years. So this suggestion could be ignored. However, two beat houses for the lower hotspot areas are suggested in each of 2011, 2012, and 2013 (Figures 9.20, 9.21, and 9.22), so this suggestion should be examined.

Figure 9.6 Depicting the encircled "red" hotspot zone.

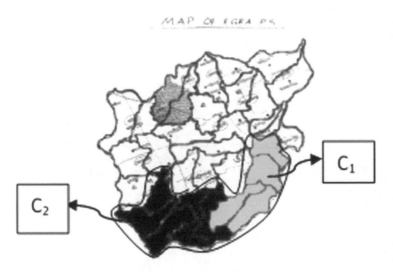

Figure 9.7 Splitting the "red" zone into two clusters, C_1 and C_2.

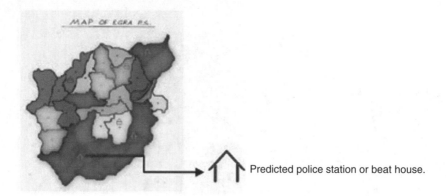

Predicted police station or beat house.

Figure 9.8 Depicting suitable locations for construction of beat houses.

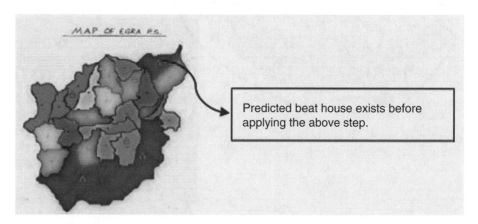

Predicted beat house exists before applying the above step.

Figure 9.9 Before.

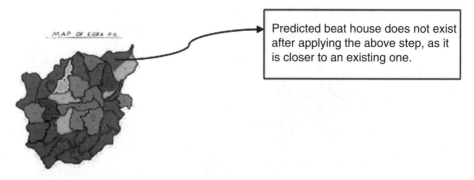

Predicted beat house does not exist after applying the above step, as it is closer to an existing one.

Figure 9.10 After.

9.4 Implementation and Results

In this section, results of implementation, operations, and outputs are exhibited. To fulfill the intended tasks, NetBeans (JAVA) [6, 7] was used. To carry out this implementation job, a flat-file system is selected for data storage; no database is taken up for such purposes, to increase portability.

The acts of culling up "Create New Profile" or "Open an Existing Profile," creating a new profile, and recessing an existing profile are portrayed in the Figures 9.11, 9.12, and 9.13, respectively.

The technique has an inbuilt user-friendly digitization tool. For digitization of a newly fed map, at first a suitable name should be given. It is suggested to use the Geographic-Name of the area being digitized (Figure 9.12). By simply mouse dragging through the border line of the raster map, the digitization [1] task of the present raster map [1] is fulfilled. By the virtue of the blue trailing line, how much portion has already

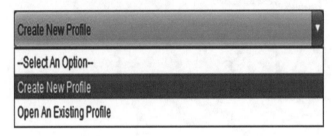

Figure 9.11 Choosing options.

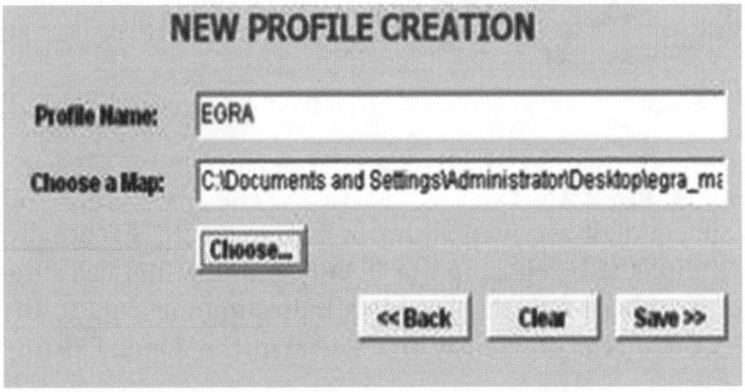

Figure 9.12 Creating a new profile.

Figure 9.13 Opening an existing profile.

Figure 9.14 Digitization of a raster map.

digitized can be manifested. Such a digitized map is shown in Figure 9.14. Eventually, the digitization procedure is completed by clicking the "SAVE" button. This saves the digitization action carried out, with a proper display of an indication message to the user. Later, when one wants to work on the previously digitized map, the "Open Existing Profile" option is chosen.

After digitization of the raster map, imparting associated data to each region is performed. To do so, the region requiring data association is clicked, and then the "Enter Data" button is clicked. A window, requiring name and population of the clicked region, pops out. Finally, the "OK" button of the data association window is clicked, after insinuating suitable information related to the considered region. The associated data is then saved in a file. Now, if the user requires fetching data with respect to any specific region, he or she has to click on the "Enter Data" button, after clicking on the considered region. After doing so, a message displaying the name and population of the clicked region pops out, as depicted in Figure 9.15.

The next task is to enter the number of cases received with respect to each crime type (shown in Table 9.3), related to a particular region in a specific year. To do so, the

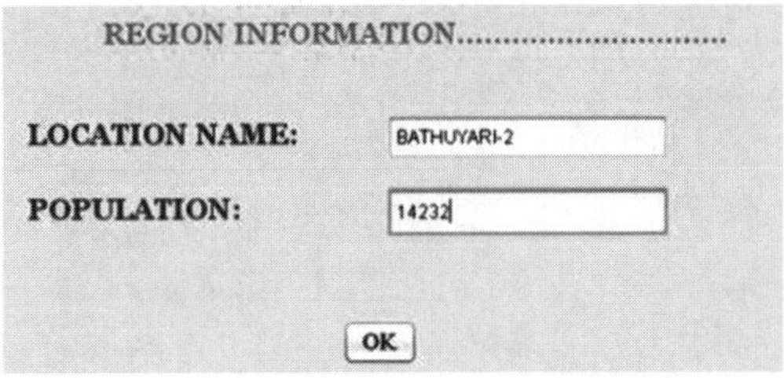

Figure 9.15 Data association.

REGION INFORMATION............................

LOCATION NAME: BATHUYARI-2

POPULATION: 14232

OK

Figure 9.16 "Click Here" button.

Figure 9.17 Crime-Info window.

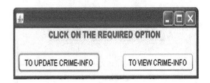

region for which crime information is needed is clicked. The user is then required to click on the "Click Here" button, under the "Crime Info" label (shown in Figure 9.16). A window comprising two buttons, namely, "TO UPDATE CRIME-INFO" and "TO VIEW CRIME-INFO," pops out (shown in Figure 9.17) and the user can click on the desired button. For updating the desired values with respect to each crime type need to be entered (Figure 9.18). After entering suitable data, the "OK" button of the crime update window is clicked for storing the imparted information in a file (Figure 9.18). Now, if the user needs to retrieve crime information with respect to a particular region, he or she just has to click on the "TO VIEW CRIME-INFO" button, after clicking on that particular region. In order to take a glance at the existing or predictive police stations, the user needs to click on the "Existing PS" or "Predictive PS", under the label "Police Station," as shown in Figure 9.19.

The proposed technique finally conveys a quick look at the crime-sensitive areas of the map considered (here, Egra, situated at Purba Medinipur in West Bengal, India) for the desired years (here, for 2011, 2012, and 2013).

It is very clear from the above three outcomes that construction of two new police stations is being suggested for the 3 consecutive years, whereas, in addition to these two, one more police station was suggested only on the basis of the crime report for 2011. So the concerned higher authority should consider constructing the two new police stations in areas as nearby as possible to those suggested, and the suggestion based on only one year's crime report could be ignored.

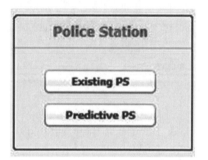

CRIME STATEMENT OF EGRA POLICE STATION FOR THE YEAR 2011 (against the clicked region)		
RANKING:	**CRIME HEAD:**	**TOTAL CASE RECEIVED:**
1	MURDER	0
2	DACOITY	0
3	ARMS ACT,376 IPC	1
4	498A/302 IPC	0
5	363A/366A IPC, 498A/304 IPC, RIOTING, O.A.W	5
6	ROBBERY	2
7	498A/323 IPC, 399/402 IPC, NDPC Act	0
8	BURGLARY	0
9	THEFT	3
10	354 IPC, Others	0

OK

Figure 9.18 Number of cases received against respective crime types.

Figure 9.19 Buttons under "Police Station" label.

Police Station

Existing PS

Predictive PS

9.5   Analysis and Comparisons

This section constitutes a number of tables, outlining a comparative study between the proposed technique and a number of other clustering techniques and existing hotspot detection methods.

- *Comparison with the K-means clustering technique* [8]: In K-means clustering technique, one needs to specify the value of K (i.e., the number of clusters), which is not always easy to determine beforehand. Defining the number of clusters can be advantageous or disadvantageous with respect to the purpose of use. When the area of application is hotspot detection, it is never possible to mention the value of K beforehand. Thus, in gist, a comparison can be drawn between the proposed technique and the K-means clustering technique, illustrated in Table 9.10.
- *Comparison with the fuzzy clustering method* [14]: The fuzzy clustering method is used to generate clusters based on the distance only. All regions have probabilities to be included in a cluster. A region may even be a member of more than one cluster, which can prove to be a disadvantage while performing hotspot detection tasks using

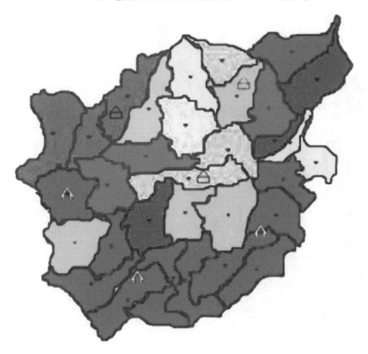

Figure 9.20 In the year 2011 (3 suggested police stations).

Figure 9.21 In the year 2012 (2 suggested police stations).

Figure 9.22 In the year 2013 (2 suggested police stations).

Table 9.10 Depicting the comparative study between *K*-means and the proposed method

Characteristics	Techniques	
	K-means	**Proposed method**
Automatic determination of clusters	No	Yes
Basis of cluster formation	Based on minimum distance (find the closest centroid)	Mainly based on associated data (crime reports) as well as distance
Prediction of new beat house or police station	No	Yes

FCM. Table 9.11 depicts the summarized version of the comparative study drawn between the proposed work and FCM clustering technique.

- *Comparison with ISODATA clustering method for hotspot detection* [10, 11]: The ISODATA clustering method can also be used to group the regions that are affected by crime. To detect hotspots using this method, the TNTmips 6.4 software is used, which is problematic for native users. Besides, to run ISODATA, it is needed to

Table 9.11 Portraying the comparative study between the fuzzy clustering method and proposed method

Characteristics	Techniques	
	Fuzzy c-means	**Proposed method**
Automatic determination of clusters	No, since the number of clusters to be formed needs to be given as input	Yes
Basis of cluster formation	Based on minimum distance (find the closest centroid)	Mainly based on associated data (crime reports) as well as distance
Prediction of new beat house or police station	No	Yes

Table 9.12 Illustrating the comparative study between ISODATA and the proposed method

Characteristics	Techniques	
	ISODATA	**Proposed method**
Automatic determination of clusters	Yes	Yes
Software dependency of cluster	Yes (to detect hotspots using this method, we need to use TNTmips 6.4 software)	Yes
Prediction of new beat house or police station	No	Yes
Complexity	More complex	Less complex
Initial mandatory parameters	Yes and must be specified to run the method	No

specify various mandatory parameters like initial cluster means, splitting parameters, lumping parameters, the minimum number of pixels in a cluster, and the number of iterations, which are actually problematic for novices. Moreover, another disadvantage of the software is that the result does not always reflect the same properties and selected options. All the settings should be tried and balanced to reach the desired number of groups. Thus, a summary can be drawn out, as depicted in Table 9.12.

- *Comparison with the STAC clustering method of hotspot detection* [9]: Another method is the STAC clustering method, which is also a crime hotspot detection technique. But the problem of using this method is that valuable information about the densely populated crime areas in the map is overlooked. Also, a fixed distance gives too big clusters, which are unable to attach useful data. Since it was a mainframe program, this had never been tested in real life (i.e., a police department). Besides all this, in this technique, clusters that are formed as hotspots may overlap; as such, the same incidents may be shared by different clusters. Another issue is that, as STAC illustrates the clusters as circles, some regions can be properly described by this; however, as the actual shapes of regions are irregular, they cannot be well described by a circle. Table 9.13 shows a comparative study in gist.

Table 9.13 The comparative study between STAC and proposed method is drawn out

Characteristics	Techniques	
	STAC	Proposed method
Automatic determination of clusters	No (the program reports how many hotspot areas it has found, and then the user has to choose how many of these to map)	Yes
Basis of cluster formation	Based on current, local-level law enforcement and community information	Mainly based on associated data (crime reports) as well as distance
Software dependency	Yes	No
Prediction of new beat house or police station	No	Yes
Complexity	More complex	Less complex

Table 9.14 Illustrating the advantages of the proposed method over RADIUS methodology

Characteristics	Techniques	
	RADIUS	Proposed method
Software dependency	Yes (implemented using MS Excel)	No
Output generation	Not properly displayed and gives very general approximation	Output comes with a detail graphical representation
Software dependency	Yes	No
Prediction of new beat house or police station	No	Yes
Complexity	More complex	Less complex

As illustrated in section 9.1, the present method is well applicable for different hotspot detections other than crime, such as earthquake-prone areas, landslide-prone areas, and so on. The present method thus could be easily compared to some existing clustering techniques in those fields, illustrated here:

1. *Comparison with RADIUS – a technique for hotspot detection related to earthquakes* [12, 13]: Table 9.14 depicts a comparative study between the RADIUS methodology and the proposed technique.
2. *Comparison with MCE – a technique for hotspot detection related to landslides* [15]: MCE is a tool utilized for landslide detection works. Table 9.15 portrays the advantages of the proposed method and landslide detection methods utilizing MCE.

Table 9.15 Comparison with MCE – a technique for hotspot detection related to landslides

Characteristics	Techniques	
	MCE	Proposed method
Basis	Based on many analytical hierarchy processes (AHPs)	Based on factors like proximity of earthquake, exposure of rock material, etc.
Software dependency	AHP method is accessible as a built-in tool inside IDRISI Andes Software	No
Prediction of new beat house or police station	No	Yes
Complexity	More complex	Less complex

9.6 Conclusions

In a country like India, depiction of crime takes place in a number of abhorrent ways, like rape, murder, robbery, kidnapping, and so on. These crimes do not require any introduction since these sights appear frequently in everyday newspapers. The reason behind such ruthless professions can be many. However, the main reason lies in poverty. Aristotle once said, "Poverty is the parent of revolution and crime." However, who is facing the malefic consequences of crime? It is the common and innocent people. As such, the common people are perpetually in search of slugs that can readily extend a supportive hand for expelling crime and its vicious employees, the criminals. Initially, police prevent common and innocent individuals from coming in contact with the harmful consequences of crime. Thus, police stations are required to be set up after going through a proper survey of crime-sensitive areas.

The intended work plays a decisive role in stretching out various measures in order to get rid of crime. It not only points out the crime-sensitive zones, to make the common people aware, but also predicts the most suitable locations for setting up beat houses for eliminating crime from such sensitive zones. By just having a quick view on the outcome of the intended work, anyone can understand how crime flourishes as well as the efforts to stop such flourishing. This would definitely torment criminals' mind, since they would be forced to think twice before giving birth to such villainous actions. As such, the proposed work would surely play a key role in discouraging crime in society.

Acknowledgments

The authors expresses a $1695.01 deep sense of hearty gratitude to the police personnel of each of the seven police stations taken into consideration, for supplying all the needful information that is the backbone of the proposed works. The authors are thankful to the Department of Computer Science, Barrackpore Rastraguru Surendranath

College, Kolkata, India, for providing all the infrastructural support to carry out the intended work.

References

1 Anirban Chakraborty, J.K. Mandal, Arun Kumar Chakraborti (2011) A File base GIS Anchored Information Retrieval Scheme (FBGISIRS) through vectorization of raster map. *International Journal of Advanced Research in Computer Science*, **2** (4), 132–138.

2 S.A. Mollah. *Numerical analysis and computational procedure: including computer fundamentals and programming in Fortran77*, 2nd ed. Books and Allied, New Delhi.

3 FEMA. (2005). *Earthquake hazard mitigation for nonstructural elements: field manual*. FEMA 74-FM. FEMA, Washington, DC. Available from http://mitigation .eeri.org/files/FEMA74_FieldManual.pdf

4 Wikipedia. (N.d.) Earthquake prediction. Available from www.en.wikipedia.org/wiki/ Earthquake-prediction

5 Shilpi Chakraborty https://en.wikipedia.org/wiki/Earthquake_prediction and Ratika Pradhan. (2012) Devlopment of GIS based landslide information system for the region of East Sikkim. *International Journal of Computer Application*, **49** (7).

6 Zetcode. (N.d.) Java Swing tutorial. Available from www.zetcode.com/tutorials/ javaswingtutorial

7 Tutorials Point. (N.d.) Java tutorials. Available from www.tutorialspoint.com/java/ index.htm

8 Anirban Chakraborty, J.K. Mandal, S.B. Chandrabanshi, and S. Sarkar. (2013) A GIS anchored system for selection of utility service stations through k-means method of clustering, *Conference Proceedings Second International Conference on Computing and Systems (ICCS-2013)"* McGraw-Hill Education (India), Noida, pp 244–251.

9 Illinois Criminal Justice Information Authority (ICJIA). *STAC user manual*. ICJIA, Chicago. Available from: http://www.icjia.org/public/pdf/stac/hotspot.pdf

10 Liu, W., Hung, C.-C., Kuo, B.C., and Coleman, T. (2008) *An adaptive clustering algorithm based on the possibility clustering and ISODATA for multispectral image classification*. Available from http://www.academia.edu/2873708/AN-ADAPTIVE-CLUSTERING-ALGORITHM-BASED-ON-THE-POSSIBILITY-CLUSTERING-AND-ISODATA-FOR-MULTISPECTRAL-IMAGE-CLASSIFICATION

11 Ball, G.H., and Hall, D.J. (1965) *Isodata: a novel method of data analysis and pattern classification*. Available from www.dtic.mil/cgi-bin/GetTRDoc?Location=U2& doc=GetTRDoc.pdf&AD=AD0699616

12 van Westen, C., Slob, S., Montoya de Horn, L., and Boerboom, L. (N.d.) *Application of GIS for earthquake hazard and risk assessment: Kathmandu, Nepal*. International Institute for Geo-Information Science and Earth Observation, Enschede, the Netherlands. Available from http://adpc.net/casita/Case_studies/Earthquake %20hazard%20assessment/Application%20of%20GIS%20for%20earthquake%20hazard %20assessment%20Kathmandu%20%20Nepal/Data_preparation_image_files.pdf.

13 Ravi Sinha, K.S.P. Aditya, and Achin Gupta. (2000), GIS-based urban seismic risk assessment using Risk. IITB, ISET. *Journal of Earthquake Technology*, **45** (3–4), 41–63.

14 Nikhil R. Pal, Kuhu Pal, James M. Keller, and James C. Bezdek. (2005) A possibilistic fuzzy c-means clustering algorithm. *IEEE Transaction on Fuzzy Systems*, **13** (4).

15 Ulrich Kamp, Benjamin J. Growley, Ghazanfar A. Khattak, and Lewis A. Owen. (2008) GIS-based landslide susceptibility mapping for the 2005 Kashmir earthquake region. *Geomorphology*, **101**, 631–642.

10

Hyperspectral Data Processing: Spectral Unmixing, Classification, and Target Identification

Vaibhav Lodhi, Debashish Chakravarty, and Pabitra Mitra

Indian Institute of Technology, Kharagpur, West Bengal, India

10.1   Introduction

Hyperspectral imaging is also termed as imaging spectroscopy because of the convergence of spectroscopy and imaging. It is an emerging and widely used remote-sensing technique that has been used in laboratory and earth observation applications in different domains like food inspection [1], vegetation [2], water resources [3], mining and mineralogical operations [4], agricultural and aquacultural [5], biomedical [6], and industrial engineering [7, 8]. In earth observation, data is normally collected using spaceborne, airborne, and unmanned aerial system (UAS) platforms. A hyperspectral sensor collects a stack of images with a spectral resolution of a few nanometers, and generates a continuous spectral signature in terms of reflectance/radiance value as a function of wavelength. On the other hand, multispectral imaging, a subset of hyperspectral, generates discrete spectral profiles due to its low spectral resolution. Hyperspectral data consists of two spatial dimensions along X and Y along with a third spectral dimension (n) to construct a 3D data ($X * Y * n$). Due to this, hyperspectral data is also known as hyperspectral data cubes. A hyperspectral imaging concept, where a spectral signature at location (X_i, Y_j) is a function of the X spectral dimension, is shown in Figure 10.1.

Hyperspectral data refers to a hypercube, due to its 3D nature (two spatial and one spectral dimension). In hyperspectral imaging, a reflectance spectrum for each pixel of the object under study is collected. During data collection from spaceborne and airborne platforms, hyperspectral data suffers from atmospheric interactions, internal system effects, dark current, system noise, and so on. In order to minimize these effects and for postprocessing operations of the hyperspectral data, one requires preprocessing operations. On the contrary, for characterization and identification of materials, classification and analysis, parameters retrieval, and so on, postprocessing operations are required. Eventually, hyperspectral image processing is divided into two broad categories – preprocessing and postprocessing. Atmospheric correction, calibration, denoising, dimension reduction, and so on come under the category of preprocessing, while spectral unmixing, classification, target identification, and the like are in postprocessing operations. Numerous algorithms are available to perform

Hybrid Intelligence for Image Analysis and Understanding, First Edition.
Edited by Siddhartha Bhattacharyya, Indrajit Pan, Anirban Mukherjee, and Paramartha Dutta.
© 2017 John Wiley & Sons Ltd. Published 2017 by John Wiley & Sons Ltd.
Companion Website: www.wiley.com/go/bhattacharyya/hybridintelligence

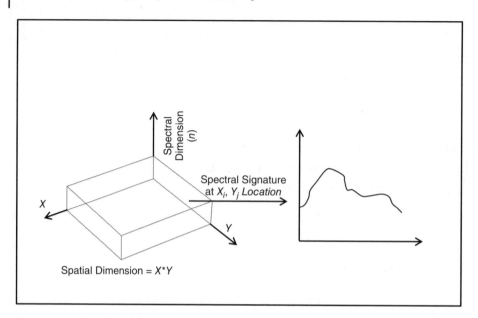

Figure 10.1 Hyperspectral imaging concept.

pre- and postprocessing operations over hyperspectral data. Each algorithm has its own advantages, disadvantages, and limitations to perform the designated processing operations. Hence, in order to exploit the advantages for overcoming the limitations and disadvantages, a combination of algorithms (i.e., hybridization) is generally followed. Hybridization of algorithms in hyperspectral imaging improves image analysis, accuracy, and understanding, leading to better interpretation. The objective of this chapter is to present understanding of hyperspectral imaging, processing algorithms, and discussion of hybrid approaches under spectral unmixing, classification, and target identification sections.

This chapter is organized in the following way. In Sections 10.2 and 10.3, we discuss background, a brief hardware description, and steps of image processing for hyperspectral imaging. In Section 10.4, "Spectral Unmixing," we review the unmixing process chain, mixing model, and methods used in unmixing. In Section 10.5, supervised classification methods are discussed in hyperspectral image classification. In Section 10.6, some methods under target identification are discussed. Hybrid techniques are discussed under spectral unmixing, classification, and target identification sections. Finally, a conclusion is presented in Section 10.7.

10.2 Background and Hyperspectral Imaging System

The term spectroscopy, first used in the late 19th century, provided the empirical foundation for atomic and molecular physics. Subsequently, astronomers began to use it for finding radial velocities of cluster, stars, and galaxies and stellar compositions. Advancement in technology of spectroscopy since the 1960s and the potentiality of the same led to the development of initial research-level imaging spectrometer [9]. Subsequently, significant improvement has been achieved in the development of airborne imaging spectrometry. The limitations of airborne spectrometry led to the development of spaceborne imaging spectrometry [10]. Usability of hyperspectral imaging is not limited to earth observations only but is growing in Ximportance in industrial and research-oriented usages also. It has been used in numerous applications such as non-destructive food inspection, mineral mapping and identification, coastal ocean studies, solid and gaseous target detection, aviation fuel inspection, and biomedical applications.

Generally, a hyperspectral imaging system (HIS) contains illumination source (light source), dispersive unit, charge-coupled device (CCD)/ complementary metal oxide semiconductor (CMOS) cameras, and image acquisition system, as shown in Figure 10.2. A brief description of the HIS components is provided here.

Light sources illuminate the material or sample under study. The tungsten halogen lamp is most commonly used in HIS; it consists of a halogen element (F / Cl / Br / I) in the quartz tubes. It has higher luminous efficiency and a longer lifetime compared to normal light bulbs.

The wavelength dispersive unit plays a vital role in the HIS. It disperses the broad-band wavelengths into narrow-band wavelengths. Normally, any one of the components [i.e., filter wheels, tunable filter, linear variable filter (LVF), grating, or prism] is used to achieve spectral dispersion. In general, filter wheels carries a set of discrete band pass filters. A band pass filter allows light of a particular wavelength while rejecting other

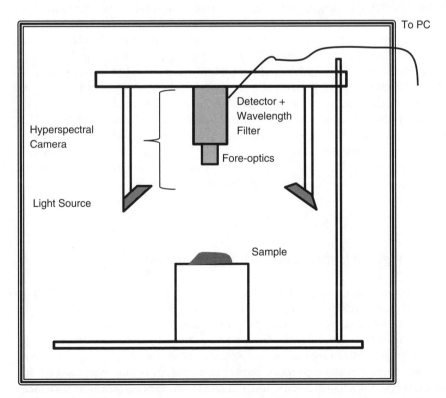

Figure 10.2 Diagrammatic representation of the schematics for a commonly used hyperspectral imaging system.

wavelengths of light. Available filter wheels operate over a wide range, starting from the ultraviolet (UV) to infrared (IR) range.

A tunable filter scans the area under investigation electronically, wavelength by wavelength, as per the full-width half maximum (FWHM) criterion. Liquid crystal tunable filter (LCTF) and acousto-optic tunable filter (AOTF) are the most widely used electronically controlled tunable filters. The LCTF-based HIS are relatively slower but have better imaging performance than the systems using AOTFs. LCTFs have relatively wider field of view (WFOV), larger apertures, and low wavefront distortions compared to the AOTF-based ones.

LVF is also known as a wedge filter due to its wedge-shaped geometry at the top-view end. Coating of the filter is done in such a manner that it passes narrow bands from one end to another that vary linearly along the length of LVF.

Prism and grating are the most commonly used components for the dispersion in HIS.

An image sensor placed within a camera is a 2D focal plane array (FPA) that collects the spatial and spectral information simultaneously. CCD and CMOS are the two most frequently used image sensors in cameras. Basic components of CCD and CMOS are photodiodes, made up of light-sensitive materials to convert light energy into electrical energy. The main difference between the CCD and CMOS image sensors is that the photo detector and readout amplifier are included for each pixel in a CMOS sensor, while this is absent in the case of a CCD sensor. Silicon (Si), mercury cadmium tellurium (HgCdTe or MCT), and indium gallium arsenide (InGaAs) are the normally used materials in detector arrays for hyperspectral imaging to work in different wavelength ranges. The spectral response of the silicon is in visible-near-infrared (VNIR) range, while InGaAs and MCT operate in the short-wave infrared (SWIR) range and long-wave infrared (LWIR) range. CCD and CMOS have their own associated advantages and disadvantages. Advantages of CCD are low cost, high pixel count and sensitivity, wide spectrum response, and high integration capability. There have been four different architectures of CCD that are available, and they are interline CCD, frame interline CCD, frame transfer CCD, and full frame CCD to improve the performance and usability of the image sensor. On the other hand, advantages of CMOS are low power consumption, on-chip functionality, and low operating voltage. An image acquisition system (frame grabber) is an electronic device to trigger the camera and acquire the digital frame successively.

10.3 Overview of Hyperspectral Image Processing

Hyperspectral image processing refers to the combined execution, of complex algorithms for storage, extraction, and manipulation of the hypercube data to perform tasks such as classification, analysis, target identification, and parameter retrieval. [11–13]. Hyperspectral image-processing work flow is quite different from that of color image processing; it includes data calibration, atmospheric correction, pixel purity with endmember selection, dimension reduction, and data processing. To compare, color image processing involves steps like preprocessing (linearization, dark current compensation, etc.), white balance, demosaicking, color transformation, postprocessing, and display. Color image postprocessing steps involve image enhancement, which leads to loss of

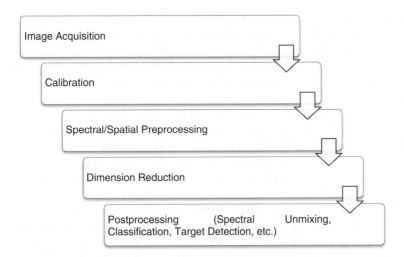

Figure 10.3 General processing steps involved for hyperspectral data.

spectral information in hyperspectral imaging. General processing steps involved for hyperspectral data have been presented in Figure 10.3.

10.3.1 Image Acquisition

Hyperspectral data is 3D data where two dimensions are spatial and the third one is spectral. The detector used in a hyperspectral imager is suitable to capture data in a 2D framework. So, in order to acquire the third dimension, either a scanning mechanism or tunable filter is used. In case of the use of a scanning mechanism, the dispersive components like prism, grating, and wedge filter are normally used. Generally, AOTF [14] and LCTF [15] are the most commonly available tunable filters that have been used in real-life imagers. Three types of data formats are obtained during acquisition: BSQ (band sequential), BIL (band interleaved by line), and BIP (band interleaved by pixel). In BSQ, each image belongs to one narrow-wavelength band. The BSQ format remains suitable for spatial processing in order to create regions of interest (ROIs) or to extract spatial features. In the BIL format, the first row of all spectral bands is gathered in the order of their wavelengths, and followed by the second row, the third row, and so on. This is suitable for both spectral and spatial processing operations. In BIP, the first pixels of all spectral bands are stored as per the respective wavelength bands, and followed by the second pixels, the third pixels, and so on. This is suitable for spectral processing. The choice of the optimal data format is one of the most desirable features for the processing of hyperspectral images. However, the BIL format is a good compromise for carrying out better processing of images.

10.3.2 Calibration

Raw data that is obtained as the output of the acquisition setup is related to the radiance as a function of spatial position and spectral band at the sensor–pupil plane. Radiance depends on the properties of the object of interest and is affected by factors like dark current, responsivity at the FPA, optical aberrations and transmission, and so on. Extracting quantitative information from the raw data requires a method to determine the system response along three important dimensions, namely, spectral, spatial, and radiometric [16]. This method is referred to as hyperspectral sensor calibration. Calibration of the data is required to ascertain the accurateness and repeatability of the results generated by the hyperspectral imaging system. In this section, spectral, spatial, and radiometric calibrations are discussed.

In spectral calibration, band numbers are linked with the corresponding wavelength-related data [16]. Generally, monochromatic laser sources and pencil-type calibration lamps are used to calibrate the wavelength. Linear and nonlinear regression analysis is used to guess the wavelength at unknown bands. Spectral calibration of the instrument has to be done at the manufacturing end before commercializing in the market for the customers to use.

Spatial calibration is also known as geometric correction, and its function is to assign every image pixel to a known unit such as meters or known features in the object space [16]. Spatial calibration gives information about the spatial dimensions of every sensor pixel on the surface of the object. It reduces the optical aberrations, like smile and keystone effects, which distort the imaging geometry and thereby introduce the spatial error components. Due to advancements, in recent years, such sensors are available that minimize the smile and keystone well below the tolerance limits.

A digital imaging sensor produces images with inherent artifacts called camera shading, which arises due to nonuniform sensitivity at the sensor plane with respect to the properties of received signals obtained from the whole of the object space. Camera-shading corrections apply to the image in case of variations in sensitivity of the imaging sensor. In radiometric calibration, one has to recalculate the digital numbers in an image based on several factors such as dark current, exposure time, camera shading correction, and so on to obtain obtaining the radiometrically corrected image. Standard radiometric calibration methods are the lamp-integrating sphere, detector-integrating sphere, detector-diffuser [17], and so on.

10.3.3 Spatial and Spectral Preprocessing

Spatial preprocessing is to manipulate or enhance the image information in the spatial dimension. Normal digital image-processing techniques like enhancement and filtering can be applied here. In the case of hyperspectral images, spatial preprocessing techniques for denoising and sharpening are not applied to raw and calibrated hyperspectral images, generally because they affect the spectra of the data [16].

Roughly, spectral preprocessing can be grouped into two parts: (1) endmember extraction and (2) chemometric analysis. *Endmember* refers to the pure spectra and has importance in remote-sensing applications. It can be acquired in field or on the ground and in laboratories by spectro-radiometry in order to build a spectral library of different materials. Endmembers can be extracted from the hyperspectral images by spectral preprocessing algorithms like the Pixel Purity Index (PPI), N-finder (N-FINDR) [11], and so on. Obtained endmembers can be used in target identification, spectral unmixing, and classification. A chemometrics approach refers to the extraction of information from a system by data-driven means. Spectral data obtained from chemometrics is used for multivariate analysis techniques like partial least squares (PLS), principal component analysis (PCA), and so on.

10.3.4 Dimension Reduction

Hyperspectral data contains a significant amount of redundancy. In order to process the data efficiently, one needs to reduce or compress the dimension of the hypercube data. The objective of dimension reduction is to map high-dimensional data into low-dimensional space without distorting the major features of the original data. Dimensionality reduction techniques can be either transformation-based or selection-based approaches; the difference between these two techniques is whether they transform or preserve the original data sets during the dimensional reduction process.

10.3.4.1 Transformation-Based Approaches

Transformation-based approaches are commonly used where originality of the data sets is not needed for further processing, but these preserve the important features under consideration. Under these approaches, many methods like PCA [18], discrete wavelet transform (DWT) [19], Fisher's linear discriminant analysis (FLDA) [20], minimum noise fraction, and others have been used for dimensional reduction. Some of the methods are discussed in brief here:

Principal component analysis: In hyperspectral data, adjacent bands are correlated and approximately contain the same information as the sample under study. PCA reduces the correlation among the adjacent bands and converts the original data into lower dimensional space. PCA analysis is based on the statistics to determine the correlation among adjacent bands of hyperspectral data.

Discrete wavelet transform: Wavelet transform is one of the powerful tools in signal processing and has been used in remote-sensing applications like feature extraction, data compression and fusion, and so on. It is applied to single pixel locations in the spectral domain while transforming the data into lower dimensional space of the object under study. During transformation, it preserves the characteristics of the spectral signature.

Fisher's linear discriminant analysis: In pattern recognition, FLDA is used for dimension reduction, which projects original data into low-dimensional space. It is a parametric feature extraction method and well suitable for normally distributed data.

10.3.4.2 Selection-Based Approaches

Selection-based approaches attempt to find a minimal subset of the original data set without missing their physical meaning. This approach is based on the band selection method to reduce the dimension of the data. In band selection, selecting the subset bands contains class information from the original bands. Categorization of band selection approach based on class information is divided into two types: 1) Supervised band selection methods and 2) Unsupervised band selection methods.

When class information is available, supervised band selection is the preferred one for dimension reduction of data. It involves two steps: subset generation and subset evaluation. In case of non-availability of class information, unsupervised band selection is used. There are a number of methods available for unsupervised band selection, like methods based on information theory, Constrained Band Selection (CBS) [21], and so on.

10.3.5 Postprocessing

The hyperspectral imagery consists of a lot of information. Interpreting the information requires pre- and postprocessing, analysis, and accordingly making conclusions. Spectral unmixing, classification, target identification, and so on. come under the category of hyperspectral postprocessing. In spectral unmixing, this involves unmixing of mixed pixels into sets of endmembers and their abundance fractions. Mixing may be linear or nonlinear. Classification assigns each pixel of hypercube to a particular class depending on its spatial and spectral information; while, in target identification, a hyperspectral sensor measures the reflected/emitted light from the object under observation in the working wavelength range. Processing of these obtained data using algorithms/techniques assists in identifing the target of interest. The theme of this chapter is to discuss spectral unmixing, classification, and target identification in postprocessing, which will be discussed further in this chapter.

10.4   Spectral Unmixing

Hyperspectral imaging systems record the scenes consisting of numerous different materials that contribute to the formation of spectra at a given pixel (single pixel). This is generally referred to as a mixed pixel in a hypercube. In such a mixed pixel, one needs to identify the individual pixel spectra and their proportions present. Spectral unmixing is the process by which identification of individual pure pixel spectras or pure spectral signatures, called endmembers, and their abundance values (i.e., proportion present in the mixed pixel) are characterized and analyzed. Generally, endmembers represent the pure pixel spectrum present in the image, and percentages of endmember at each pixel represent the abundance of materials. Mixed pixels contain a mixture of two or more different materials and in general occur due to two main reasons. The first reason is that spatial resolution of the hyperspectral sensing devices is quite low, leading to a condition where adjacent endmembers of the scene jointly occupy the projected location of a single pixel, target, or area of interest that is smaller compared to the pixel size. Generally, it occurs in the spaceborne and airborne platforms. The second reason is because two different materials are mixed in such a way as to generate a homogeneous–heterogenous material composition (e.g., moist soil, mineral, etc.). In a broad way, spectral unmixing is a special case of an inverse problem in which estimating constituents and abundances at each pixel location is carried out by using one or more observations of the received mixed signals.

10.4.1   Unmixing Processing Chain

Steps involved in the processing chain of hyperspectral unmixing have been shown in Figure 10.4. These steps are atmospheric correction, dimension reduction, and unmixing (endmembers and inversion). Radiant light at the sensor coming from the scene attenuates and gets scattered by the atmospheric particulate materials. Atmospheric correction compensates these effects by transforming radiance data into reflectance data. To improve the performance of the algorithms for analyzing, one needs to execute the dimension reduction step. This step is optional and is used by some algorithms

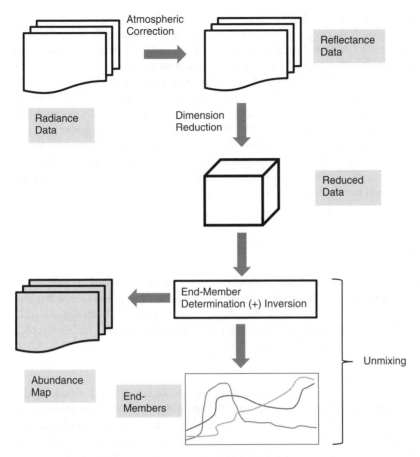

Figure 10.4 Hyperspectral unmixing processing chain.

in order to reduce complexity problems as well as the execution time of the same. In the endmember determination step, estimation of the endmembers present in the mixed pixel of the scene is carried out. At last, in the inversion step, abundance maps are generated in order to find the partial fractions of the components present in each mixed pixel by using the observed and determined endmembers spectra.

10.4.2 Mixing Model

Implementation of any approach for hyperspectral unmixing depends upon the model chosen, which describes how individual materials combine in a pixel-producing mixed spectrum at the sensor end. Mixing models try to describe the fundamental physics that are the basis of hyperspectral imaging. On the other hand, unmixing uses these models for endmembers determination and inverse operation in order to find the endmembers and their fractional abundance present in each of the mixed pixels. Mixing models have been categorized into two parts, linear mixing models and nonlinear mixing models, which are described in brief in this section.

10.4.2.1 Linear Mixing Model (LMM)

In LMMs, the mixing scale is macroscopic and incident light supposedly interacts with just one substance. The reflected light beams from all the different materials, which are completely and exclusively separated linearly, get mixed when measured at the sensor end. In this way, a linear relationship is assumed to exist between the fractional abundance of the materials comprising the area and spectra of the reflected radiation. In unmixing, LMMs have been used broadly in the past few decades due to their simplicity and ability to provide approximate solutions. A very simple conceptual model of application of LMMs for three different materials is illustrated in Figure 10.5.

If L indicates the number of spectral bands, s_i represents the endmember spectra of the ith endmember, and α_i represents the ith endmember abundance value, then the observed spectra y of any pixel in the scene can be represented as Equation (10.1):

$$y = \alpha_1 s_1 + \alpha_2 s_2 \ldots \alpha_M s_M + w \tag{10.1}$$

$$= \sum_{i=1}^{M} \alpha_i s_i + w = S\alpha + w \tag{10.2}$$

where M refers to the number of endmembers; S refers to the matrix of endmembers; and w accounts for error terms for additive noise and other model inadequacies. The component α represents the fractional abundance, which then satisfy the constraints $\alpha_i \geq 0$, for $i = 1, \ldots, M$ and are named the abundance non-negativity constraint (ANC) and abundance sum constraint (ASC), respectively.

10.4.2.2 Nonlinear mixing model

Normally, formation of an intimate mixture of materials is obtained when all the components of the material are uniformly mixed in a random manner that leads to the generation of mixed ground representative cells producing mixed hyperspectral image pixels. Due to this, incoming radiations falling on the sensor experience reflections from multiple materials (i.e., multiple locations), and the resultant spectra of reflected radiation may no longer hold the linear proportion of constituent material spectra. The nonlinearity in the aggregate spectra makes LMMs not suitable for use to analyze this condition, and hence a different approach called the nonlinear mixing model is used for

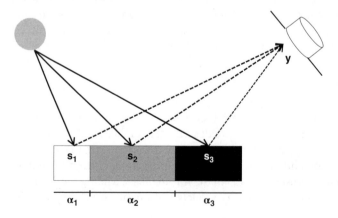

Figure 10.5 Conceptual diagram of a simple linear mixture model geometry.

Figure 10.6 Nonlinear mixture model.

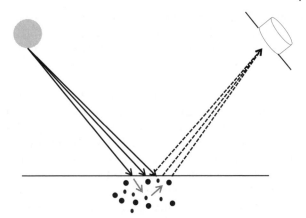

analyzing the mixed spectra. One simple representation of a nonlinear mixing model is illustrated in Figure 10.6.

10.4.3 Geometrical-Based Approaches to Linear Spectral Unmixing

The underlying assumption of geometrical-based approaches under LMMs is that the spectral vectors come under a simplex, whose vertices correspond to pure reference materials (endmembers). Geometrical-based approaches discuss algorithms for the presence or absence of pure pixels in the data set. In the case of the pure pixels, algorithms are much easier than with the nonpure pixels case [22]. In general, the geometrical-based techniques are divided into two broad parts: pure pixel-based and minimum volume (MV) based on [23]. Brief discussions on pixel-based and MV-based techniques are given in this section.

10.4.3.1 Pure Pixel-Based Techniques

These algorithms presume the presence of at least one pure pixel per endmember component. This assumption enables the design of effective algorithms from the computational view and is a prerequisite that may not hold in many cases. In any case, these techniques find the set of the most pure pixels in the hypercube. In hyperspectral linear unmixing applications, they have presumably been the most frequently used because of their clear conceptual meaning and easygoing computational complexity. Some of the techniques are discussed here:

Pixel purity index (PPI): In the PPI [24], data are subjected to dimensional reduction and improvement of the signal-to-noise ratio (SNR, or noise whitening) using the minimum noise fraction (MNF) transform process. The PPI projects data repeatedly onto a number of random vectors. The points corresponding to the extreme values in each of the projections are recorded. A cumulative account records the number of times every pixel (i.e., a given spectral vector) is found to be an element of the extremes. The points with the highest extreme score are the purest ones.

N-FINDR: N-FINDR [25] is an autonomous algorithm to find the pure pixels in the image. It works by "inflating" a simplex inside the data and starts with a random set of image pixels. For every pixel and every endmember, the endmember is replaced with the spectra of the pixel, followed by recalculation of the volume. If it increases, the spectra of the new pixel replace that endmember. This process is repeated until no more replacements are possible.

Vertex component analysis (VCA): VCA [26] is an unsupervised endmember extraction method to unmix any given linear mixture. This method exploits two basic facts: (1) the endmembers belong to the set of vertices of simplex, and (2) the affine transformation of simplex is also simplex. VCA works with projected and unprojected data. It iteratively projects data onto the orthogonal direction of the subspace spanned by the endmembers already identified and determined. The new endmember signature(s) corresponds to the extreme(s) of the projection. The algorithm keeps iterating until all the endmembers are exhausted. VCA performs much better than PPI and better than or comparable to N-FINDR.

Iterative error analysis (IEA): IEA is a linear-constrained endmember extraction algorithm based on LMM. This algorithm selects endmembers one by one based on the previously extracted endmembers. In an unmixed image, pixels that minimize the remaining errors are selected as endmembers in every iteration. This technique look for endmembers that reduce the remaining error in an abundance map by involving a series of linear-constrained spectral-unmixing steps.

Simplex growing algorithm (SGA): SGA [27] finds the vertices corresponding to the maximum volume by growing simplex, iteratively. Initially, it starts with two vertices and begins to grow simplex by increasing its vertices one at a time. This algorithm terminates when the number of vertices reaches the number of endmembers, estimated by virtual dimensionality (VD; i.e., the number of spectrally distinguishable signatures in hyperspectral imagery).

10.4.3.2 Minimum Volume-Based Techniques

Minimum volume-based techniques are used in the case of non-availability of pure pixels in the object under study. In other words, it violates the assumption of pure pixels. Some of the minimum volume-based techniques and their extensions are minimum volume simplex analysis (MVSA) [28], simplex identification via split augmented lagrangian (SISAL) [30], minimum volume enclosing simplex (MVES) [29], and robust minimum volume enclosing simplex (RMVES) [31], among others.

10.4.4 Statistics-Based Approaches

Geometrical-based approaches yield poor results in the case of huge mixing of spectral mixtures. In these cases, statistical methods are efficient as compared to geometrical-based approaches. In statistical methods, unmixing algorithms process the mixed pixel using statistical representations. The illustrations can be analytical expressions that denote the probability density function (parametric). It comes with the price of higher complexity and computation.

Normally, the number of materials and their reflectance values (in terms of both the magnitude and the proportional composition) remain unknown. In such cases, hyperspectral unmixing comes into the section of the blind source separation problem. The

Independent Component Analysis (ICA) [32] is a well-known technique for solving blindly unmixing problems. It assumes that the spectral vectors of different substances are non-Gaussian in nature and are independent of each other. Due to this assumption, ICA uses all the higher order statistics-based parameters for the mixed signal rather than first-order (mean, etc.) and second-order statistics (covariance, etc.). Algorithms like PCA and the like use only second-order statistics. ICA is able to provide correct unmixing in the case of independent sources, since the minimum of the mutual information corresponds only to the independent sources. In case of dependence fractional abundances, ICA is not able to provide the correct unmixing. To overcome the limitations imposed by the ICA, a new unmixing method has been proposed named dependent component analysis (DECA) [33]. This technique models the abundance fractions as mixtures of Dirichlet densities, thus enforcing the constraints on abundance fractions imposed by the acquisition process, namely, non-negativity and constant sum. The mixing matrix is inferred by a generalized expectation-maximization (GEM)-type algorithm.

Finally, some methods have been used for spatial and spectral information that combined the order to improve the endmembers extraction. Automated morphological endmember extraction (AMEE) [34] is a mathematical morphology-based unsupervised pixel purity determination and endmember extraction algorithm. It uses both the spectral and spatial information in a combined manner. Spatial-spectral endmember extraction (SSEE) [35] is a projection-based unmixing algorithm to analyze the scene into subsets. As a result, the spectral contrasts of low-contrast endmembers increase while improving the potential for these extracted endmembers to be selected for discrimination.

10.4.5 Sparse Regression-Based Approach

A semi-supervised approach, the sparse regression-based approach, is also used to find the endmembers in a spectral library (S) containing many spectra (p) in which only a few of them remain present in the pixel and $p << L$ (L refers to the number of spectral bands). This means that the vector of fractional abundances is sparse in nature. From Section 10.4.2.1, the equation for LMM can be rewritten as given in Equation (10.3) as:

$$y = Sf + w \tag{10.3}$$

where f represents fractional abundances ($L - by - 1$ vector) of the members contained in S. In unmixing, the mixing matrix (M) is to find the vector of fractional abundances f, given y and S. As the number of actual endmembers (q) is much lesser than the number of spectra (p) contained in S, the vector of fractional abundances f is sparse.

10.4.5.1 Moore–Penrose Pseudoinverse (MPP)

S is not a square matrix (i.e., not normally invertible); it is not possible to determine the $\hat{f}$ of f by multiplying the inverse of S with y:$\hat{f} = S^{-1}R$. As per S, the product of $S^T S$ is square and invertible. An estimate of f can be found using Equation (10.4):

$$\hat{f} = (S^T S)^{-1} S^T R = S^{\sharp} R \tag{10.4}$$

where $(S^{\sharp} = (S^T S)^{-1} S^T)$ is the moore–penrose pseudoinverse code of the matrix S; and $\hat{f} = f$ only in case of no error term (w). In case of a bad conditioned case of S, the error term becomes more important.

10.4.5.2 Orthogonal Matching Pursuit (OMP)

OMP is an alternative to matching pursuit. It is an iterative technique that finds the spectral signatures from the spectral library that best explain a predetermined residual at each iteration. At the first iteration, the initial residual is equal to the observed spectrum of the pixel, the vector of fractional abundances is null, and the matrix of the indices of selected endmembers is empty. At each iteration, OMP searches the member from the spectral library that best explains the actual residual, adds this member to the endmembers matrix, updates the residual, and calculates the estimate of fractional abundance using the selected endmembers.

10.4.5.3 Iterative Spectral Mixture Analysis (ISMA)

ISMA is an iterative technique extracted from the spectral mixture analysis. It searches the optimal endmember set by investigating the variation in the root mean squared (RMS) error along the iterations. This algorithm branches into two parts. Initially, it calculates an unconstrained solution of the unmixing problem using all the available spectra in the spectral library; then, it removes the spectra with the lowest estimated fractional abundance and repeats the process with the remaining endmembers, until one endmember remains. The second part of ISMA consists in finding the critical iteration, which corresponds to the iteration of the first abrupt variation in the RMS error, calculated as given in Equation (10.5):

$$\Delta RMS = 1 - \frac{RMS_{j-1}}{RMS_j} \tag{10.5}$$

where RMS_j corresponds to the j_{th} iteration RMS error. The critical iteration corresponds to the optimal set of endmembers. The notion of retrieving the true endmember set by analyzing the change in the RMSE is based on the fact that before extracting the optimal set of endmembers, the RMSE changes within a certain (small) range and limits, and it has bigger changes when one endmember from the optimal set is removed, as the remaining endmembers are not sufficient to generate a model with good accuracy with respect to the actual observations. ISMA calculates, at every iteration, an unconstrained solution in place of a constrained one, as it is expectable that when the endmember set approaches the optimal one, the abundance fractions will approach the true ones.

10.4.6 Hybrid Techniques

Spectral unmixing is one of the mandate tasks of spectra-based imaging. Unmixing determines the pure pixels and their abundance value. Mixing may be linear or nonlinear based. Numerous algorithms or approaches are available to solve spectral unmixing. However, hybrid approaches have been researched in order to solve spectral unmixing with better accuracy in the presence or absence of endmembers, and to solve cases like unsupervised or semi-supervised unmixing or the like. Some of the hybrid techniques as proposed in spectral unmixing are explained in brief here.

In [36], a hierarchical Bayesian model is proposed for a semi-supervised hyperspectral linear spectral-unmixing process. An assumption of this model is that the pixel reflectances contain linear mixing of pure spectra along with an additive Gaussian noise. Similarly, for endmember extraction and linear unmixing for hyperspectral data, a joint Bayesian [37] model is proposed for betterment of unmixing results.

Generally, endmember extraction algorithms are applied for obtaining the presence of materials that are either known or to be determined. However, [38] a proposed technique called genetic orthogonal projection (GOP) for endmember extraction is a fully unsupervised approach. This proposed technique is a hybridization of orthogonal projection and genetic algorithms for endmember extraction.

Conventional hyperspectral image processing performs denoising and spectral unmixing separately. However, in [39], a sparse representation framework is proposed to improve both the denoising and spectral unmixing components iteratively in a feedback manner. In a proposed hybrid-combined representation framework, sparse coding for denoising and sparsity for spectral unmixing are used in an iterative manner for improving the performance of both denoising and unmixing results.

In [40], a neural network-based hybrid mixture model is proposed to obtain the information from nonlinearly mixed pixels. The proposed model is composed of three stages or processes, namely, (1) extracting endmembers using the N-FINDR algorithm, (2) using endmembers by LMM for abundance estimation, and (3) feeding training samples with abundance value to neural network-based multilayer perceptron (MLP) architecture as an input, to get the refined output in case of the nonlinear mixture class as well. Here, the hybridization using the MLP helps in better segregation of nonlinear unmixing approaches.

10.5 Classification

Hyperspectral image classification is one of the dominant research areas in recent times, in classification for a given data cube to assign a unique class for given pixel vectors. High spatial resolution is quite important to assume the data content of pure pixels (single spectral signature). In contrast, for the case of mixed pixels, spectral-unmixing methods have been used for analysis. In this part, we discuss some of the key methods of a supervised classification approach. Supervised classification has been mostly used in [41], but it has certain limitations like high dimensionality of data and limited training samples [42], among others. To solve these problems, feature mining [23], semi-supervised learning, and sub space-based approaches [43] have also been developed and used. Feature mining and supervised learning are discussed briefly in this section.

10.5.1 Feature Mining

Hyperspectral data contains rich amounts of information for a variety of applications in different fields of study. It provides opportunity for a wide range of applications, not focused to solve a given single problem. Each band may or may not reveal the unique absorption feature(s) for the object under study in a given problem. Thus, a given band may be able to solve one problem but not the other sufficiently. Therefore, the original hyperspectral bands are essentially the candidate features for a particular application. Identifying efficient features is one of the critical preprocessing steps for image classification. Generally, it is not a good procedure to use the entire spectral band's information to solve a particular problem. As a result, it increases the computational load and the algorithms face significant challenges. Feature mining is one of the important tasks included in the feature selection (FS) and feature extraction (FE) stages. Finding effective subspace that contains a minimum number of attributes is needed to describe selected properties of given data for hyperspectral image analysis. Feature mining is discussed in two sections here, feature selection and feature extraction.

10.5.1.1 Feature Selection (FS)

Suppose X indicates the original hyperspectral data, that is, $X = [x_i]$, where $x_i \in \Re^L$; $i = 1, 2, \ldots, d$; and d and L represent the number of pixels and spectral bands. The objective of the FS is to find the good subset p from the total features $L, p < L$. There are two main types of approaches for FS: the filter approach and wrapper approach. For FS, the filter approach is classifier independent, while the wrapper approach is classifier dependent. Here, filter-based approaches (supervised and unsupervised) will be discussed.

The parametric supervised FS involves class modeling of the data using a training set. Jeffries-Matusita (JM) distance has been broadly used to measure the separability of two class density functions and performs fairly when the Gaussian distribution presumption is applicable. Other distance measures such as Euclidean distance, spectral angle, Mahalanobis distance, and so on. have also been used; in the case of nonparametric supervised FS, the use of training data directly without modeling class data is reported. Information theory can also be employed to execute nonparametric supervised FS. Mutual information (MI) provides a degree of linear and nonlinear dependency between two variables and can be used in both FS and FE. In case of unsupervised FS, without considering any specific applications, the basis for the selection criteria of FS is SNR. MI is used to find the subset bands that have minimum dependency [44].

10.5.1.2 Feature Extraction

Numerous approaches have been proposed in hyperspectral imaging to execute feature extraction prior to data classification. Feature extraction exploits the entire available spectral band in order to extract relevant features. The most available approach for the generation of features is based on extractions obtained from PCA and MNF algorithms. In these approaches, data is projected onto new space in which the first few components contain most of the complete information of the data, and therefore only a few features are retained. Segmented PCA is better than conventional PCA because it reduces the computational load. DWT has been used for generation of feature space that separates high- and low-frequency components. This allows a form of derivative analysis that has been also used to generate features prior to hyperspectral image classification [45]. Other broadly used techniques are decision boundary feature extraction (DBFE) and nonparametric weighted feature extraction (NWFE) [42]. Another strategy for feature extraction has been based on the grouping of neighboring bands, using methods such as the weighted sum or average of each group [46]; this approach also performs equally well for most data sets.

10.5.2 Supervised Classification

In supervised classification, the target of interest in the data is identified using available training samples. In hyperspectral data, each voxel has its own spectra compared with the available training samples, that is, a spectral signature, which is used for the classification of images accordingly.

10.5.2.1 Minimum Distance Classifier

This classifier is used to classify unknown image data into classes that minimize the distance between the image data and the class in case of multifeature space. In this, a minimum distance indicates the index of similarity. First-order and second-order statistics-based classifiers are described briefly here.

First-order statistics classifiers: First-order statistics classifiers provide local statistical information (e.g., mean), which is given here:

> *Euclidean distance (ED):* ED is one of the most broadly used features to measure the distance between spectral signatures. Suppose that two spectral signature vectors are represented as $S_j = (S_{j1}, S_{j2}, \ldots, S_{jL})^T$ and $S_k = (S_{k1}, S_{k2}, \ldots, S_{kL})^T$, and let L be the number of spectral bands; then, the ED is given by Equation (10.6):

$$ED(S_j, S_k) = \|S_j - S_k\| = \sqrt{\sum_{l=1}^{L} (S_{jl} - S_{kl})^2} \tag{10.6}$$

> *City block distance (CBD):* CBD is given in Equation (10.7):

$$CBD(S_j, S_k) = \sum_{l=1}^{L} |S_{jl} - S_{kl}| \tag{10.7}$$

> where $S_j = (S_{j1}, S_{j2}, \ldots, S_{jL})^T$ and $S_k = (S_{k1}, S_{k2}, \ldots, S_{kL})^T$ are the two spectral signature vectors; and L is the number of spectral bands.

> *Tchebyshev (maximum) distance (TD):* TD is represented and computed as provided in Equation (10.8):

$$TD(S_j, S_k) = \max_{1 \le l \le L} |S_{jl} - S_{kl}| \tag{10.8}$$

> where $S_j = (S_{j1}, S_{j2}, \ldots, S_{jL})^T$ and $S_k = (S_{k1}, S_{k2}, \ldots, S_{kL})^T$ are two spectral signature vectors; and L is the number of spectral bands.

Second-order statistics classifiers: Covariance-based classifiers are known as second-order statistics classifiers, which are given here:

> *Mahalanobis distance (MD) classifier:* MD has been used for different purposes, such as observing the difference between two data sets and detecting outliers. It calculates the distance using variance–covariance values of two variables, which are given in Equation (10.9):

$$MD = \sqrt{(x_i - \bar{x}) C_x^{-1} (x_i - \bar{x})^T} \tag{10.9}$$

> where C_x is the variance–covariance matrix of two variables, $x1$ and $x2$. The ellipse, formed based on the distance values, and it represents the equal MDs toward the center point of data. Prior knowledge of threshold is needed in order to distinguish the two classes.

> *Bhattacharyya distance (BD) classifier:* The BD is given in Equation (10.10)

$$BD_{i,j} = \frac{1}{8}(S_i - S_j)^T \left(\frac{\Sigma_i + \Sigma_j}{2} \right)^{-1} (S_i - S_j) + \frac{1}{2} \ln \left(\frac{|(\Sigma_i + \Sigma_j)/2|}{\sqrt{|\Sigma_i||\Sigma_j|}} \right) \tag{10.10}$$

where $\sum_i$ and $\sum_j$ are class sample covariance matrices that correspond to the two spectral signatures, namely, S_i and S_j.

10.5.2.2 Maximum Likelihood Classifier (MLC)

MLC is a pixel-based method, that assumes that the members of each class are normally distributed in the feature space. A pixel with an associated observed feature vector X is assigned to class c_i if:

$$X \in c_i \quad \text{if} \quad g_i(X) > g_j(X) \quad \text{for all} \quad i \neq j, i, j = 1, 2..N$$

For multivariate Gaussian distributions, $g_j(X)$ is given in Equation (10.11):

$$g_j(X) = \ln(p(c_i)) - \frac{1}{2} \ln \left| \sum_j \right| - \frac{1}{2}(X - M_j)^T \sum_j^{-1}(X - M_j) \tag{10.11}$$

where M_j and $\sum_j$ are the sample mean vector and covariance matrix of class j, respectively; and g_j is the discriminating function. It involves the estimation of class mean vectors and covariance matrices.

10.5.2.3 Support Vector Machines (SVMs)

SVM are used to solve supervised classification and regression problems; they were introduced by Boser, Guyon, and Vapnik [47, 48]. Numerous classification algorithms have been proposed to classify remote-sensing images. Other classification algorithms are MLCs, neural network classifiers [49], decision tree classifiers [49], and so on. MLC is a parametric classifier based on statistical theory and has some limitations. Neural network classifiers use a nonparametric approach to avoid the problems of MLC. MLP is the most broadly used neural network classification algorithm in remotely sensed images. A decision tree classifier uses different approaches for classification. It breaks a complex problem into a number of simple decision-making processes. SVM is one of the competitive methods in classifying high-dimensional data sets and employs the optimization techniques in order to obtain the boundaries between the classes. SVMs have been used in multispectral and hyperspectral data classification tasks [50, 51]. SVMs may be classified into linear SVMs and nonlinear SVMs. A linear SVM is based upon the statistical learning theory that divides linearly separable feature space into two classes with the maximum margin. Nonlinear SVMs can be used in case the feature space is not linearly separable. In nonlinear SVMs, the underlying concept is to represent the data in higher dimensional feature space such that nonlinearly mapped data is separable in higher dimensional space and linear SVMs can be applied to the newly mapped data. Nonlinear mapping via dot products into high-dimensional space is computationally complex. Nonlinear SVM using a kernel becomes computationally feasible, providing the same output as nonlinear mapping.

10.5.3 Hybrid Techniques

In classification, one labels the group(s) of pixels to a particular class. A number of approaches or algorithms are available to classify the data. However, hybrid approaches for hyperspectral image classification have been proposed in order to solve problems in a different way and/or to improve certain factors like output accuracy, efficient framework, and so on. Some of the hybrid approaches are explained in brief in this section.

In [52], the authors proposed a hybrid approach for semi-supervised classification for analysis of hyperspectral data. In the proposed approach, the multinomial logistic regression (MLR) integrates with different spectral-unmixing chains in order to exploit the spectral-unmixing and classification approaches together. In another work [53], the researchers proposed a hybrid approach to improve the classification performance of hyperspectral data. The proposed hybrid approach consists of rotation forests that integrate with Markov random fields (MRFs).

For hyperspectral image classification, several authors are using genetic algorithms and SVMs, as effective feature selection strategies [54]. To enhance the classification accuracy, several authors fused supervised and unsupervised learning techniques [55]. In particular, authors fused the strengths of the SVM classifier and the fuzzy c-means clustering algorithms.

In [56], deep convolutional neural networks (DCNNs) are introduced for spectral-spatial hyperspectral image classification. A proposed framework is hybridization of PCA, logistic regression (LR), and DCNNs. Similarly, to solve nonlinear classification problems, a new framework [57] is proposed based on SVMs using multicenter models and MRFs in probabilistic decision frameworks.

10.6 Target Detection

In reconnaissance and surveillance applications, hyperspectral imaging has been used to identify targets of interest. Target detection is considered as a binary classification whose aim is to label each pixel in the data as a target or background. The general aim of target detection is to identify rare target classes embedded in broadly populated and heterogeneous background classes. The process of identifying and detecting the target of interest in hyperspectral imagery is normally a two-stage process. The first stage is an anomaly detector [58, 59] that detects the spectral vectors that are significantly different from the background pixels spectra. The second stage is to detect and confirm that the detected anomaly is a target or natural clutter. This can be accomplished by using spectra from the spectral library or training sets. Spectral signature–based target detectors (SSTDs) have been used broadly due to the availability of spectral signatures of materials in spectral libraries. This assumes that the target signature is known and tries to detect all the pixels that have a high degree of correlation with the target signature. This method depends upon the reference target spectra, which have been collected by using a handheld spectro-radiometer. However, hyperspectral data collected using spaceborne or airborne platforms need atmospheric correction; its accuracy is vital for identification of the target. The performance of SSTDs is limited due to factors like atmospheric compensation, spatial-spectral and radiometric calibrations, and so on. This section discusses some of the basic algorithms of target detection.

10.6.1 Anomaly Detection

Anomaly detectors are also called novelty detectors or outlier detectors, and they are used to identify the target isolated from the background pixels. This method is perceived as a special case of target detection in which no prior information is provided about the spectra of targets of interest. The aim is to find anomalous objects in the image with reference to the background. Anomalies vary according to the particular applications. They can be as varied as mineral identification, crop stress determination, and manmade object detection. Unlike SSTDs, anomaly detection can be applied to raw radiance image data, and it does not require atmospheric compensation to be considered. This is due to the intrinsic nature of anomaly detection, which does not need prior information about the target spectra and can focus on the data for exploring those pixels whose spectra are notably different from the background. Anomaly detection is the first step for image analysis to provide ROIs that may consist of potential targets to be explored using spectral matching techniques to finally determine the target of interest [60]. Lots of techniques have been proposed for anomaly detection, but an Reed–Xiaoli (RX) algorithm is considered as the benchmark of anomaly detection for hyperspectral data. In this subsection, an overall review of anomaly detection is discussed.

10.6.1.1 RX Anomaly Detection

The RX anomaly detection algorithm has been developed for spectral detection of anomaly, by identifying the targets of unknown spectral characteristics against a background clutter with an unknown spectral covariance relationship. This technique has been applied often in hyperspectral image analysis-based applications and is viewed as a standard anomaly detection technique in hyperspectral and multispectral applications. The RX algorithm is a constant false alarm rate (CFAR) adaptive anomaly detector that is extracted from the generalized likelihood ratio test (GLRT) [61]. CFAR allows the detector to use a single threshold value to maintain the desired false alarm rate regardless of the background variation at different positions in the scene. Suppose that a pixel $x = [x_1, x_2, \ldots, x_L]^T \in \mathfrak{R}^L$ is the observation test vector consisting of L number of bands; the output of the RX algorithm will be given as shown in Equation (10.12):

$$RX(x) = (x - \mu_b)^T C_b^{-1} (x - \mu_b) \tag{10.12}$$

where μ_b is the the the global sample mean and C_b is the sample covariance matrix of the image. The RX algorithm applies the square of the MD between the text pixel and local background mean. The background mean and covariance matrix can be estimated globally from the whole hyperspectral data and locally using a double concentric sliding window approach. To estimate the covariance globally, one needs more advanced methods or techniques like a linear or stochastic mixture model or some clustering techniques. On the other hand, a double-sliding window approach can be used to estimate the covariance matrix locally to consist of test pixels, an inner window region (IWR), a guard band, and an outer window region (OWR), as shown in Figure 10.7. Local mean vector and covariance matrix are calculated from the pixel(s) falling within the OWR. Inner window size depends upon the size of target of interest in the image. Guard band size is marginally larger than the IWR and smaller than the OWR. It is used to denote the probability that some target spectra will inhabit the OWR and hence affect the background model [62]. A local RX algorithm calculation is computationally quite intensive compared to the global RX algorithm due to the need of estimation of a covariance matrix at each location of test pixel following the double concentric sliding window approach. Several variations of the RX detector approach have been proposed to surpass some of the limitations.

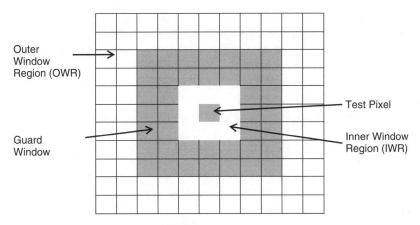

Figure 10.7 Double concentric sliding window.

10.6.1.2 Subspace-Based Anomaly Detection

In a subspace anomaly detection algorithm, projection vectors are generated using the OWR and IWR covariance matrices [63]. Denoting a spectral vector for the IWR centered at test pixel $x_k = (x_k(1), x_k(2), \ldots, x_k(L))$, where L refers to the number of spectral bands; and $k = 1, 2, \ldots, N_{iw}$, that is, there are a total of N_{iw} number of pixels in the IWR, matrix $X = [x_1, x_2, \ldots, x_{N_{iw}}]_{L*N_{iw}}$. In the same way, spectral vector for OWR $y_l = (y_l(1), y_l(2), \ldots, y_l(L))$, where L indicates the number of spectral bands and $l = 1, 2 \ldots, N_{ow}$ (i.e., the total number of pixels in the OWR matrix $Y = [y_1, y_2, \ldots, y_{N_{ow}}]_{L*N_{ow}}$). The covariance matrices for the IWR and OWR are given by Equations (10.13) and (10.14):

$$C_x = \frac{1}{N_{iw} - 1}(X - \hat{\mu}_X)(X - \hat{\mu}_X)^T \tag{10.13}$$

$$C_y = \frac{1}{N_{ow} - 1}(Y - \hat{\mu}_Y)(Y - \hat{\mu}_Y)^T \tag{10.14}$$

where $\hat{\mu}_X$ and $\hat{\mu}_Y$ are the statistical means of the IWR and OWR (back-ground), respectively. The projection separation statistics for test pixel r is represented as Equation (10.15):

$$s' = (r - \hat{\mu}_Y)^T W W^T (r - \hat{\mu}_Y) \tag{10.15}$$

where $W = [w_1, w_2 \ldots w_m]$ is the matrix with m projection vectors; $W W^T$ is known as projection vectors and indicates a subspace characterizing the spectra used to generate the projection vectors w_i; and, if the projection separation statistics s' is greater than some threshold value, anomaly is detected.

10.6.2 Signature-Based Target Detection

The performance of the anomaly detection is limited by no prior knowledge of target spectra in hyperspectral imagery. With signature-based detection, the reference spectral signature of the target becomes available. A spectral fingerprint is collected in the form of reflectance from the laboratory or field. For some target detection applications, a spectral characteristic of the target is known *a priori*. Spectral characteristics of targets can be defined by a single target spectrum or target subspace. In this manner, the background gets modeled as a statistically Gaussian distribution or with subspace that represents the local or whole background statistics. In this subsection, focus is on the classical detection algorithm of a spectral angle mapper, spectral matched filter, and matched subspace detector, which are briefly explained here.

10.6.2.1 Euclidean distance

The ED for signature-based target detection is given in Equation (10.16).

$$ED(S_j, S_k) = \|S_j - S_k\| = \sqrt{\sum_{l=1}^{L} (S_{jl} - S_{kl})^2} \tag{10.16}$$

where $S_j = (S_{j1}, S_{j2}, \ldots, S_{jL})^T$ and $S_k = (S_{k1}, S_{k2}, \ldots, S_{kL})^T$ are two spectral signature vectors; and L is the number of spectral bands.

10.6.2.2 Spectral Angle Mapper (SAM)

SAM is one of the most frequently used methods under the supervised hyperspectral imaging classification approach. Normally, it compares the image spectral signature with the known spectral signature (collected in the laboratory or in the field with a spectro-radiometer) or an endmember. In SAM, a pixel with zero or minimum spectral angle compared to the reference spectra is assigned to the class that is defined by the reference vector. On modifying the threshold for classification using SAM, the probability for target identification may increase. This algorithm is based on the assumption that every pixel contains only one ground material and can be uniquely assigned to one class. Spectral similarity is obtained by considering the spectral signature as a vector in L-dimensional space, where L refers to the number of spectral bands. It measures the spectral similarity by determining angles between two spectra, treating them as vectors in space with dimensionality equal to the number of bands. SAM determines the spectral similarity between the test spectra and reference spectra by using Equation (10.17) [64]:

$$\alpha = \cos^{-1} \left(\frac{\sum_{i=1}^{L} t_i r_i}{\left(\sum_{i=1}^{L} t_i^2 \right)^{1/2} \left(\sum_{i=1}^{L} r_i^2 \right)^{1/2}} \right) \tag{10.17}$$

where α = spectral angle between the test and reference spectra; L = number of bands; t_i = test spectra; and r_i = reference spectra.

10.6.2.3 Spectral Matched Vilter (SMF)

In SMFs, hyperspectral target detection can be expressed using two complementary hypotheses. The assumption for the spectral matched filter is:

$H_0 = X : m$, Target absent

$H_1 = X : as + m$, Target present

where a represents an unknown target abundance measure (i.e., a=0 in case of no target and a>0 in case of target); $s = [s_1, s_2, \ldots, s_n]^T$ is target spectra; and m is the zero-mean Gaussian random additive background clutter noise. This technique is based on the presumption that Gaussian distributions for background clutter noise $N(0, \hat{C}_b)$ and target $N(as, \hat{C}_b)$ have the same covariance matrices (where a represents the scalar abundance value indicating target strength). The output of this method for a test pixel x using GLRT is given [65] by:

$$D_{SMF}(x) = \frac{s_T \hat{C}_b^{-1} x}{\sqrt{s_T \hat{C}_b^{-1} s}} \gtrless_{H_0}^{H_1} \eta \qquad (10.18)$$

where $\hat{C}_b$ represents the estimated covariance matrix for the centered observation data; and η represents any threshold value.

10.6.2.4 Matched Subspace Detector (MSD)

In MSD [66], target pixel vectors are represented as a linear combination of target and background spectra that are presented by subspace target and background spectra, respectively. The target identification subject is represented as two competing hypotheses, which are given as:

$$H_0 = X : B\xi + m, \text{ Target absent}$$

$$H_1 = X : S\Theta + B\xi + m, \text{ Target present}$$

where S and B indicate target and background subspaces, respectively; Θ and ξ are the coefficients that indicate the abundance in S and B; and m represents Gaussian random noise. The MSD [66] model using GLRT has been given in Equation (10.19):

$$D_{\eta MSD}(x) = \frac{X_T(1 - P_b)X}{X_T(1 - P_{tb})X} \gtrless_{H_0}^{H_1} \eta MSD \qquad (10.19)$$

where $P_b = BB^\sharp$ is a projection matrix associated with the background subspace; $P_{tb} = [SB][SB]^\sharp$ is a projection matrix associated with the target-and-background subspace; and $D_{\eta MSD}(x)$ is compared with threshold ηMSD in order to find which hypothesis is related to x.

10.6.3 Hybrid Techniques

In the workflow of hyperspectral image processing and analysis, target identification and verification tasks are one of the vital tasks of postprocessing and analysis. There are a number of algorithms available in order to identify the target of interest in applications like search and rescue, defense and intelligence, and so on. However, algorithms face some critical challenges (e.g., no prior information of target, etc.) in order to identify the objects of interest using approaches like subpixel target detection, anomaly detection, and so on. In order to overcome these critical challenges, combinations and/or frameworks of collectively applied algorithms are proposed (i.e., hybridization of algorithms) for target identification.

Normally, spatial resolution is limited in hyperspectral image and subpixel targets that occupy some part of the pixel. Therefore, subpixel target detection is a challenging task in which the target on ground itself is smaller than the pixel size in the image. Hybrid algorithms and/or frameworks are proposed in order to identify these subpixel targets. Adaptive matched subspace detector (AMSD) is used to identify targets that are different from background image pixels based on the statistical characteristics. In [67], the authors proposed a hybrid detector system that uses the advantages of both detectors [i.e., fully constrained least squares (FCLS) and AMSD] for increasing the accuracy of a subpixel target detection operation. In another work [68], the authors proposed two hybrid detectors to exploit the advantages of these two approaches by modeling the background using the characteristics of both physics- and statistics-based approaches in order to accurately identify the target(s). Other different hybrid detectors systems and approaches [69–71] are proposed for subpixel target identification utilizing existing and/or new techniques in a varied manner.

In situations and applications like surveillance, no prior knowledge related to the object of interest is available. In such situations, one needs to perform object detection and identification in conditions with no *a priori* knowledge. Hybridization and/or frameworks of algorithms have also been proposed for efficient identification of such anomalies. The RX detector is one of the most frequently used anomaly detection algorithms. Several authors have proposed hybrid and/or modified versions of RX [72, 73] detector systems in order to increase the efficiency of the anomaly detection step for further processing the outputs.

10.7 Conclusions

Hyperspectral imaging involves multiple disciplines of knowledge adapting vast concepts from signal and image processing, statistics, and machine learning to hybrid computing techniques. The importance of the hyperspectral imaging has been increasing due to the large number of spaceborne and airborne sensors systems available, commercially available sensors, and the number of applications supported by this technique. This imposes some challenges like high dimension, spectral unmixing (linear and nonlinear), atmospheric effect elimination, and so on, which require sophisticated and hybrid algorithms for data processing, analysis, and interpretation. In this chapter, we have presented some major data analysis methods and algorithms for spectral unmixing, classification, and target detection along with a critical discussion of the hybrid approaches for systems and algorithms. As a conclusion, hybrid hyperspectral image and analyses comprise a prominent and emerging field, contributing actively with frontier cross-disciplinary research activities. It is expected to mature in the coming years from all aspects of the hybrid approaches in sensor development, algorithms, and applications.

References

1 Feng, Y.Z. and Sun, D.W. (2012) Application of hyperspectral imaging in food safety inspection and control: a review. *Critical Reviews in Food Science and Nutrition*, **52** (11), 1039–1058.

2 Adam, E., Mutanga, O., and Rugege, D. (2010) Multispectral and hyperspectral remote sensing for identification and mapping of wetland vegetation: a review. *Wetlands Ecology and Management*, **18** (3), 281–296.

3 Olmanson, L.G., Brezonik, P.L., and Bauer, M.E. (2013) Airborne hyperspectral remote sensing to assess spatial distribution of water quality characteristics in large rivers: the Mississippi River and its tributaries in Minnesota. *Remote Sensing of Environment*, **130**, 254–265.

4 Kruse, F.A., Boardman, J.W., and Huntington, J.F. (2003) Comparison of airborne hyperspectral data and eo-1 hyperion for mineral mapping. *IEEE Transactions on Geoscience and Remote Sensing*, **41** (6), 1388–1400.

5 Dale, L.M., Thewis, A., Boudry, C., Rotar, I., Dardenne, P., Baeten, V., and Pierna, J.A.F. (2013) Hyperspectral imaging applications in agriculture and agro-food product quality and safety control: a review. *Applied Spectroscopy Reviews*, **48** (2), 142–159.

6 Lu, G. and Fei, B. (2014) Medical hyperspectral imaging: a review. *Journal of Biomedical Optics*, **19** (1), 010 901–010 901.

7 ElMasry, G., Wang, N., ElSayed, A., and Ngadi, M. (2007) Hyperspectral imaging for nondestructive determination of some quality attributes for strawberry. *Journal of Food Engineering*, **81** (1), 98–107.

8 Qiao, J., Ngadi, M.O., Wang, N., Gariépy, C., and Prasher, S.O. (2007) Pork quality and marbling level assessment using a hyperspectral imaging system. *Journal of Food Engineering*, **83** (1), 10–16.

9 Solomon, J. and Rock, B. (1985) Imaging spectrometry for earth remote sensing. *Science*, **228** (4704), 1147–1152.

10 Pearlman, J.S., Barry, P.S., Segal, C.C., Shepanski, J., Beiso, D., and Carman, S.L. (2003) Hyperion, a space-based imaging spectrometer. *Geoscience and Remote Sensing, IEEE Transactions*, **41** (6), 1160–1173.

11 Chang, C.I. (2013) *Hyperspectral data processing: algorithm design and analysis*, John Wiley & Sons, Hoboken, NJ.

12 Eismann, M.T. (2012) *Hyperspectral remote sensing*. SPIE, Bellingham, WA.

13 Sun, D.W. (2010) *Hyperspectral imaging for food quality analysis and control*. Elsevier, Amsterdam.

14 Inoue, Y. and Penuelas, J. (2001) An AOTF-based hyperspectral imaging system for field use in ecophysiological and agricultural applications. *International Journal of Remote Sensing*, **22** (18), 3883–3888.

15 Wang, W., Li, C., Tollner, E.W., Rains, G.C., and Gitaitis, R.D. (2012) A liquid crystal tunable filter based shortwave infrared spectral imaging system: design and integration. *Computers and Electronics in Agriculture*, **80**, 126–134.

16 Park, B. and Lu, R. (2015) *Hyperspectral imaging technology in food and agriculture*, Springer, New York.

17 Yu, X., Sun, Y., Fang, A., Qi, W., and Liu, C. (2014) Laboratory spectral calibration and radiometric calibration of hyper-spectral imaging spectrometer, in *Systems and Informatics (ICSAI), 2014 2nd International Conference*, IEEE, pp. 871–875.

18 Rodarmel, C. and Shan, J. (2002) Principal component analysis for hyperspectral image classification. *Surveying and Land Information Science*, **62** (2), 115.

19 Pu, R. and Gong, P. (2004) Wavelet transform applied to eo-1 hyperspectral data for forest lai and crown closure mapping. *Remote Sensing of Environment*, **91** (2), 212–224.

20 Du, Q. (2007) Modified Fisher's linear discriminant analysis for hyperspectral imagery. *IEEE Geoscience and Remote Sensing Letters*, **4** (4), 503.

21 Chang, C.I. and Wang, S. (2006) Constrained band selection for hyperspectral imagery. *IEEE Transactions on Geoscience and Remote Sensing*, **44** (6), 1575–1585.

22 Bioucas-Dias, J.M. and Plaza, A. (2010) Hyperspectral unmixing: geometrical, statistical, and sparse regression-based approaches, in *Remote sensing*. SPIE, Bellingham, WA, pp. 78 300A–78 300A.

23 Bioucas-Dias, J.M., Plaza, A., Dobigeon, N., Parente, M., Du, Q., Gader, P., and Chanussot, J. (2012) Hyperspectral unmixing overview: geometrical, statistical, and sparse regression-based approaches. *Selected Topics in Applied Earth Observations and Remote Sensing, IEEE Journal*, **5** (2), 354–379.

24 Boardman, J.W., Kruse, F.A., and Green, R.O. (1995) Mapping target signatures via partial unmixing of AVIRIS data, in *Proceedings of Annual JPL Airborne Geoscience Workshop, Pasadena, CA*, pp. 23–26.

25 Winter, M.E. (1999) N-FINDR: an algorithm for fast autonomous spectral end-member determination in hyperspectral data, in *SPIE's International Symposium on Optical Science, Engineering, and Instrumentation*. SPIE, Bellingham, WA, pp. 266–275.

26 Nascimento, J.M. and Dias, J.M.B. (2005) Vertex component analysis: a fast algorithm to unmix hyperspectral data. *Geoscience and Remote Sensing, IEEE Transactions*, **43** (4), 898–910.

27 Chang, C.I., Wu, C.C., Liu, W., and Ouyang, Y.C. (2006) A new growing method for simplex-based endmember extraction algorithm. *IEEE Transactions on Geoscience and Remote Sensing*, **44** (10), 2804–2819.

28 Li, J. and Bioucas-Dias, J.M. (2008) Minimum volume simplex analysis: a fast algorithm to unmix hyperspectral data, in *IGARSS 2008-2008 IEEE International Geoscience and Remote Sensing Symposium*, vol. 3, IEEE, pp. III–250.

29 Chan, T.H., Chi, C.Y., Huang, Y.M., and Ma, W.K. (2009) A convex analysis-based minimum-volume enclosing simplex algorithm for hyperspectral unmixing. *IEEE Transactions on Signal Processing*, **57** (11), 4418–4432.

30 Bioucas-Dias, J.M. (2009) A variable splitting augmented Lagrangian approach to linear spectral unmixing, in *2009 First Workshop on Hyperspectral Image and Signal Processing: Evolution in Remote Sensing*, IEEE, pp. 1–4.

31 Ambikapathi, A., Chan, T.H., Ma, W.K., and Chi, C.Y. (2011) Chance-constrained robust minimum-volume enclosing simplex algorithm for hyperspectral unmixing. *IEEE Transactions on Geoscience and Remote Sensing*, **49** (11), 4194–4209.

32 Bayliss, J.D., Gualtieri, J.A., and Cromp, R.F. (1998) Analyzing hyperspectral data with independent component analysis, in *26th AIPR Workshop: Exploiting New Image Sources and Sensors*. SPIE, Bellingham, WA, pp. 133–143.

33 Nascimento, J.M. and Bioucas-Dias, J.M. (2007) Dependent component analysis: a hyperspectral unmixing algorithm, in *Pattern Recognition and Image Analysis*, Springer, Berlin, pp. 612–619.

34 Plaza, A., Martinez, P., Pérez, R., and Plaza, J. (2002) Spatial/spectral endmember extraction by multidimensional morphological operations. *Geoscience and Remote Sensing, IEEE Transactions*, **40** (9), 2025–2041.

35 Rogge, D., Rivard, B., Zhang, J., Sanchez, A., Harris, J., and Feng, J. (2007) Integration of spatial–spectral information for the improved extraction of endmembers. *Remote Sensing of Environment*, **110** (3), 287–303.

36 Dobigeon, N., Tourneret, J.Y., and Chang, C.I. (2008) Semi-supervised linear spectral unmixing using a hierarchical bayesian model for hyperspectral imagery. *IEEE Transactions on Signal Processing*, **56** (7), 2684–2695.

37 Dobigeon, N., Moussaoui, S., Coulon, M., Tourneret, J.Y., and Hero, A.O. (2009) Joint Bayesian endmember extraction and linear unmixing for hyperspectral imagery. *IEEE Transactions on Signal Processing*, **57** (11), 4355–4368.

38 Rezaei, Y., Mobasheri, M.R., Zoej, M.V., and Schaepman, M.E. (2012) Endmember extraction using a combination of orthogonal projection and genetic algorithm. *IEEE Geoscience and Remote Sensing Letters*, **9** (2), 161–165.

39 Yang, J., Zhao, Y.Q., Chan, J.C.W., and Kong, S.G. (2016) Coupled sparse denoising and unmixing with low-rank constraint for hyperspectral image. *IEEE Transactions on Geoscience and Remote Sensing*, **54** (3), 1818–1833.

40 Kumar, U., Raja, K.S., Mukhopadhyay, C., and Ramachandra, T. (2012) A neural network based hybrid mixture model to extract information from non-linear mixed pixels. *Information*, **3** (3), 420–441.

41 Plaza, A., Benediktsson, J.A., Boardman, J.W., Brazile, J., Bruzzone, L., Camps-Valls, G., Chanussot, J., Fauvel, M., Gamba, P., Gualtieri, A., *et al.* (2009) Recent advances in techniques for hyperspectral image processing. *Remote Sensing of Environment*, **113**, S110–S122.

42 Landgrebe, D.A. (2005) *Signal theory methods in multispectral remote sensing*, vol. 29. John Wiley & Sons, Hoboken, NJ.

43 Bioucas-Dias, J.M. and Nascimento, J.M. (2008) Hyperspectral subspace identification. *Geoscience and Remote Sensing, IEEE Transactions*, **46** (8), 2435–2445.

44 Hossain, M.A., Pickering, M., and Jia, X. (2011) Unsupervised feature extraction based on a mutual information measure for hyperspectral image classification, in *Geoscience and Remote Sensing Symposium (IGARSS), 2011 IEEE International*, IEEE, pp. 1720–1723.

45 Bruce, L.M. and Li, J. (2001) Wavelets for computationally efficient hyperspectral derivative analysis. *Geoscience and Remote Sensing, IEEE Transactions*, **39** (7), 1540–1546.

46 Kumar, S., Ghosh, J., and Crawford, M.M. (2001) Best-bases feature extraction algorithms for classification of hyperspectral data. *Geoscience and Remote Sensing, IEEE Transactions*, **39** (7), 1368–1379.

47 Boser, B.E., Guyon, I.M., and Vapnik, V.N. (1992) A training algorithm for optimal margin classifiers, in *Proceedings of the Fifth Annual Workshop on Computational Learning Theory*. ACM, New York, pp. 144–152.

48 Vapnik, V. (2013) *The nature of statistical learning theory*. Springer Science & Business Media, New York.

49 Goel, P., Prasher, S., Patel, R., Landry, J., Bonnell, R., and Viau, A. (2003) Classification of hyperspectral data by decision trees and artificial neural networks to identify weed stress and nitrogen status of corn. *Computers and Electronics in Agriculture*, **39** (2), 67–93.

50 Huang, C., Davis, L., and Townshend, J. (2002) An assessment of support vector machines for land cover classification. *International Journal of Remote Sensing*, **23** (4), 725–749.

51 Gualtieri, J.A. and Cromp, R.F. (1999) Support vector machines for hyperspectral remote sensing classification, in *The 27th AIPR Workshop: Advances in Computer-Assisted Recognition*. SPIE, Bellingham, WA, pp. 221–232.

52 Dópido, I., Li, J., Gamba, P., and Plaza, A. (2014) A new hybrid strategy combining semisupervised classification and unmixing of hyperspectral data. *IEEE*

Journal of Selected Topics in Applied Earth Observations and Remote Sensing, **7** (8), 3619–3629.

53 Xia, J., Chanussot, J., Du, P., and He, X. (2015) Spectral–spatial classification for hyperspectral data using rotation forests with local feature extraction and markov random fields. *IEEE Transactions on Geoscience and Remote Sensing*, **53** (5), 2532–2546.

54 Li, S., Wu, H., Wan, D., and Zhu, J. (2011) An effective feature selection method for hyperspectral image classification based on genetic algorithm and support vector machine. *Knowledge-Based Systems*, **24** (1), 40–48.

55 Alajlan, N., Bazi, Y., Melgani, F., and Yager, R.R. (2012) Fusion of supervised and unsupervised learning for improved classification of hyperspectral images. *Information Sciences*, **217**, 39–55.

56 Yue, J., Zhao, W., Mao, S., and Liu, H. (2015) Spectral–spatial classification of hyperspectral images using deep convolutional neural networks. *Remote Sensing Letters*, **6** (6), 468–477.

57 Tang, B., Liu, Z., Xiao, X., Nie, M., Chang, J., Jiang, W., Li, X., and Zheng, C. (2015) Spectral–spatial hyperspectral classification based on multi-center SAM and MRF. *Optical Review*, **22** (6), 911–918.

58 Stein, D.W., Beaven, S.G., Hoff, L.E., Winter, E.M., Schaum, A.P., and Stocker, A.D. (2002) Anomaly detection from hyperspectral imagery. *Signal Processing Magazine, IEEE*, **19** (1), 58–69.

59 Matteoli, S., Diani, M., and Corsini, G. (2010) A tutorial overview of anomaly detection in hyperspectral images. *Aerospace and Electronic Systems Magazine, IEEE*, **25** (7), 5–28.

60 Chang, C.I. and Chiang, S.S. (2002) Anomaly detection and classification for hyperspectral imagery. *Geoscience and Remote Sensing, IEEE Transactions*, **40** (6), 1314–1325.

61 Reed, I.S. and Yu, X. (1990) Adaptive multiple-band CFAR detection of an optical pattern with unknown spectral distribution. *Acoustics, Speech and Signal Processing, IEEE Transactions*, **38** (10), 1760–1770.

62 Ranney, K.I. and Soumekh, M. (2006) Hyperspectral anomaly detection within the signal subspace. *Geoscience and Remote Sensing Letters, IEEE*, **3** (3), 312–316.

63 Goldberg, H. and Nasrabadi, N.M. (2007) A comparative study of linear and nonlinear anomaly detectors for hyperspectral imagery, in *Defense and Security Symposium*. SPIE, Bellingham, WA, pp. 656 504–656 504.

64 Kruse, F., Lefkoff, A., Boardman, J., Heidebrecht, K., Shapiro, A., Barloon, P., and Goetz, A. (1993) The spectral image processing system (SIPS)–interactive visualization and analysis of imaging spectrometer data. *Remote Sensing of Environment*, **44** (2), 145–163.

65 Fuhrmann, D.R., Kelly, E.J., and Nitzberg, R. (1992) A CFAR adaptive matched filter detector. *Transactions on Aerospace and Electronic Systems*, **28** (1), 208–216.

66 Scharf, L.L. and Friedlander, B. (1994) Matched subspace detectors. *Signal Processing, IEEE Transactions*, **42** (8), 2146–2157.

67 Broadwater, J., Meth, R., and Chellappa, R. (2004) A hybrid algorithm for subpixel detection in hyperspectral imagery, in *Geoscience and Remote Sensing Symposium, 2004. IGARSS'04. Proceedings. 2004 IEEE International*, vol. 3, IEEE, pp. 1601–1604.

68 Broadwater, J. and Chellappa, R. (2007) Hybrid detectors for subpixel targets. *IEEE Transactions on Pattern Analysis and Machine Intelligence*, **29** (11), 1891–1903.

69 Zhang, L., Du, B., and Zhong, Y. (2010) Hybrid detectors based on selective endmembers. *IEEE Transactions on Geoscience and Remote Sensing*, **48** (6), 2633–2646.

70 Zhou, L., Zhang, X., Guan, B., and Zhao, Z. (2014) Dual force hybrid detector for hyperspectral images. *Remote Sensing Letters*, **5** (4), 377–385.

71 Xu, Y., Wu, Z., Li, J., Plaza, A., and Wei, Z. (2016) Anomaly detection in hyperspectral images based on low-rank and sparse representation. *IEEE Transactions on Geoscience and Remote Sensing*, **54** (4), 1990–2000.

72 Zare-Baghbidi, M., Homayouni, S., and Jamshidi, K. (2015) Improving the RX anomaly detection algorithm for hyperspectral images using FFT. *Modeling and Simulation in Electrical and Electronics Engineering*, **1** (2), 33–39.

73 Khazai, S. and Mojaradi, B. (2015) A modified kernel-RX algorithm for anomaly detection in hyperspectral images. *Arabian Journal of Geosciences*, **8** (3), 1487–1495.

11

A Hybrid Approach for Band Selection of Hyperspectral Images

Aditi Roy Chowdhury[1], Joydev Hazra[2], and Paramartha Dutta[3]

[1] Department of Computer Science and Technology, Women's Polytechnic, Jodhpur Park, Kolkata, West Bengal, India
[2] Department of Information Technology, Heritage Institute of Technology, Kolkata, West Bengal, India
[3] Department of Computer and System Sciences, Visva-Bharati University, Santiniketan, West Bengal, India

11.1 Introduction

Image processing is an emerging field of research. Image classification, analysis, and processing are very useful in different areas of research like pattern recognition [1], image registration [2–4], image forensics, and so on. Another form of image processing is hyperspectral images. The human eye can observe wavelengths between 380 nm to 760 nm, and they are classified as the visible spectrum. Other forms of radiation are perceived as infrared and ultraviolet (UV) light, as given in Figure 11.1. Most multispectral images (Landsat, AVHRR, etc.) measure radiation reflected from wide and separate wavelength bands. But most hyperspectral images measure radiation as narrow and contiguous wavelength bands. A hyperspectral image consists of different spectral bands with a very fine spectral resolution. The Airborne Visible/Infrared Imaging Spectrometer (AVIRIS) from NASA Jet Propulsion Laboratory (NASA/JPL) and the Hyperspectral Digital Imagery Collection Experiment (HYDICE) from Naval Research Laboratory are examples of two sensors that can collect image information with hundreds of spectral bands. For accurate object identification, spectral information provides an important role. Hyperspectral images have sufficient spectral information to identify and distinguish different materials uniquely. The image spectra of the hyperspectral image can be compared with some laboratory reflectance spectra in order to recognize and to prepare a ground truth map of surface materials, such as particular types of land, vegetation, or minerals. The hyperspectral image itself is an image cube as represented in Figure 11.2, where the z-axis represents the band. Here, each pixel is represented by a vector consisting of the reflectance of the object of a band. The length of each vector is equal to the number of bands of any particular data set.

Thus, the data volume to be processed for a hyperspectral image is huge. This vast amount of data creates problem in data transmission and storage. For image analysis, the computational complexity is also very high. Particularly, traditional image analysis techniques face challenges to handle this huge volume of data. From this purview, band selection is necessary. It is obvious to perform this selection process without hampering

Hybrid Intelligence for Image Analysis and Understanding, First Edition.
Edited by Siddhartha Bhattacharyya, Indrajit Pan, Anirban Mukherjee, and Paramartha Dutta.
© 2017 John Wiley & Sons Ltd. Published 2017 by John Wiley & Sons Ltd.
Companion Website: www.wiley.com/go/bhattacharyya/hybridintelligence

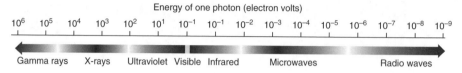

Figure 11.1 Electromagnetic spectrum [5].

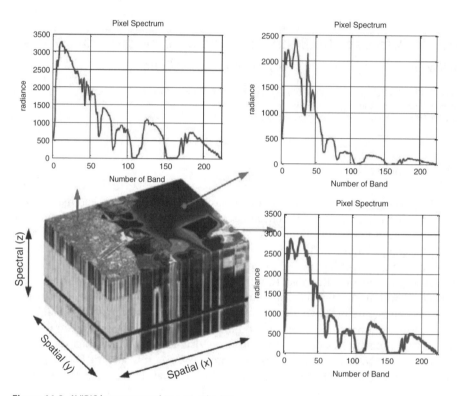

Figure 11.2 AVIRIS hyperspectral image cube [6].

classification accuracy. For the hyperspectral image, some bands are affected by various atmospheric effects like noise. These bands are irrelevant. Some redundant bands are also there to increase the complexity of the analysis system and produce an incorrect prediction. Again, some brands contain vast information about the scene, but their predictions are incorrect. So, minute investigation is needed to detect the relationship between the bands and the scene.

Data dimensionality can be reduced by mapping the high-dimensional data onto a low-dimensional space by using certain techniques. Principal component analysis (PCA), maximum noise fraction transform (MNF) [7, 8], and linear discriminant analysis (LDA) have been used as dimension reduction techniques. PCA is used to maximize the variance of the transformed data (or to minimize the reconstruction error), whereas Fisher's LDA is used to maximize the class separability. PCA and noise-adjusted PCA were proposed for unsupervised band [9] selection. However, the main problem with these methods is that they do not recover the original bands that correspond to wavelengths in the main data sets. Thus, some crucial information may be lost or distorted.

The orthogonal subspace approach was also used to transform the data [10]. Merging different adjacent bands is also used in dimensionality reduction [11, 12]. Another dimensionality reduction approach is band selection. In band selection, we can reduce the dimensionality of hyperspectral images by selecting only the relevant bands without losing their physical meaning. Hence, it is advantageous over dimensionality reduction since it preserves the original data of each band. Band selection can be done based on feature selection methodology or by extracting new bands containing the maximal information. Hyperspectral image analysis deals with the detection or classification of objects. The main aim of band selection is to select a combination of bands from the original one so that the accuracy is unchanged or tolerably degraded, and the computational complexity is significantly reduced. Hence, band selection is popular among researchers.

Recently, different band selection methods have been proposed. Basically, band selection techniques can be broadly classified into two groups, supervised and unsupervised, depending upon the availability of image data. Supervised methods are basically a predictive technique. They use training data to establish a model for prediction. These methods preserve a priori knowledge of image data. Among different supervised techniques, canonical analysis was employed in [9]. In MVPCA (maximum variance-based PCA), the bands are sorted depending on the importance of individual bands as well as the correlation with other bands. Key problems of supervised band selection are the number of selected bands and the criteria to choose the band. Different measurement criteria have been proposed like consistency measure [13], mutual information [14], and information measure [15]. An exhaustive search strategy can give an optimal solution. Popular search strategies are like maximum influence feature selection (MIFS), minimum redundancy–maximum relevance (mRMR), and so on. Most recently, a clonal selection algorithm (CSA) was used as an efficient search strategy. In [16], a CSA based on Jeffris–Matusita distance measure was used for dimensionality reduction. The trivariate mutual information (TMI) and semisupervised TMI-based (STMI) methods [17] are used as band selection methods incorporating aa priori CSA as a search strategy.

Due to unavailability of the required prior knowledge, the supervised technique is sometimes not useful. Therefore, unsupervised techniques that can offer better performance over supervised techniques regardless of the type of objects to be detected are more useful. Band index (BI) and mutual information (MI) have been used to measure the statistical dependence between different bands [18]. The linear prediction (LP)-based band similarity metric performs well for a small number of pixels [19–21]. Another unsupervised technique is band clustering, where the initial spectral bands are split into disjoint clusters or subbands [22].

Another approach in band selection is feature selection. Different feature extraction techniques like locality preserving projection (LPP) [23], PCA [24], and locally linear embedding (LLE) [25] transform or map data from high dimensions to low dimensions. However, feature selection reduces the dimensionality by selecting some important features among a set of original features. Feature ranking [26, 9] and feature clustering

[27, 28] are feature selection techniques. Mitra *et al.* [29] propose an unsupervised feature selection algorithm based on similarity between features.

In this chapter, we used a hybrid technique to select important bands. Here, we used CSA to identify the best band combinations based on the fuzzy k-nearest neighbors (KNN) technique, incorporating the features extracted by 2D PCA on each class separately. Classification accuracy of different band combinations selects the best set of bands.

The organization of this chapter is as follows: Section 11.2 revisits a brief description of different concepts of relevant techniques used in the proposed work. The proposed methodology and underlying algorithms are described in Section 11.3. Section 11.4 presents a brief description of data sets used in the experiment as well as the experimental values and result analysis. Section 11.5 contains conclusive remarks.

11.2 Relevant Concept Revisit

In this section, a brief description of feature selection using 2D PCA, CSA, and fuzzy KNN is provided.

11.2.1 Feature Extraction

The term *feature* is very important in image processing and analysis. It can be defined as a function of some quantifiable properties of an image. Feature extraction and selection are mechanisms to highlight some important and significant characteristics of an object. There are many feature extraction techniques. Features can be broadly classified into two groups: application specific and application independent. Application-specific features are like human faces, fingerprints, veins and some other conceptual features. Application-independent features can be divided into: local features (features calculated on image segmentation, clustering, and edge identification), global features (features calculated on the entire image or some prespecified regular sub-image), and pixel-level features (features calculated on texture, color, and pixel coordinates). Another form of classification is low level versus high level. Low-level features are directly related with the original image. Some feature extraction techniques are directly used on the original image to extract low-level features. Low-level feature extraction techniques include the Canny, Sobel, and Prewitt edge detection techniques. High-level features are basically based on these low-level features, and they concern finding shapes in an image object.

11.2.2 Feature Selection Using 2D PCA

PCA is one of the popular feature extraction techniques. It is used as both a classical feature extraction and data representation [30] technique. It determines the covariance structure of a set of variables. Basically, it allows us to identify the principal directions along which the data vary. In dimensionality reduction, one of the well-known methods is PCA. The first N projections consist of the principal component (i.e., they contain the most variance in an N-dimensional subspace). Hence, PCA maps N-dimensional data to 1D subspace. However, the main difficulty related to PCA is that it ignores spatial information (i.e., it considers only the spectral information of the image for reduction).

According to Wiskott *et al.* [31], if the training data set does not provide explicit information regarding invariance, then PCA could not capture even the simplest invariance.

Here, the 2D image is converted to the 1D vector, and then the covariance matrix is calculated. But, according to Wiskott *et al.* [31], PCA cannot capture any small invariance in the training data unless it is explicitly provided. To overcome this problem, Yang *et al.* [30] propose an updated version named 2D PCA. It is operated on 2D matrices rather than 1D vectors. An image covariance matrix can be directly constructed from the original image data.

Let I be the digital image of size $m \times n$ and Y be the n-dimensional unitary column vector. Using the linear transformation given here:

$$X = IV \tag{11.1}$$

we can project image I onto V. A covariance matrix can be calculated on the projected feature vector of the training sample and can be represented as:

$$C_i = E(X - EX)(X - EX)^T$$

or:

$$= E[(I - EI)V][(I - EI)V]^T \tag{11.2}$$

Now, the total scatter of the projected vector can be characterized by trace.

So the trace of C_i is:

$$t_r(C_i) = V^T(E(I - EI)^T(I - EI))V \tag{11.3}$$

Let the covariance matrix of the image be:

$$I_c = E[(I - EI)^T(I - EI)] \tag{11.4}$$

The discriminatory power of the projection vector V can be measured using the following formula:

$$t_r(C_i) = V^T I_c V \tag{11.5}$$

A set of optimal projection axes $V_1, V_2, \ldots, V_k$ can be obtained by maximizing equation (11.4) (i.e., the eigenvector of I_c). Actually, $V_1, V_2, \ldots, V_k$ are the orthonormal eigenvectors of I_c corresponding to the first k largest eigenvalues. For these first k values, we have a set of projected feature vectors $X_1, X_2, \ldots, X_k$. This feature vector set of size $m \times n$ represents the feature matrix of the original image I.

11.2.3 Immune Clonal System

An evolutionary search algorithm is a random search technique used in different optimization problems. Immune clonal systems are inspired by biological immune systems and belong to the evolutionary search strategy [32]. Nature is an important source of inspiration for researchers to develop various computational algorithms. The biological immune system is a robust technique that defends the body from foreign pathogens. The ultimate target of all immune response is to identify and destroy an antigen (Ag), which is usually a foreign molecule. Illness-causing microorganism and viruses are called pathogens. Living organisms like animals, plants, and so on are exposed to different pathogens. Pathogens are generally antigens. The immune system is capable of remembering each infection or antigen so that the next exposure to the same antigen can be more efficiently and effectively dealt with. The adaptive or acquired immune system consists of lymphocytes, mainly B- and T-cells. These cells are responsible for identifying or destroying any specific substances. like antigens. After stimulation, the immune system generates antibodies to protect the body from the foreign antigens. The basic mechanism of the biological immune system is a complex process. At first after detecting some antigen, antigen presenting cells (APCs) fragment into antigenic peptide and are joined with major histocompatibility complex (MHC) molecules. Then T-cells can recognize them as a different peptide–MHC combination and secrete some chemical signals. B-cells respond to this signal and separate the antigens from the MHC molecule. B-cells then secrete antibodies, and these antibodies can neutralize or destroy the antigens they found in the MHC molecule. Some B-cells and T-cells are memory cells to protect the body from any second attempt of the same antigens. Antibodies on the B-cell perform affinity maturation (i.e., mutation and editing) to improve their response against antigens. The biological immune system is briefly depicted in Figure 11.3, and a corresponding flowchart is given in Figure 11.4. An immune clonal system or AIS can be used as an unsupervised learning technique. It is useful since it can maintain diversity in population and uses a global search mechanism.

The step-by-step procedure of the CLONALG algorithm is described in Algorithm 11.1.

11.2.4 Fuzzy KNN

Classification of objects is an important part of much research work. KNN is a popular classification technique. One of the problems with this technique is that each sample of the sample space gives equal importance to identify the class memberships of any pattern or object, irrespective of their importance [34]. Thus, the fuzzy version of the KNN is introduced to overcome this shortcoming. The basic idea of the fuzzy KNN (FKNN) is to assign membership as a function of the object distance from its k-nearest neighbors. It is a sophisticated pattern recognition procedure. The procedure is as follows:

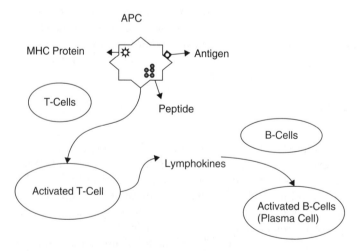

Figure 11.3 Basic biological immune system [33].

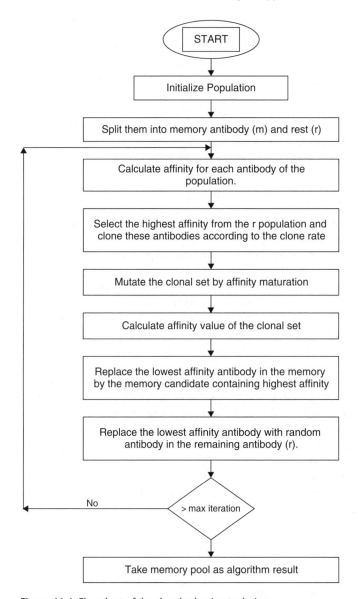

Figure 11.4 Flowchart of the clonal selection technique.

Let $V = v_1, v_2, \ldots, v_l$ be a set of data of length l. Each object v_i is defined by t characteristics $v_i = (v_{i1}, v_{i2}, \ldots, v_{it})$. Let y be the unclassified elements, N be the set of n-nearest neighbors, and $m_i(y)$ be the membership of y in class i. Let the data set contain c classes. Hence, the algorithm of FKNN is given in Algorithm 11.2.

Here, class membership can be calculated as:

$$m_i(y) = \sum_{j=1}^{n} m_{ij} * (1/\|y - v_i\|^{2/(m-1)}) / \sum_{j=1}^{n} (1/\|y - v_i\|^{2/(m-1)}) \tag{11.6}$$

Algorithm 11.1   Clonal selection algorithm

---

1. Initialization: In the CLONALG technique, the first step is initialization [i.e., preparation of an antibody population (P) of size N]. This population consists of two components, a memory part (m) that becomes the solution of the problem and a remaining part (r) used for different operations of the CLONALG technique.
2. Loop: Each iteration is termed a *generation*. The algorithm then proceeds by executing different generations to the antibodies of the predefined population. The number of generations or stopping criteria is problem specific.
   a. Select antibody: A single antibody is selected from the pool of antibody for the current generation.
   b. Affinity calculation: Affinity values are calculated for all antibodies. Affinity calculation is problem dependent.
   c. Selection: A set of n antibodies that have the highest affinity value are selected from the entire antibody pool.
   d. Cloning: The set of selected antibodies is then cloned in proportion to their affinity value (highest affinity = more cloning).
   e. Mutation: In affinity maturation or mutation, there is modification of a bit string where the flipping of a bit is achieved by an affinity proportionate probability distribution. Here, the degree of maturation is inversely proportional to their parent's affinity (i.e., for the highest affinity, the mutation rate is lower).
   f. Clone exposure: Affinity value is calculated for clones, and they are ranked accordingly.
   g. Candidature: Among the clone antibody, the highest affinity value is selected as a candidate memory antibody. If the affinity value of the candidate memory antibody is greater than the affinity value of the antibody in the memory (m) pool, then it replaces the said antibody.
   h. Replacement: Finally, the d individuals in the remaining r antibody pool with the lowest affinity are replaced.
3. Finish: After reaching the stopping criteria, the m component of the antibody pool is taken as a solution. Solutions may be single or multiple based on the problem domain.

Algorithm 11.2   Fuzzy KNN algorithm

---

1. Initialize: The number of closest element n.
2. For $i = 1$:
3. Calculate the distance among y and v_i.
4. Check if $i <= k$, then add v_i to N; otherwise, check whether v_i is closer to y than any previously recorded nearest neighbor, and, if true, then delete the farthest neighbor and include v_i in the set N.
5. For each class c, calculate the membership value given in equation (11.6).
6. Repeat steps 2 to 5 until $i = t$.

11.3   Proposed Algorithm

In hyperspectral images, selecting the band combination that gives the best result is the most challenging task. Among a set of B bands, one has to select b number of bands (B). The proposed semisupervised band selection method can be divided into the following steps: Extract features of each class of the image using 2D PCA, perform the artificial clonal selection algorithm on a randomly chosen population, identify the most promising bands using classification accuracy, and then rank the bands and obtain the results. In the proposed band selection technique, at first, the 2D PCA technique was used to extract features from the hyperspectral image cube. A ground truth table is used to obtain the detailed information about each class. Using 2D PCA, we can extract features of each class separately. In the second step (i.e., the clonal selection step), a pool of antibody population was initialized. Algorithm 11.3 clearly depicts the basic mechanism of the proposed technique. Figure 11.5 represents the flowchart of the proposed algorithm. From this pool of antibody, n number of antibody is selected for m-cells and the rest in r-cells. Classification accuracy for each class is performed by the FKNN technique. This accuracy value is considered as an affinity value of the CSA, and antibodies with the highest accuracy are selected for cloning and mutation. In two successive iterations, no change in m-cells indicates the result of the algorithm.

Algorithm 11.3   Algorithm of the Proposed technique

---

1. Read the hyperspectral image of P number of different bands and C number of classes.
2. Extract features of each class using 2D PCA.
3. Initialize the antibody population of size N as given in the CSA.
4. Calculate the classification accuracy of each antibody by fuzzy KNN, and best sets are separately stored in m-cells.
5. Antibodies in m-cells are selected for clones and perform mutations on the newly generated clone set.
6. Calculate the affinity of the newly generated antibodies.
7. Antibodies with the highest affinity value are selected for m-cells.
8. Steps 4 to 7 are repeated until stopping criteria are met.

The advantages and disadvantages of different feature selection and classification techniques of hyperspectral images are unevenly distributed. Hence, a combined or hybrid approach may often provide the best performance.

11.4   Experiment and Result

Performance of the proposed method can be evaluated by performing several experiments on the Indian Pines data set.

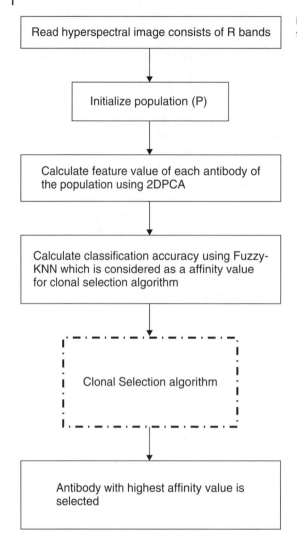

Figure 11.5 Flowchart of the proposed technique.

11.4.1 Description of the Data Set

Indian Pines data set: The most commonly used data set is the Indian Pines data set obtained by the AVIRIS. In 1992, it was instrumented over the agricultural area of northwestern Indiana's Indian pine. The spatial dimension of Indian pine is 145 × 145, and it has 224 spectral bands. The data has been captured within the spectral range of 400 to 2500 nm, and the spatial and spectral resolution are 20 m per pixel and 10 nm respectively. Irrelevant bands like four zero bands and the 35 lower SNR bands affected by atmospheric absorption are discarded. Among the lower SNR bands, water absorption bands (104–108, 150–163, 220) and noisy bands (1–3, 103, 109–112, 148–149, 159, 164–165, 217–219) are there. The remaining 185 bands are preserved. The data set contains 16 classes and also 10,366 labeled pixels, as depicted in Table 11.1. The ground truth map as represented in Figure 11.6 contains 16 mutually exclusive land-cover classes.

Table 11.1 Indian Pines data set: classes with number of samples

Class	Land type	No. of samples
C1	Alfalfa	52
C2	Corn	224
C3	Buildings-grass-trees-drives	380
C4	Grass-pasture-mowed	20
C5	Corn-min till	734
C6	Corn-no till	1234
C7	Hay-windrowed	486
C8	Grass/pasture	495
C9	Grass/trees	746
C10	Soybean-min till	2408
C11	Oats	19
C12	Soybean-no till	898
C13	Soybean-clean till	610
C14	Woods	1290
C15	Stone-steel-towers	90
C16	Wheat	210

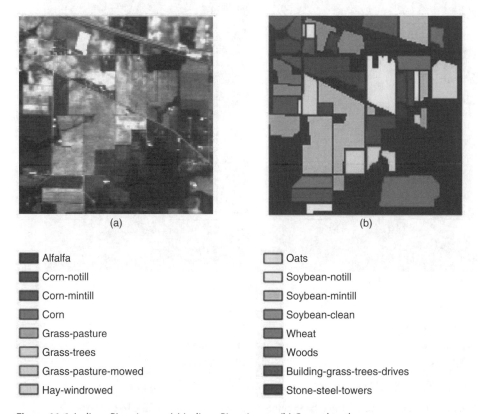

(a)

(b)

◼ Alfalfa

◼ Corn-notill

◼ Corn-mintill

◻ Corn

◻ Grass-pasture

◻ Grass-trees

◻ Grass-pasture-mowed

◻ Hay-windrowed

◻ Oats

◻ Soybean-notill

◻ Soybean-mintill

◻ Soybean-clean

◼ Wheat

◼ Woods

◼ Building-grass-trees-drives

◼ Stone-steel-towers

Figure 11.6 Indiana Pines image. (a) Indiana Pines image. (b) Ground truth.

Pavia Center data set: The Pavia data set was acquired by the Reflective Optics Spectrographic Imaging System 03 (ROSIS-03) sensor. It was instrumented over the center of Pavia, Italy. The spatial dimension of Pavia is 1096 × 1096 pixels, and it has 115 spectral bands. Removing a 381-pixel-wide black stripe in the left part of the image resulted in an image with 1,096,715 pixels. Irrelevant bands like the four lower SNR bands affected by atmospheric absorption are discarded. Among the lower SNR bands, 13 noisy bands are there. The remaining 102 bands are preserved. The data set contains nine classes. The ground truth map as represented in Figure 11.7 contains nine land-cover classes. Table 11.2 lists the land-cover classes with the corresponding number of samples.

11.4.2 Experimental Details

Different experiments are conducted on the Indian Pines data set, as described in the next section. Details of the data set are given in the previous section. A 2D PCA

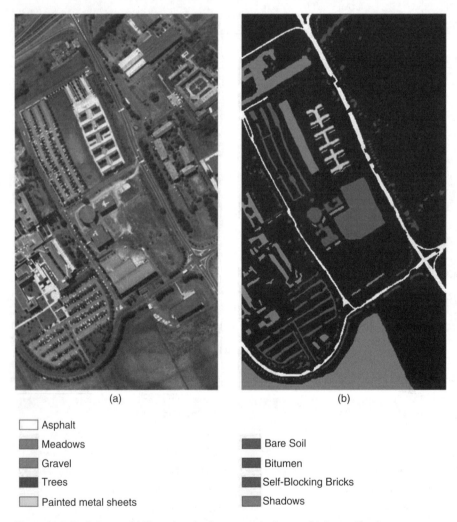

(a)   (b)

☐ Asphalt

▨ Meadows   ▨ Bare Soil

▨ Gravel   ▨ Bitumen

▨ Trees   ▨ Self-Blocking Bricks

☐ Painted metal sheets   ▨ Shadows

Figure 11.7 Pavia image. (a) Three-band color composite image. (b) Ground truth.

Table 11.2 Pavia data set: classes with number of samples

Class	Land type	No. of samples
C1	Water	65,971
C2	Trees	7598
C3	Asphalt	3090
C4	Self-blocking bricks	2685
C5	Bitumen	6584
C6	Tiles	9248
C7	Shadows	7287
C8	Meadows	42,826
C9	Bare soil	2863

technique was used to extract features from each band of the original image. The total number of the antibody population is N, and the total number of selected antibodies in each population is $n(n = 5$ to $20)$. We apply 2D PCA on the selected antibody and calculate the classification accuracy using FKNN. This value becomes the affinity value. Now, based on these affinity values, select the antibody with the highest value. Perform the clonal selection technique to find the best antibody. The result of the proposed antibody is reported in Table 11.3 and 11.4. In our experiment, we varied the number of selected bands from 5 to 30, and overall accuracy is reported in Figure 11.8. However, the optimum number of selected bands will be addressed by the future scope of our work. Comparisons of different methods with the proposed method for different data sets are represented in Figure 11.9 and 11.10, which depict the overall accuracy (in percentage) with the number of selected bands. Experimental results show that the proposed method provides convincing results in terms of accuracy for different numbers of selected bands. Table 11.4 shows the list of selected bands with different k values using our proposed method.

11.4.3 Analysis of Results

The performance of the proposed method for band selection is compared with other well-known methods. Among them, the paper presented in [35] is a competitive

Table 11.3 Selected bands for Indian Pines data set obtained by the proposed method

Band number	Selected band
$k = 5$	16, 42, 71, 138, 181
$k = 10$	4, 9, 19, 54, 55, 102, 138, 143, 172, 184
$k = 15$	8, 17, 19, 30, 31, 44, 58, 69, 85, 100, 124, 127, 128, 135, 171
$k = 20$	4, 8, 16, 18, 24, 31, 32, 35, 44, 46, 67, 80, 82, 87, 117, 124, 127, 135, 171, 184

Table 11.4 Accuracy of classification using the algorithm

Band number	5	7	9	11	13	15	17	19	20
Overall accuracy	83.44	83.81	84.23	84.66	85.21	85.65	86.23	86.12	86.03

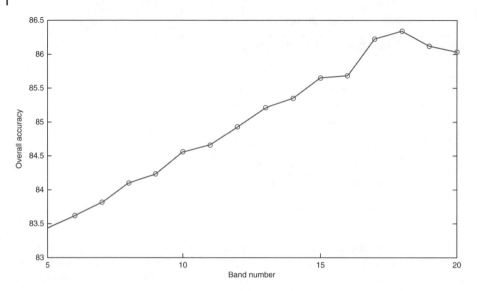

Figure 11.8 Overall accuracy of the proposed algorithm on the Indiana data set.

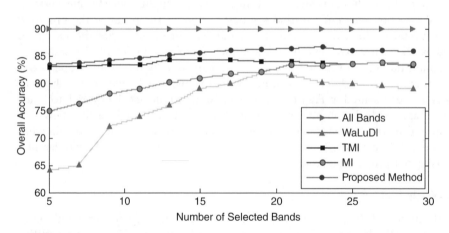

Figure 11.9 Comparison of the proposed method with MI, WaLuDi, and TMI in terms of overall accuracy for the Indiana data set.

algorithm based on the mutual information (MI) between a band and the reference image of the data set. More MI means more information content in the band with respect to the reference map. Thus, redundancy can be minimized. The next method is a semisupervised method [17]. Trivariate MI (TMI) is used to measure the trivariate correlation among two bands and the reference image. Semisupervised TMI (STMI) is also used, as desired prior information about classes in a hyperspectral image is not available sometimes. A CSA-based search strategy is used to optimize the result. Another technique is feature clustering [27]. The wards linkage strategy using divergence (WaLuDi) is a feature-clustering technique where the finalized cluster is further used for band selection. The optimal number of selected bands is difficult to identify as it varies from image to image. In the present chapter, the comparisons carried

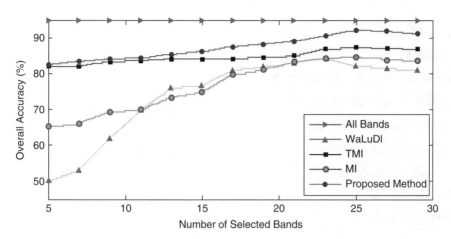

Figure 11.10 Comparison of the proposed method with MI, WaLuDi, and TMI in terms of overall accuracy for the Pavia Center data set.

out for different numbers of selected bands for each method are between $n(n = 5$ to 30). Figure 11.9 and 11.10 depict the comparison results of the above-mentioned method with our proposed method. The graph shows the average overall accuracy (in percentage) versus the number of bands. Experimental results justify the effectiveness of the proposed technique. The proposed method (i.e., the fuzzy KNN classifier) is compared with another well-known classifier named the support vector machine (SVM) [36]. SVMs are based on linear discriminant functions. Thus, the classification is based on a function of the form $w^t X + b$, where w and b are learned from training data. For a linearly separable two-class problem, the positive class can be characterized by a hyperplane $w^t X + b = 1$, and the negative class can be characterized by $w^t X + b = -1$. The decision boundary for the separating plane is characterized by $w^t X + b = 0$. Figure 11.11 depicts this idea clearly. The graph presented in Figure 11.12 shows the result obtained using the proposed method with a SVM. The x-axis represents the overall accuracy of the different classifier obtained by the Indiana image, and the y-axis represents a different number of bands. The proposed method is also compared

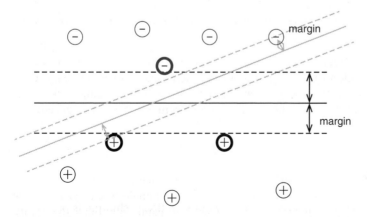

Figure 11.11 Basics of a SVM.

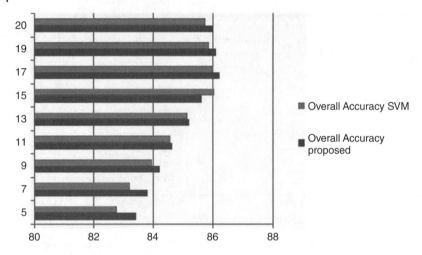

Figure 11.12 Comparison of the proposed method (2D PCA and fuzzy KNN) with 2D PCA and SVM in terms of overall accuracy for the Indiana data set.

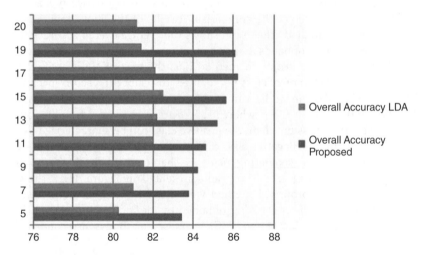

Figure 11.13 Comparison of the proposed method (2D PCA) and LDA in terms of overall accuracy for the Indiana data set.

with another feature extraction technique, LDA. Fisher's linear discriminant is used in supervised classification. It is used to map n-dimensional data to 1D data using a projection vector. Classification is then carried out in this 1D data. The graph represented in Figure 11.13 shows the result obtained from a 2D PCA (proposed) and LDA classifier.

11.5 Conclusion

In this chapter, a novel semisupervised hybrid band selection technique is presented. This hybrid technique is a combination of 2D PCA and clonal selection algorithm (CSA).

Basically, to remove redundant bands, this technique calculates the feature information on each band and selects the most informative bands using a CSA. The CSA uses the fuzzy KNN technique to identify the most informative bands. The overall accuracy measure has been incorporated to measure the performance of the proposed method. This method shows a convincing result in comparison with other well-known methods like MI, WaLuDi, and TMI for different data sets of hyperspectral images, namely, Indiana and Pavia Center. A hybrid combination of 2D PCA with fuzzy KNN gives slightly better performance than a 2D PCA and SVM combination. Comparison of the proposed method with the Bayesian belief network and hidden Markov model (HMM) in place of fuzzy KNN is left for future investigation. In the proposed algorithm, the number of selected bands varies from 5 to 30, but the number of selected bands for optimum results is left for future research to determine.

References

1 Ayadi, M.F. and Kamel, M.S. (2011) Survey on speech emotion recognition features,classification schemes and databases. *Pattern Recognition*, **44** (3), 572–587.
2 Hazra, J. and Roy Chowdhury, A (2013) An approach for determining angle of rotation of a gray image using weighted statistical regression. *International Journal of Scientific and Engineering Research*, **4** (8), 1006–1013.
3 Hazra, J. and Roy Chowdhury, A. (2014) Statistical regression based rotation estimation technique of color image. *International Journal of Computer Applications*, **102** (15), 1–4.
4 Hazra, J. and Roy Chowdhury, A. (2015) Transformation parameter estimation using parallel output based neural network. *Applied Soft Computing*, **46**, 868–874.
5 Gonzalez, R.E. and Woods, R.F. (2002) Digital image processing, 2nd ed. Prentice Hall, Upper Saddle River, NJ.
6 Aviris–Airborne Visible/Infrared Imaging Spectrometer. Available from http://aviris.jpl.nasa.gov.
7 Green, A. A., Berman, M. and Craig, M.D. (Jan. 1988) A transformation for ordering multispectral data in terms of image quality with implications for noise removal. *IEEE Trans. Geosci. and Remote Sens.*, **26** (1), 65–73.
8 Lee, J. B. and Berman, M. (May 1990) Enhancement of high spectral resolution remote-sensing data by a noise-adjusted principal components transform. *IEEE Trans. Geosci. and Remote Sens*, **28** (3), 295–304.
9 Chang, C.I, Du, T.L.S. and Althouse, M.L.G. (Jun. 1999) A joint band prioritization and band decorrelation approach to band selection for hyperspectral image classification. *IEEE Trans. Geosci. and Remote Sens.*, **37** (6), 2631–2641.
10 Harsanyi, J.C. and Chang, C.I. (Jul. 1994) Hyperspectral image classification and dimensionality reduction: an orthogonal subspace projection approach. *IEEE Trans. Geosci. and Remote Sens*, **32** (4), 779–785.
11 De Backer, S., Kempeneers, P. and Scheunders, P. (Jul. 2005) A band selection technique for spectral classification. *IEEE Geosci. and Remote Sens. Lett.*, **2** (3), 319–323.
12 Kumar, S. and Crawford, M.M. (Jul. 2001) Best-bases feature extraction algorithms for classification of hyperspectral data. *IEEE Trans. Geosci. and Remote Sens.*, **39** (7), 1368–1379.

13 Lashkia, G. and Anthony, L. (Apr. 2004) Relevant, irredundant feature selection and noisy example elimination. *IEEE Trans. Syst., Man, Cybern. B, Cybern.,* **34** (2), 888–897.

14 Shannon, E. (Jul.-Oct. 1948) A mathematical theory of communication. *Bell Syst. Tech. J.,* **27** (3), 379–423.

15 Huang, R. and He, M. (Apr. 2005) Band selection based on feature weighting for classification of hyperspectral data. *IEEE Geosci. Remote Sens. Lett.,* **2** (2), 156–159.

16 Zhang, L. and Zhong, Y. (Dec. 2007) Dimensionality reduction based on clonal selection for hyperspectral imagery. *IEEE Geosci. Remote Sens. Lett.,* **45** (12), 4172–4186.

17 Jie Feng, L.C., Jiao, X.Z. and Sun, T. (July 2014) Hyperspectral band selection based on trivariate mutual information and clonal selection. *IEEE Trans. Geosci. and Remote Sens.,* **52** (7), 4092–4105.

18 Bajcsy, P. and Groves, P. (2004) Methodology for hyperspectral band selection. *Photogrammetric Engineering and Remote Sensing,* **70**, 793–802.

19 Tan, K. and Du, P. (2011) Combined multi-kernel support vector machine and wavelet analysis for hyperspectral remote sensing image classification. *Chinese Optics Letters,* **9** (1), 011 003–110 006.

20 Du, Q. and Yang, H. (2008) Similarity-based unsupervised band selection for hyperspectral image analysis. *IEEE Geosci. Remote Sens. Lett.,* **5** (4), 564–568.

21 Yang, H. and Chen, G. (2011) Unsupervised hyperspectral band selection using graphics processing units. *IEEE J. Sel. Topics Appl. Earth Observ. Remote Sens.,* **4** (3), 660–668.

22 Cariou, C. and Chehdi, K. (2011) Bandclust: An unsupervised band reduction method for hyperspectral remote sensing. *IEEE Geosci. Remote Sens. Lett.,* **8** (3), 565–569.

23 Lei, L., Prasad, S. and Bruce, L.M. (Apr. 2012) Localitypreserving dimensionality reduction and classification for hyperspectral image analysis. *IEEE Trans. Geosci. Remote Sens.,* **50** (4), 1185–1198.

24 Agarwal, A., El-Ghazawi, T. and Le-Moigne, J. (2007) Efficient hierarchical-PCA dimension reduction for hyperspectral imagery. *Proc. IEEE Int. Symp. Signal Process. Inf. Technol.,* 353–356.

25 Roweis, S.T. and Saul, L.K. (Dec. 2000) Nonlinear dimensionality reduction by locally linear embedding. *Science,* **290** (22), 2323–2326.

26 Chang, C.I. and Wang, S. (Jun. 2006) Constrained band selection for hyperspectral imagery. *IEEE Trans. Geosci. Remote Sens.,* **44** (6), 1575–1585.

27 Martínez-Usó, A., Pla, F. and García-Sevilla, P. (Dec. 2007) Clustering based hyperspectral band selection using information measures. *IEEE Trans. Geosci. Remote Sens.,* **45** (12), 4158–4171.

28 Martínez-Usó, A., Pla, F. and García-Sevilla, P. (Apr. 2012) Clustering based hyperspectral band selection using information measures. *IEEE Trans. Geosci. Remote Sens.,* **5** (2), 531–543.

29 Mitra, P. and Pal, S. (2002) Unsupervised feature selection using feature similarity. *IEEE Transactions on Pattern Analysis and Machine Intelligence,* **24** (3), 301–312.

30 Yang, J., Zhang, D. and Yu Yang, J. (2004) Two-dimensional PCA: a new approach to appearance-based face representation and recognition. *IEEE Transaction on Pattern Analysis and Machine Intelligence,* **26** (1), 131–137.

31 Wiskott, L., Fellous, J.M. and von der Malsburg, C. (1997) Face recognition by elastic bunch graph matching. *IEEE Transaction on Pattern Analysis and Machine Intelligence*, **19** (7), 775–779.

32 D. Castro, L.N. and von Zuben, F.J. (2000) The clonal selection algorithm with engineering application. *GECCO'00-Workshop Proceedings*, pp. 36–37.

33 Hazra, J. and Roy Chowdhury, A. (Nov. 2015) Immune based feature selection in rigid medical image registration using supervised neural network. *Handbook of Research on Advanced Hybrid Intelligent Techniques and Applications*, pp. 551–581.

34 Keller, J. and Gray, M. (1985) A fuzzy k-nearest neighbor algorithm. *IEEE Transaction on System, Man, Cybernetics*, **15** (4), 580–585.

35 Guo, B. and Gunn, R.D.J.N. (2006) Band selection for hyperspectral image classification using mutual information. *IEEE Geoscience and Remote Sensing Letters*, **3** (4), 522–526.

36 Kolekar, M.H. and Dash, D.P. (2015) A nonlinear feature based epileptic seizure detection using least square support vector machine classifier. *IEEE Region 10 Conference TENCON*.

12

Uncertainty-Based Clustering Algorithms for Medical Image Analysis

Deepthi P. Hudedagaddi and B.K. Tripathy

School of Computer Science and Engineering (SCOPE), VIT University, Vellore, Tamil Nadu, India

12.1   Introduction

Image segmentation frames the basic and crucial step in pattern recognition and analysis of the image. It includes segregating the image in accordance with a few characteristics like intensity and texture [1]. In recent years, the fuzzy c-means (FCM) clustering algorithm and its variations have been extensively used [2, 3]. In comparison with the hard c-means algorithm, FCM has been proved to give better and optimum results. Procuring several segmentation results of the image helps in analyzing an image in several perspectives. In this regard, techniques of parallel processing have to be exploited.

Images can be defined as 2D coordinate representations of pixels. Image processing is a technique used in the early part of the 20th century. The study of image segmentation plays a vital role in image analysis and has attracted several researchers. Imprecision in data occurs due to various reasons. One has to pay a heavy cost if he or she tries to remove imprecision before any kind of processing. This is true for any kind of data, and image data is no exception. It has been established of late that hybrid techniques are more efficient than the individual ones. So, while considering segmentation, we focus on hybrid techniques obtained from uncertainty-based models. Several techniques, like Otsu threshold and performance metrics such as statistical methods, PSNR and RMSE, and the David Bouldin (DB) index and Dunn (D) index, are used for measuring the efficiency of different image segmentation algorithms.

12.2   Uncertainty-Based Clustering Algorithms

Taking Zadeh's fuzzy sets and Pawlak's rough sets [4], Attanassov [5] came up with intuitionistic fuzzy sets; Dubois and Prade came up with rough fuzzy sets. Accordingly, clustering algorithms have found an evolution in this direction. FCM [6], rough c-means (RCM) [7, 8], hybrid variations like rough FCM (RFCM) [9–11], and intuitionistic FCM (IFCM) [5] have been developed. From literature, it can be observed that hybrid algorithms that deal with vagueness and uncertainty have better performance when compared with several indices [12].

Hybrid Intelligence for Image Analysis and Understanding, First Edition.
Edited by Siddhartha Bhattacharyya, Indrajit Pan, Anirban Mukherjee, and Paramartha Dutta.
© 2017 John Wiley & Sons Ltd. Published 2017 by John Wiley & Sons Ltd.
Companion Website: www.wiley.com/go/bhattacharyya/hybridintelligence

Segmenting images into regions is a clustering or classifying process. It simplifies image representation and makes it more understandable for analysis. The process can be conventional techniques (crisp) and uncertainty-based models (soft). Some of the conventional techniques used are Canny, Sobel, Prewitt, Robert, and hard c-means, where vagueness and uncertainty are not distinguished, and which have been well considered by uncertainty models in the artificial intelligence domain.

As vagueness and uncertainty concepts are the most suitable that have not been distinguished properly, they need to be controlled for image analysis.

Cluster analysis is a major aspect in data mining. It has found applications in fields such as web mining, biology, image processing, and market segmentation. Clustering involves segregating samples in a way that the samples are extremely similar to one another in the same cluster and unidentical to ones in the other.

C-means is a traditional crisp clustering algorithm where every object belongs to a single cluster, thereby providing a quick rate of convergence. However, it fails in handling overlapping. Since most real-world data is uncertain and has boundaries overlapping, it paved the way for developing uncertainty-based clustering algorithms. These have been successful in meeting the needs of a particular scenario by being open to minor modifications.

12.2.1 Fuzzy C-Means

FCM was introduced by James C. Bezdek. Fuzzy, or soft, clustering allows an object to belong to more than one cluster. These come with a set of membership values that provide an insight to the level of membership to a particular cluster.

1. Allocate initial cluster centers.
2. Compute Euclidean distance d_{ik} from data elements x_k and centroids v_i:

$$d(x, y) = \sqrt{(x_1 - y_1)^2 + (x_2 - y_2)^2 + \cdots + (x_n - y_n)^2} \tag{12.1}$$

3. Generate membership matrix U:
   If $d_{ij} > 0$, then

$$\mu_{ik} = \frac{1}{\sum_{j=1}^{C} \left(\frac{d_{ik}}{d_{jk}} \right)^{\frac{2}{m-1}}} \tag{12.2}$$

   Else

$$\mu_{ik} = 1$$

4. Centroids are computed with

$$V_i = \frac{\sum_{j=1}^{N} (\mu_{ij})^m x_j}{\sum_{j=1}^{N} (\mu_{ij})^m} \tag{12.3}$$

5. Compute new U using steps 2 and 3.
6. If $\|U^{(r)} - U^{(r+1)}\| < \epsilon$, then stop. If not, redo from step 4.

12.2.2 Rough Fuzzy C-Means

RFCM [13] was developed by S. Mitra and P. Maji; with the combination of fuzzy and rough set concepts. The property of rough sets to deal with uncertainty, incompleteness, and vagueness along with a fuzzy set concept of membership that evaluates overlapping clusters have been used in RFCM.

1. Assign v_i for c clusters.
2. Calculate μ_{ik} using equation (12.2).
3. Let μ_{ik} be maximum and μ_{jk} be next-to-maximum membership values of object x_k.

   If $\mu_{ik} - \mu_{jk} < \epsilon$, then
   $x_k \in \overline{BU}_i$ and $x_k \in \overline{BU}_j$, and x_k shall not belong to lower approximation.
   Else

   $\qquad x_k \in \underline{BU}_i$

4. Compute new cluster means with:

$$
V_i = \begin{cases}
w_{low} \dfrac{\sum_{x_k \in \underline{BU}_i} x_k}{|\underline{BU}_i|} + w_{up} \dfrac{\sum_{x_k \in \overline{BU}_i - \underline{BU}_i} \mu_{ik}{}^m x_k}{\sum_{x_k \in \overline{BU}_i - \underline{BU}_i} \mu_{ik}{}^m} \\
\qquad if\, |\underline{BU}_i| \neq \emptyset\ and\ |\overline{BU}_i - \underline{BU}_i| \neq \emptyset \\[2ex]
\dfrac{\sum_{x_k \in \overline{BU}_i - \underline{BU}_i} \mu_{ik}{}^m x_k}{\sum_{x_k \in \overline{BU}_i - \underline{BU}_i} \mu_{ik}{}^m} \\
\qquad if\, |\underline{BU}_i| = \emptyset\ and\ |\overline{BU}_i - \underline{BU}_i| \neq \emptyset \\[2ex]
\dfrac{\sum_{x_k \in \underline{BU}_i} x_k}{|\underline{BU}_i|} \qquad\qquad ELSE
\end{cases}
\tag{12.4}
$$

5. Redo from step 2 till termination criteria are obtained.

   Note: Range for values of *m* is [1.5, 2.5]. However, for all practical purposes, it is taken to be 2.

12.2.3 Intuitionistic Fuzzy C-Means

The IFCM proposed by T. Chaira includes a component called hesitation value (π), which increases clustering accuracy.

1. Initial cluster centers are assigned.
2. Compute Euclidean distance d_{ik} between data objects x_k and centroids v_i with equation (12.1).
3. Generate membership matrix U:

   If $d_{ij} > 0$, then compute μ_{ik} using equation (12.2)
   Else

   $\qquad \mu_{ik} = 1$

4. Compute the hesitation matrix π.

5. Compute the modified matrix U′ with

$$\mu'_{ik} = \mu_{ik} + \pi_{ik} \tag{12.5}$$

6. Centroids are computed using

$$V_i = \frac{\sum_{j=1}^{N} (\mu'_{ij})^m x_j}{\sum_{j=1}^{N} (\mu'_{ij})^m} \tag{12.6}$$

7. Calculate new U using steps 2 to 5.
8. If $\|U'^{(r)} - U'^{(r+1)}\| < \epsilon$, then stop. If not, redo from step 4.

12.2.4 Rough Intuitionistic Fuzzy C-Means

1. Initial c clusters are assigned [14].
2. Calculate Euclidean distance d_{ik} using equation (12.1).
3. Compute U matrix.
   If $d_{ik} = 0$ *or* $x_j \in \underline{B}U_i$, then

   $$\mu_{ik} = 1$$

   Else compute μ_{ik} using equation (12.2)
4. Compute π_{ik}.
5. Compute μ'_{ik}.

   $$\mu'_{ik} = \mu_{ik} + \pi_{ik}$$

6. Let μ'_{ik} be maximum and μ'_{jk} be next-to-maximum membership values of object x_k.
   If $\mu'_{ik} - \mu'_{jk} < \epsilon$, then

   $x_k \in \overline{B}U_i$ and $x_k \in \overline{B}U_j$ and x_k shall not be member of any lower approximation.
   Else

   $$x_k \in \underline{B}U_i$$

7. Compute new cluster means with Equation (12.4).

12.3 Image Processing

Image processing involves performing operations on an image by converting it into digital format. This is a rapidly growing field finding its applications in various aspects of diagnosis of disease, weather forecasting, classification of areas, and so on. It has been a core research area and has witnessed large-scale developments in varied fields. It is characterized by three important steps: obtaining the image, image manipulation, and recognizing patterns.

The medical field is facing challenges in diagnosis of diseases. It has become a difficult and time-consuming process. Image-processing techniques have been applied to preprocess any digital image, and soft computing methods are applied to cluster images to distinguish between vagueness and uncertainty that exist in real-time environments. Thereby, investigations are carried out to process digital images for enhancement of quality, which is segmented based on uncertainty algorithms for reconstruction of images. In turn, the reconstructed image is visualized in 3D map generation for interpretation of images for various purposes of medical and satellite applications with minimum time complexity and improvised performance.

12.4   Medical Image Analysis with Uncertainty-Based Clustering Algorithms

Several clustering algorithms presume distinction among clusters such that one pattern belongs to one cluster. With the numerous medical images available, handling and analyzing them in computational ways have become the need of the hour. Image segmentation algorithms have a major role to play in biomedical applications like tissue volume quantifying, diagnosis, anatomy, and computer-integrated surgeries.

With the advancement and advantages of magnetic resonance imaging (MRI) over other diagnostic imaging, it has been a practice of several researchers to use them for their study [15–18]. Of these, fuzzy techniques have better performance as they retain much information from the original image. FCM assigns pixels to clusters without labeling. But, due to the spatial intensity inhomogeneity induced by the radiofrequency coil in MR images, conventional FCM has not been efficient. The inhomogeneity issue of the images is easily dealt with by modeling the image and with the use of multiplier fields. Of late, researchers have begun using spatial information to obtain better segmentation results. Tolias and Panas [19] developed a rule-based fuzzy technique to impose spatial continuity, and in [20], a small constant was used to change membership of the cluster center. Pham *et al.* [21] made changes to the objective function of FCM by introducing a multiplier field. In the same manner, Ahmed *et al.* developed an algorithm to label a pixel by considering its immediate neighbor. Of late, Pham made FCM's objective function to constrain the behavior of the membership functions.

Several clustering algorithms like FCM, RCM, IFCM, RFCM, RIFCM, spatial FCM, and spatial IFCM [12–14, 22] have been developed to date, which have been further extended to both kernel-based and possibilistic versions such that their efficiency is increased and they can be applied to a greater number of applications. FCM with respect to adaptive clustering has already been developed. Though FCM is powerful in several ways, it fails to determine the right number of clusters for pattern classification. It needs the user to specify the number of clusters.

12.4.1   FCM with Spatial Information for Image Segmentation

Chuang *et al.* [23] introduced a FCM segmentation technique that reduced noise effect and made the clustering more homogeneous. Usually, in conventional FCM, a noisy pixel is misclassified. They incorporated spatial information that recomputed membership value, and hence the degree of a pixel belonging to a cluster was altered.

They used the concept of correlation among neighboring pixels. This means the probability that neighboring pixels belong to the same cluster is great. This concept has not been used in standard FCM algorithms. A spatial function is defined in equation (12.1).

The clustering process is a two-step process. The initial step is similar to FCM in calculating membership function. The second step includes the membership value of each pixel computed from spatial domain and spatial function. FCM continues with new values that are used with spatial function. When the difference between two cluster centers with two iterations is less than threshold, iteration can be stopped.

They evaluated their results based on cluster validity functions. Partition coefficient V_{pc} and partition entropy V_{pe} signify the fuzzy partitions. The partitions that have less fuzziness indicate better performance. Higher V_{pc} value and lesser V_{pe} value indicate better clustering.

$$V_{pc} = \frac{\sum_j^N \sum_i^c u_{ij}^2}{N} \tag{12.8}$$

and

$$V_{pe} = -\frac{\sum_j^N \sum_i^c [u_{ij} \log u_{ij}]}{N} \tag{12.9}$$

A good clustering result provides pixels that are compact within a single cluster and pixels that are different between different clusters. Minimizing V_{xb} leads to good clustering. Application of sFCM on MRI image is shown in Figure 12.1.

$$V_{xb} = \frac{-\sum_j^N \sum_i^c u_{ij}?x_j - v_i?^2}{N(\min_{i \neq k}\{\|v_k - v_i\|^2\})} \tag{12.10}$$

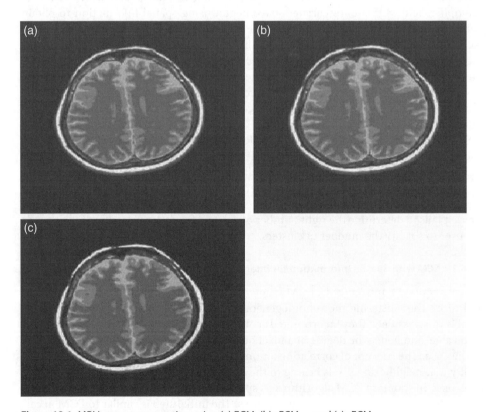

Figure 12.1 MRI image segmentation using (a) FCM, (b) sFCM$_{1,1}$, and (c) sFCM$_{1,2}$.

Figure 12.2 Segmentation results on (a) original image, (b) same image with mixed noise, results of (c) FCM_S1, (d) FCM_S2, (e) EnFCM, (f) FGFCM_S1, (g) FGFCM_S2, and (h) FGFCM.

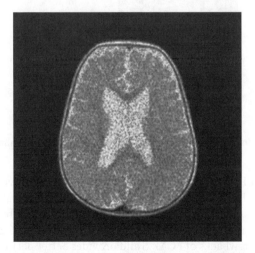

Figure 12.3 MRI image – speckle noise.

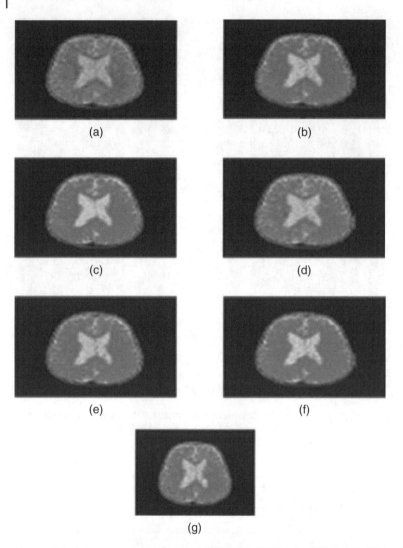

Figure 12.4 Noisy image segmentation. (a) FCM, (b) sFCM1,1, (c) sFCM$_{1,2}$, (d) sFCM$_{2,1}$, (e) sIFCM$_{1,1}$, (f) sIFCM$_{1,2}$, and (g) sIFCM$_{2,1}$.

12.4.2 Fast and Robust FCM Incorporating Local Information for Image Segmentation

Cai *et al.* [24] found FCM with spatial constraints (FCM_S) provide better results for image segmentation. They also found the disadvantages that come along, like: insensitivity to noise, robustness to noise, and time for segmentation. They have worked on using local spatial and gray information together to frame a fast generalized FCM (FGFCM). This overcomes the drawbacks of FCM_S and provides better performance with respect to time and segmentation. They have also introduced special cases such as FGFCM_S1 and FGFCM_S2. The characteristics of these are: (1) factor S_{ij} as both a spatial and gray similarity measure to guarantee noise immunity for images, and (2) fast clustering, that is, segmenting time is dependent on the number of gray levels q rather than the size $N(>q)$ of the image. Also, computational complexity is reduced from $O(NcI_1)$ to $O(qcI_2)$, where c indicates the number of clusters, and I_1 and I_2 indicate the number of iterations. The experiments of FGFCM on synthetic and real-world images in Figure 12.2 prove it to be effective and efficient [25, 26].

12.4.3 Image Segmentation Using Spatial IFCM

Tripathy *et al.* [27] developed the IFCM with spatial information (sIFCM).This was an extension to Chuang's work. The algorithm is provided in detail in [27].

They used DB and D indices to measure the cluster quality in addition to the evaluation metrics used by Chuang *et al.* [8, 9]. Speckle noise of mean 0 and variance 0.04 was induced in the image. FCM and sFCM were applied to the image. V_{pc} and V_{pe} are calculated.

The DB index is defined as the ratio of the sum of within-cluster distance to between-cluster distance.

$$DB = \frac{1}{c} \sum_{i=1}^{c} \max_{k \neq i} \left\{ \frac{S(v_i) + S(v_k)}{d(v_i, v_k)} \right\} \ for \ 1 < i, k < c \tag{12.11}$$

It aims to minimize within-cluster distance and maximize intercluster separation. Therefore, a good clustering procedure should have low DB value.

The D index is used for identification of clusters that are compact and separated.

$$Dunn = \min_i \left\{ \min_{k \neq i} \left\{ \frac{d(v_i, v_k)}{\max_l S(v_l)} \right\} \right\} \ for \ 1 < k, i, l < c \tag{12.12}$$

It aims at maximizing the intercluster distance and minimizing the intracluster distance. Hence, a higher D value proves to be more efficient.

A brain MRI image of 225×225 dimensions was used for proving their results. The number of clusters $c = 3$. Results of MRI imaging with speckle noise is shown in Figures 12.3 and 12.4.

The results are distinct in images with speckle noise. Conventional FCM misclassifies spurious blobs and spots. Increasing the parameter q, which is the degree of the spatial function, modifies the membership function to accommodate spatial information to a greater degree, and produces better results. The table 12.1 shows the performance of the different techniques applied on the noisy image.

Table 12.1 Cluster evaluation results on speckle noise image

Method	Results on the noisy image			
	V_{pc}	V_{pe}	DB	D
FCM	0.6975	2.8195×10^{-4}	0.4517	3.4183
sFCM$_{1,1}$	0.7101	5.9541×10^{-9}	0.4239	3.6734
sFCM$_{2,1}$	0.6922	7.7585×10^{-12}	0.4326	3.4607
sFCM$_{1,2}$	0.6874	4.2711×10^{-12}	0.4412	3.6144
sIFCM$_{1,1}$	0.7077	1.1515×10^{-08}	0.4254	3.6446
sIFCM$_{2,1}$	0.7135	8.1312×10^{-13}	0.4276	3.7472
sIFCM$_{1,2}$	0.713	4.6770×10^{-13}	0.4393	3.4968

Table 12.2 Performance indices of sFCM on leukemia image

Index	FCM	sFCM$_{2,1}$	sFCM$_{1,1}$	sFCM$_{1,2}$
V_{pc}	0.2118	0.2300	0.2219	0.4963
V_{pe}	0.0178	2.05×10^{-6}	0.0003	7.69×10^{-8}
V_{xb}	0.0168	0.0060	0.0074	0.0522
DB	0.4670	0.4185	0.4197	0.4173
D	2.2254	2.9951	2.9318	3.4340

Table 12.3 Performance indices of sIFCM on leukemia image

Index	FCM	IFCM	sIFCM$_{1,1}$	sIFCM$_{1,2}$	sIFCM$_{2,1}$
V_{pc}	0.2118	0.2074	0.4762	0.2241	0.2281
V_{pe}	0.0178	0.02585	3.56E-005	1.68E-005	5.90E-006
V_{xb}	0.0168	0.02096	0.05929	0.0069	0.0063
DB	0.467	0.4906	0.4126	0.4208	0.4188
D	2.2254	2.0529	3.5866	3.0203	2.9655

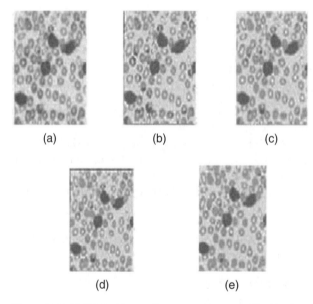

(a) (b) (c)

(d) (e)

Figure 12.5 (a) Original image. Segmented images of leukemia using (b) FCM, (c) sFCM$_{2,1}$, (d) sFCM$_{1,1}$, and (e) sFCM$_{1,2}$.

12.4.3.1 Applications of Spatial FCM and Spatial IFCM on Leukemia Images

Deepthi *et al.* [28, 29] have applied the spatial clustering algorithms to leukemia images and have found the following results.

The results in Table 12.2 show that the sFCM succeeds in providing better results than conventional FCM.

Figure 12.5 shows the segmented images with the application sFCM provide better clarity and understanding of the presence of leukemia cells than images with conventional FCM.

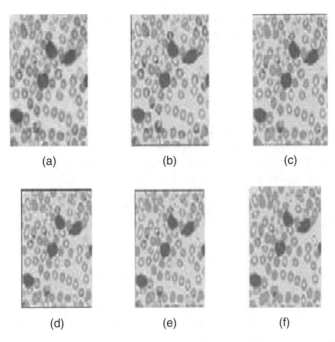

(a)   (b)   (c)

(d)   (e)   (f)

Figure 12.6 (a) Original image. Segmented images of leukemia using (b) FCM, (c) IFCM, (d) sFCM$_{2,1}$, (e) sFCM$_{1,1}$, and (f) sFCM$_{1,2}$.

The results in Table 12.3 show that the sIFCM succeeds in providing better results than conventional FCM.

The segmented images in Figure 12.6 with the application sIFCM provide better clarity and understanding of the presence of leukemia cells than images with conventional FCM and IFCM.

12.5 Conclusions

The field of uncertainty-based clustering has varied applications. One major application is image segmentation, which is a vital process in image analysis. These algorithms provide more realistic information and hence help in diagnosis of diseases. Different modifications of FCM and IFCM are applied on medical images. Similarly extensive work is happening in the application of rough and fuzzy-based hybrid algorithms for analyzing medical images.

References

1 Cheng, H.D., Jiang, X., Sun, Y., and Wang, J. (2001) Color image segmentation: advances and prospects. *Pattern Recognition*, **34** (12), 2259–2281.
2 Pham, D.L. and Prince, J.L. (1999) An adaptive fuzzy c-means algorithm for image segmentation in the presence of intensity inhomogeneities. *Pattern Recognition Letters*, **20** (1), 57–68.

3 Chen, W., Giger, M.L., and Bick, U. (2006) A fuzzy c-means (FCM)-based approach for computerized segmentation of breast lesions in dynamic contrast-enhanced MRimages 1. *Academic Radiology*, **13** (1), 63–72.

4 Pawlak, Z. (1982) Rough sets. *International Journal of Computer & Information Sciences*, **11** (5), 341–356.

5 Atanassov, K.T. (1986) Intuitionistic fuzzy sets. *Fuzzy Sets and Systems*, **20** (1), 87–96.

6 Maji, P. and Pal, S.K. (2007) RFCM: a hybrid clustering algorithm using rough and fuzzy sets. *Fundamenta Informaticae*, **80** (4), 475–496.

7 Lingras, P. and West, C. (2004) Interval set clustering of web users with rough k-means. *Journal of Intelligent Information Systems*, **23** (1), 5–16.

8 Maji, P. and Pal, S.K. (2007) Rough set based generalized fuzzy-means algorithm and quantitative indices. *IEEE Transactions on Systems, Man, and Cybernetics, Part B (Cybernetics)*, **37** (6), 1529–1540.

9 Michalopoulos, M., Dounias, G., Thomaidis, N., and Tselentis, G. (2002), Decision making using fuzzy c-means and inductive machine learning for managing bank branches performance. Paper presented at the European Symposium on Intelligent Techniques (ESIT'99), Crete, Greece, March.

10 Mitra, S., Banka, H., and Pedrycz, W. (2006) Rough&# 8211; fuzzy collaborative clustering. *IEEE Transactions on Systems, Man, and Cybernetics, Part B (Cybernetics)*, **36** (4), 795–805.

11 Tripathy, B., Tripathy, A., Govindarajulu, K., and Bhargav, R. (2014) On kernel based rough intuitionistic fuzzy c-means algorithm and a comparative analysis, in Advanced Computing, Networking and Informatics-Volume 1, Springer, pp. 349–359.

12 Tripathy, B., Tripathy, A., and Rajulu, K.G. (2014) Possibilistic rough fuzzy c-means algorithm in data clustering and image segmentation, in *Computational Intelligence and Computing Research (ICCIC), 2014 IEEE International Conference*, IEEE, pp. 1–6.

13 Bhargava, R. and Tripathy, B. (2013) Kernel based rough-fuzzy c-means, in International Conference on Pattern Recognition and Machine Intelligence, Springer, pp. 148–155.

14 Bhargava, R., Tripathy, B., Tripathy, A., Dhull, R., Verma, E., and Swarnalatha, P. (2013) Rough intuitionistic fuzzy c-means algorithm and a comparative analysis, in *Proceedings of the 6th ACM India Computing Convention*, ACM, p. 23.

15 Wells, W.M., Grimson, W.E.L., Kikinis, R., and Jolesz, F.A. (1996) Adaptive segmentation of MRI data. *IEEE Transactions on Medical Imaging*, **15** (4), 429–442.

16 Bezdek, J.C., Hall, L., and Clarke, L. (1992) Review of MRimage segmentation techniques using pattern recognition. *Medical Physics*, **20** (4), 1033–1048.

17 Dawant, B.M., Zijdenbos, A.P., and Margolin, R.A. (1993) Correction of intensity variations in MR images for computer-aided tissue classification. *IEEE Transactions on Medical Imaging*, **12** (4), 770–781.

18 Bezdek, J.C. (2013) Pattern recognition with fuzzy objective function algorithms, Springer Science & Business Media, New York.

19 Tolias, Y.A. and Panas, S.M. (1998) On applying spatial constraints in fuzzy image clustering using a fuzzy rule-based system. *IEEE Signal Processing Letters*, **5** (10), 245–247.

20 Tolias, Y.A. and Panas, S.M. (1998) Image segmentation by a fuzzy clustering algorithm using adaptive spatially constrained membership functions. *IEEE Transactions on Systems, Man, and Cybernetics – Part A: Systems and Humans*, **28** (3), 359–369.

21 Pham, D.L., Xu, C., and Prince, J.L. (2000) Current methods in medical image segmentation 1. *Annual Review of Biomedical Engineering*, **2** (1), 315–337.

22 Tripathy, B. and Ghosh, A. (2012) Data clustering algorithms using rough sets. *Handbook of Research on Computational Intelligence for Engineering, Science, and Business*, p. 297.

23 Chuang, K.S., Tzeng, H.L., Chen, S., Wu, J., and Chen, T.J. (2006) Fuzzy c-means clustering with spatial information for image segmentation. *Computerized Medical Imaging and Graphics*, **30** (1), 9–15.

24 Cai, W., Chen, S., and Zhang, D. (2007) Fast and robust fuzzy c-means clustering algorithms incorporating local information for image segmentation. *Pattern Recognition*, **40** (3), 825–838.

25 Wang, X.Y. and Bu, J. (2010) A fast and robust image segmentation using FCM with spatial information. *Digital Signal Processing*, **20** (4), 1173–1182.

26 Wang, J., Kong, J., Lu, Y., Qi, M., and Zhang, B. (2008) A modified FCM algorithm for MRI brain image segmentation using both local and non-local spatial constraints. *Computerized Medical Imaging and Graphics*, **32** (8), 685–698.

27 Tripathy, B., Basu, A., and Govel, S. (2014) Image segmentation using spatial intuitionistic fuzzy c means clustering, in *Computational Intelligence and Computing Research (ICCIC), 2014 IEEE International Conference*, IEEE, pp. 1–5.

28 Hudedagaddi, D. and Tripathy, B. (2016) Application of spatial IFCM on leukemia images. *Soft Computing for Medical Data and Satellite Image Analysis*, **7** (5), 33–40.

29 Hudedagaddi, D. and Tripathy, B. (2015) Application of spatial FCM on cancer cell images, in *Proceedings of NCICT*, Upendranath College, India, pp. 32–36.

13

An Optimized Breast Cancer Diagnosis System Using a Cuckoo Search Algorithm and Support Vector Machine Classifier

Manoharan Prabukumar[1], Loganathan Agilandeeswari[1], and Arun Kumar Sangaiah[2]

[1] School of Information Technology & Engineering, VIT University, Vellore, Tamil Nadu, India
[2] School of Computing Science & Engineering, VIT University, Vellore, Tamil Nadu, India

13.1 Introduction

A disease that is responsible for many deaths across the world is cancer. In 2008, the percentage of deaths due to cancer was about 13%, and it is expected to increase significantly up to 12 billion in the year 2030 as per the World Health Organization (WHO) [1]. Around the world, the second most dangerous problem in health among women is breast cancer. The American Cancer Society report stated in 2009 that around 15% of people (especially female) die because of breast cancer, out of 269,800 cancer deaths approximately. In addition, as per Meselhy, out of 713,220 diagnosed cancer cases, 27% of were related to it [2]. To reduce the mortality rate of the patients, early cancer detection plays an important role. The method that is the gold standard used for such early detection is the mammography. Each mammographic examination includes four forms of images: namely, cranio caudal (CC) corresponding to the right breast and mediolateral oblique (MLO) of the left breast. The chances of identifying the non-palpable breast cancer tissues will be greater with the use of CC and MLO, since it improves the visualization. Then the radiologists use these images to determine the abnormalities present.

Due to the repetitive task of mammography, the radiologists may become confused, and it leads to failure to detect malignancy for about 10 to 30% [3]. The masses and microcalcifications are the mammographic signs of malignancy. Also, based on the quality of an image and the radiologist's expertise level, the sensitivity of screening mammography gets affected. Thus, the automatic system is required for classifying the doubtful areas in a mammogram to assist radiologists in the screening process and to avoid unnecessary biopsy. Computer-aided design (CAD) is very helpful to aid the radiologists in diagnosis. It can easily classify microcalcification, masses, and architectural distortions [4]. Several approaches are proposed by various researchers to study the mammograms and classify their images as the malignant or benign breast cancer type. For microcalcification, many research works are presented recently for automatic detection with a sensitivity of about 98%. The challenge is with mass detection; because

Hybrid Intelligence for Image Analysis and Understanding, First Edition.
Edited by Siddhartha Bhattacharyya, Indrajit Pan, Anirban Mukherjee, and Paramartha Dutta.
© 2017 John Wiley & Sons Ltd. Published 2017 by John Wiley & Sons Ltd.
Companion Website: www.wiley.com/go/bhattacharyya/hybridintelligence

masses are normally indistinguishable from their adjacent tissues, the following issues may arise: (1) image contrast will be poor, (2) speculated lesions are mostly connected to the surrounding parenchymal tissue, (3) masses are surrounded by no uniform tissue background, and (4) they have no definite size, shape, and density [5]. Hence, the accuracy of diagnosis in terms of sensitivity can be improved by double reading, which means that the reading is made by two radiologists. This may lead to increases in the operating cost. Then, the mammographic images from the same patient are compared, that is, the images from both views (i.e., CC and MLO) of the same breast are compared. Some researchers proved that the detection and diagnostic performances can be improved by this approach [6, 7]. But this approach reduces the recalls for a second inspection of the patient. The CAD system is one that operates independently on each view of mammography. Several authors have been involved in building CAD systems that can analyze two views on breast cancer detection to avoid confusion [8].

References [9] and [10] proposed the classification of mammograms by density through computation techniques. The classification presented considers local statistics and texture measures to divide the mammograms in to fat or dense tissue. They found one of the measures (local skewness in tiles) gives a good separation between fatty and dense patches.

Hadjiiski *et al.* in 1999 [11] proposed a mammographic-based breast cancer detection system using a hybrid classifier such as linear discriminant analysis (LDA) and adaptive resonance theory (ART). They utilized LDA and a neural network scheme for classification to overlook previously learned expert knowledge during classification. The uniqueness of the hybrid classifier is that the classification process is done at two levels. In the first-level classification process, the ART classifier is used for preliminary check of the classes. If the class is malignant according to the ART classifier, no further processing is done. But, if the class is labeled as mixed, then for further classification, the data is processed through a LDA classifier to classify the cancer as malignant or benign. They compared the performance of the proposed hybrid classifier with the LDA classifier and a back-propagation neural network (BPN) classifier-based system. They achieved the average area under the ROC (receiver operating characteristic) curve 0.81 as compared to 0.78 for the LDA classifier-based system and 0.80 for the BPN classifier-based system. They proved that the hybrid classifier improves the classification accuracy of the CAD system through their results.

Mousa *et al.* in 2005 [12] proposed a breast cancer diagnosis system based on wavelet analysis and fuzzy-neural. The wavelet features have been known to be used in extracting useful information from mammographic images, and fuzzy-neural classifiers are used to classify the breast cancer type. Statistical features in a multiple-view mammogram with support vector machines (SVMs) and kernel Fisher discriminant (KFD) for a digital mammographic breast image classification system were *et al.* in proposed by Liyang Wei *et al.* [13]. The proposed system achieved 85% classification accuracy. Szekely *et al.* [14] proposed a CAD system for mammographic breast cancer image classification using the texture features. It used a *et al.* in hybrid classifier such as decision trees and multiresolution Markov random models for classification of cancer as type benign or malignant. The proposed system achieved 94% classification accuracy.

To detect and classify the breast cancer in digital mammograms, Alolfe *et al.* in 2009 [15] presented a CAD system based on SVMs and LDA classifiers. The presented system has four stages: region of interest (ROI) identification, feature extraction, feature selection, and feature classification. They used a *et al.* in forward stepwise linear regression method for feature selection and a *et al.* in hybrid classifier using SVM and LDA for benign and malignant classification. They compared the classification performance of the proposed system with the SVM, LDA, and fuzzy *c*-means classifier-based system. They achieved a specificity of 90% and sensitivity of 87.5%.

Balakumaran *et al.* [16] proposed a system that proves to be a better one with the implementations of multiresolution transformations and the neural networking phenomenon. The classification approach helps in the adaptive learning pattern, which has helped in the identification of various shapes and sizes of the microcalcifications, and it differentiates between the various textures (i.e., whether it is a duct, tissue, or microcalcification). When a mass is detected, it is difficult to distinguish if it is benign or malignant, but there are differences in the features of shape and texture between them. Benign masses are typically smooth and distinct, and their shapes are similar to round. On the other hand, malignant masses are irregular, and their boundaries are usually blurry [17].

Moayedi *et al.* [18] proposed a system for automatic mass detection of mammograms based on fuzzification and SVMs. The geometrical features are extracted from the curvelet coefficients, and genetic algorithms are applied for feature weighting. Then the extracted features of the digital mammograms are applied to the fuzzy-SVM (FSVM) classifiers to classify mammogram images. The performance of the proposed was compared with the fuzzy role-based SVM, the support vector–based fuzzy neural network (SVFNN) classifier. They achieved 95.6% classification accuracy for the FSVM classifier. They proved the FSVM has a strong capability to classify the mass and nonmass images.

Balanică *et al.* [19] has proposed a system that classifies on the basis of the fuzzy system that helps in the identification of the geometrical pattern of the textures. The tumor features are extracted using the fuzzy logic technique for predicting the risk of breast cancer based on a set of chosen fuzzy rules utilizing patient age automatically.

Meselhy *et al.* [2] have proposed a method for breast cancer diagnosis in digital mammogram images using mresolution transformations, namely, wavelet or curvelet transformations. Here, mammogram images are converted into frequency coefficients and construct the feature matrix. Then, based on the statistical *t*-test method, a better feature extraction method was developed and the optimized features are extracted using the dynamic threshold method. This also helps in studying the various frequencies of the textures in the image. SVMs are used to classify benign and malignant tumors. The classification accuracy rate achieved by the proposed method using wavelet coefficients is 95.84% and using curvelet coefficients is 95.98%. The obtained results show the importance of the feature extraction step in developing a CAD system.

A innovative approach for detection of microcalcification in digital mammograms is the swarm optimization neural network (SONN) presented by Dheeba *et al.* in 2012 [20]. To capture descriptive texture information, Laws' texture features are extracted from the mammograms. These features are used to extract texture energy measures from the ROI containing microcalcification. A feedforward neural network that is used for detection of abnormal regions in breast tissue is optimally designed using the particle swarm optimization algorithm. The proposed intelligent classifier is evaluated based on the MIAS database where 51 malignant, 63 benign, and 208 normal images are utilized. It also been tested on 216 real-time clinical images with abnormalities, which showed that the results are statistically significant. Their proposed methodology achieved an area under the ROC curve (Az) of 0.9761 for the MIAS database and 0.9138 for real clinical images.

A computerized scheme for automatic detection of cancerous lesions in mammograms was examined by Dheeba *et al.* [21]. They proposed a supervised machine learning algorithm, the Differential Evolution Optimized Wavelet Neural Network (DEOWNN), to detect tumor masses in mammograms. The texture features of the abnormal breast tissues and normal breast tissues are extracted prior to classification. The DEOWNN classifier is used to determine normal or abnormal tissues in mammograms. The performance of the CAD system is evaluated using a mini-database from MIAS. They achieved sensitivity of 96.9% and specificity of 92.9%.

Prabukumar *et al.* [22] proposed a novel approach to diagnosis breast cancer using discrete wavelet transform, SVM, and neuro-fuzzy logic classifier. To optimize the number of features and achieve the maximum classification accuracy rate, dynamic thresholds are applied. The FSVM method is used to classify between normal and abnormal tissues. The proposed approach obtained classification accuracy rates of 93.9%.

Guzmán-Cabrera *et al.* [23] proposed a CAD scheme to effectively analyze digital mammograms based on texture segmentation for classifying the early-stage tumors. They used morphological operations and, clustering techniques to segment the ROI from the background of the mammogram images. The proposed algorithm was tested with the well-known digital database. It was applied for screening mammography for cancer research and diagnosis, and it was found to be absolutely suitable to distinguish masses and microcalcifications from the background tissue. The breast abnormalities in digital mammograms investigated by Dheeba *et al.* [24] involved a novel classification approach based on particle swarm optimized wavelet neural networks. The laws' texture energy measures are extracted from the mammograms to classify the suspicious regions. They used a real clinical database to test the proposed algorithm. They achieved an area under the ROC curve of 0.96853 with a sensitivity of 94.167% and specificity of 92.105%. They compared the performance of the proposed system with a SONN-and DEOWNN-based system.

To detect and segment regions in mammographic images, Pereira *et al.* [25] presented a method using a wavelet and genetic algorithm. They adopted wavelet transform and Wiener filtering for image denoising and enhancement in the preprocessing stage. They employed a genetic algorithm for suspicious regions segmentation. They achieved a sensitivity of 95% with a FP rate of 1.35 per image. Researchers proposed a system for breast cancer diagnosis using fuzzy neural networks. They used statistical features to train the classifier and achieved 83% classification accuracy.

A novel mammographic image preprocessing method was proposed by He *et al.* in 2015 [26] to improve image quality of the digital mammogram images. Instead of using an assumed correlation between smoothed pixels and breast thickness to model the breast thickness, they used a shape outline derived from MLO and CC views for the same. Then they used a selective approach to target specific mammograms more accurately. The proposed approach improved the mammographic appearances not only in the breast periphery but also across the mammograms. To facilitate a quantitative and qualitative evaluation, mammographic segmentation and risk/density classification were performed. When using the processed images, the results indicated more anatomically correct segmentation in tissue-specific areas, and subsequently better classification accuracies were achieved. Visual assessments were conducted in a clinical environment to determine the quality of the processed images and the resultant segmentation. The developed method has shown promising results. It is expected to be useful in early breast cancer detection, risk-stratified screening, and aiding radiologists in the process of decision making prior to surgery and/or treatment.

To detect abnormalities or suspicious areas in digital mammograms and classify them as malignant or nonmalignant, Sharma *et al.* in 2015 [27] proposed a CAD system. The suspicious areas in digital mammogram patches are extracted from the original large-sized digital mammograms manually. Zernike moments of different orders are computed from the extracted patches and stored as a feature vector for training and testing purposes. They tested the performance of the proposed system on benchmarked data sets such as the Image Retrieval In Medical Application (IRMA) reference data set and Digital Database for Screening Mammography (DDSM) mammogram database. Their experimental study shows that the Zernike moments of order 20 and SVM classifier give the best results compared to the other studies. The performance of the proposed CAD system is compared with the other well-known texture descriptors such as gray-level co-occurrence matrix (GLCM) and discrete cosine transform (DCT) to verify the applicability of Zernike moments as a texture descriptor. The proposed system achieved 99% sensitivity and 99% specificity on the IRMA reference data set and 97% sensitivity and 96% specificity on the DDSM mammogram database.

In this chapter, we propose an optimized breast cancer diagnosis system using a cuckoo optimization algorithm and SVM classifier. To segment the ROI, a morphological segmentation algorithm is used in our approach. It will segment the nodule of interest accurately, which leads to better classification accuracy of the proposed system. In order to increase the accuracy of classification, the optimized features are determined using the cuckoo optimization algorithm. Then these optimized features are extracted from the segmented breast cancer region, and the same is trained by SVM classifier to classify the nodules as benign or malignant.

The organization of the chapter is given as follows: Section 13.2 deals with the technical background, and architecture of the proposed system is discussed in Section 13.3. In Section 13.4, results and discussion are given. Finally, a conclusion is given in Section 13.5.

13.2   Technical Background

In this section, we describe all the methods we need to form our proposed system.

13.2.1   Morphological Segmentation

In general, morphological segmentation is performed on a thresholded image to remove unwanted regions or segment the ROI. First, the image is converted into a binary cluster by obtaining a desired threshold of an image using threshold-based segmentation. Then, perform some morphological operations on the binary cluster to obtain the desired region. The basic morphological operations that can be performed are erosion, opening, closing, and dilation. To perform a morphological operation, an appropriate structuring element is to be recognized and applied to the binary image. There are two situations for morphological operations, hit and fit. The structuring element is said to fit the image if, for each of its pixels set to 1, the corresponding image pixel is also 1. Similarly, a structuring element is said to hit, or intersect, an image if, for at least one of its pixels set to 1, the corresponding image pixel is also 1. Morphological opening opens up a gap between objects connected by bridge and hence is found to be best among all operations for segmentation. The structuring element for each magnetic resonance imaging (MRI) scan image may be different as the structure of breast tissue is different for different computed tomography (CT) slices. Hence, choosing the appropriate structuring element for each slice may be a problem. Morphological opening may even destroy brain tumor structure as the size of the structuring element increases [28, 29].

13.2.2 Cuckoo Search Optimization Algorithm

A new nature-inspired metaheuristic cuckoo search algorithm proposed by Yang and Deb (2010) [30] has been used to determine the most predominant features among the features extracted from the segmented breast cancer region. Each egg in a nest represents a solution, and a cuckoo egg represents a new solution. The aim is to employ the new and potentially better solutions (cuckoos) to replace not-so-good solutions in the nests. In the simplest form, each nest has one egg. The algorithm can be extended to more complicated cases in which each nest has multiple eggs representing a set of solutions.

The cuckoo search is based on three idealized rules: (1) each cuckoo lays one egg at a time, and dumps it in a randomly chosen nest; (2) the best nests with high-quality eggs (solutions) will carry over to the next generations; and (3) the number of available host nests is fixed, and a host can discover an alien egg with probability $pa \hat{I}$ [0,1]. In this case, the host bird can either throw the egg away or abandon the nest to build a completely new nest in a new location.

For simplicity, the last assumption can be approximated by a fraction pa of the n nests being replaced by new nests, having new random solutions. For a maximization problem, the quality or fitness of a solution can simply be proportional to the objective function. Other forms of fitness can be defined in a similar way to the fitness function in genetic algorithms. Based on the above-mentioned rules, the basic steps of the cuckoo search can be summarized as the pseudo-code, as follows:

1: Objective function $f(x), x = (x1, x2, x3, \ldots, xd)$;
2: Generate initial population of n host xi where $i = 1, 2, 3, \ldots, n$;
3: **while** $t \leq MaxGeneration$ **do**
4:     Get a cuckoo randomly by Lévy flights;
5:     Evaluate its fitness Fi;
6:     Choose a nest among $n(sayj)$ randomly
7:     **if** $Fi > Fj$ **then**
8:         Replace j by the new solution;
9:     **end if**
10:     A fraction (pa) of worse nests is abandoned and new ones are built;
11:     Keep the best solutions (or nest with quality solutions);
12:     Rank the solutions and find the current best;
13: **end while**
14: Post process results and visualization.

When the new solution $X_i(t + 1)$ is generating for the ith cuckoo, the following Levy flight is used:

$$X_i(t + 1) = X_i(t) + \alpha \oplus Levy(\lambda) \tag{13.1}$$

where α is the step size, which should be related to the scale of the problem of interest; and the product $\oplus$ means entry-wise multiplications.

13.2.3 Support Vector Machines

SVMs are effective for classification of both linear and nonlinear data. They were introduced by Vapnik *et al.* [31]. They are trained with the data set in which it is known to which class each point of that data set belongs. SVM training builds a model according to which it classifies the new data point. Each data point is viewed as a vector with p-dimensions. The data points are separated with a hyperplane.

1. *Linear separable data:* For linear separable data, data points can be separated with a $p - 1$ dimensional hyperplane, and the hyperplane that separates the two classes with the largest margin is considered.
2. *Nonlinear separable data:* Classification of nonlinear data cannot be done on that feature space. The present feature space has to be mapped with some higher dimension feature space where training sets are separable. This mapping of present feature space into higher dimension feature space is done using the kernel trick. In simple terms, SVM uses kernels for separation of nonlinear data. The linear classifier depends on the inner product between vectors. Data points are mapped into higher feature space by some transformation $K(x_i, x_j) = x_i^T x_j$ and $\phi : \rightarrow \phi(x)$ the inner product becomes.

Features that are to be classified are nonlinear in nature. They have to be mapped to some higher dimension. In our approach, we have used the radial basis function, because it provides infinite dimensional spaces.

13.3 Proposed Breast Cancer Diagnosis System

We proposed an optimized breast cancer diagnosis system using morphological segmentation, cuckoo search optimization algorithm, and multiclass SVM classifier. The proposed system involves the following phases: (1) preprocessing of input mammographic breast images using Otsu's thresholding and morphological segmentation to segment the tissues of interest; (2) from the segmented image, various features, namely statistical, texture, geometrical, and invariant feature moments, are extracted; (3) optimized feature extraction for better diagnosis is done using the cuckoo search algorithm; (4) a multiclass SVM classifier is used for better classification of mammography breast images of type benign, malignant, and noncancerous. The block diagram of the proposed optimized system is given in Figure 13.1.

13.3.1 Preprocessing of Breast Cancer Image

In this stage, digital mammographic breast cancer image quality is improved using a median filtering process. It will remove the noise present in the mammographic breast cancer image, then enhance the mammographic breast cancer image converting to a

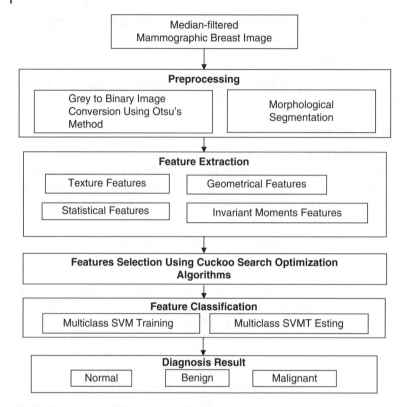

Figure 13.1 Proposed breast cancer diagnosis system.

binary image using Otsu's thresholding approach. Segmentation is the most difficult and vital process to be carried out in the diagnosis process. In order to segment the tissue of interest accurately, a morphological segmentation algorithm is used in our approach.

13.3.2 Feature Extraction

The feature extraction is the next step after the segmentation of the cancer region. In general, the features of tissues contain very crucial information about them. With the help of these features, it is possible to classify the tissues of various kinds. In our approach, we have calculated various features, namely statistical, texture, geometrical, and invariant feature moments, from each segmented tissue; and then these features are used to determine the optimum features using the cuckoo search optimization algorithm. Detailed descriptions of the various features extracted from the segmented region are given in this section.

13.3.2.1 Geometric Features

For calculating the geometric features, the segmented binary image of the breast cancer is used. The geometric features that are calculated for distinction are:

1. *Area:* This feature is calculated on the binary image of the extracted tissue. Calculation of area in binary form will not make any difference as it counts the number of pixels involved in tissue. It will make the calculation more accurate.

2. *Convex area:* This feature counts the number of pixels engaged in the tissue of interest in its convex form. Calculations are made in the ROI's binary form.
3. *Perimeter:* The number of pixels present on the boundary of tissue of interest comprises the perimeter for that tissue of interest. A binary form of an image will provide accurate results.
4. *Equivalent diameter:* It is the diameter of the circle with the same area as the tissue of interest. As it works on area, it can also be calculated on the binary form of an image.
5. *Solidity:* Solidity is the proportion of pixels present in the convex area and area. It is expressed as:

$$Solidity = Area/ConvexArea \qquad (13.2)$$

6. *Irregularity index:* Irregularity index is a scalar that tells about irregularity in shape, mainly because of their borders. It is expressed as:

$$IrregularityIndex = 4\pi. \, Area/Perimeter^2 \qquad (13.3)$$

7. *Eccentricity:* This is the ratio of distance between the foci of the tissue and its major axis length. Length contributes to a complete pixel; hence, it has to be calculated on a binary image.
8. *Size:* Calculation of the major axis length of the tissue is considered as size in our approach. Size in centimeters is determined for the extracted tissue of interest. Calculation of size will let us know that even for small-sized tissues, malignancy is possible.

13.3.2.2 Texture Features

The following texture features of the tissue of interest are calculated in the segmented gray-level image of each tissue.

9. *Energy:* Energy of tissue refers to uniformity or angular second moment. It is presented in the following way:

$$Energy = \sum_{i=1}^{M} \sum_{j=1}^{N} p^2(i,j) \qquad (13.4)$$

where $p(i,j)$ is the intensity value of the pixel at the point (i,j); and MN is the size of the image.

10. *Contrast:* It calculates contrast between a pixel's intensity and its surrounding pixels. It is expressed as:

$$Contrast = \sum_{i=1}^{M} \sum_{j=1}^{N} p^2(i,j) \qquad (13.5)$$

where $p(i,j)$ is the intensity value of the pixel at the point (i,j); and MN is the size of the image.

11. *Correlation:* It is the measure of relationship among the pixels. It calculates the extent a pixel is correlated to its neighbors.

$$Correlation = \sum_{i=1}^{M} \sum_{j=1}^{N} \frac{p(i,j) - \mu_r. \, \mu_c}{\sigma_r. \, \sigma_c} \qquad (13.6)$$

12. *Homogeneity:* It calculates how closely the elements of GLCM are distributed to the GLCM of the tissue of interest. It is expressed as:

$$Homogeneity = \sum_{i=1}^{M} \sum_{j=1}^{N} \frac{p(i,j)}{1 + |i-j|} \tag{13.7}$$

13. *Entropy:* It calculates random distribution of gray levels. As it deals with gray levels, it has to be calculated on gray-level tissue. It is expressed as:

$$Entropy = - \sum_{i=1}^{M} \sum_{j=1}^{N} p(i,j).\, logp(i,j) \tag{13.8}$$

13.3.2.3 Statistical Features

14. *Mean:* This is basically used for obtaining the mean of gray levels present in the image. It is expressed as:

$$Mean = \frac{1}{MN} \sum_{i=1}^{M} \sum_{j=1}^{N} p(i,j) \tag{13.9}$$

15. *Variance:* Variance actually observes and checks the distribution of gray levels. It is expressed as:

$$Variance = \frac{1}{MN} \sum_{i=1}^{M} \sum_{j=1}^{N} (p(i,j) - \mu) \tag{13.10}$$

16. *Standard deviation:* Standard deviation is the measure of difference or deviation existing from midpoint. It is expressed as:

$$Standard deviation = \sqrt{Variance} \tag{13.11}$$

17. *Invariant moments:*

$$1st order moments \phi_1 = \eta_{20} + \eta_{02} \tag{13.12}$$
$$2nd order moments \phi_2 = (\eta_{20} - \eta_{02})^2 + 4\eta_{11}^2 \tag{13.13}$$
$$3rd order moments \phi_3 = (\eta_{30} - 3\eta_{12})^2 + (3\eta_{21} - \mu_{03})^2 \tag{13.14}$$
$$4th order moments \phi_4 = (\eta_{30} + 3\eta_{12})^2 + (3\eta_{21} + \mu_{03})^2 \tag{13.15}$$

13.3.3 Features Selection

Some of the features extracted from the ROIs in the mammographic image are not significant when observed alone, but in combination with other features, they can be significant for classification. The best set of features for eliminating false positives and for classifying types as benign or malignant are selected using the cuckoo search optimization algorithm in the feature selection process. The best features selection procedure by the cuckoo search algorithm is given in the form of Algorithm 13.1.

Algorithm 13.1 Cuckoo search algorithm for optimal feature selection

---

1. Step 1: Initialization:

   Initialize the cuckoo search parameters as number of nest n = 50, nest size s = 25, minimum number of generations t=0, step size α = 0.30, and maximum generations Gmax = 1000.

   State fitness function (Fn) to select the optimum features using MCC as: $Fn = \frac{(TP \times TN)-(FP \times FN)}{\sqrt{(TP+FP)(TP+FN)(TN+FP)(TN+FN)}}$;

   where TP = true positive; TN = true negative; FP = false positive; and FN = false negative.

2. Step 2: Output: Choose the suitable features for determining the benign or malignant tissue using the fitness function (Fn) given in the above step.

3. Step 3: Process:

   A. While t < Gmax

      i. Get a cuckoo randomly by levy Flights i.

      ii. Evaluate the fitness function Fi using MCC

      iii. Randomly select a nest among n nests as j, and evaluate its fitness function Fj.

      iv. If Fi < Fj

         a. Replace j by new solution.

      v. Else

         a. Let i as a solution and

         b. Abandon the worst nest and build the new nest by Levy flights.

         c. Keep the current one as best

   B. Rank the fitness function of various n nests, R(Fn)

   C. Find the best Fn used as optimum feature Of.

13.3.4 Features Classification

SVMs have been successfully applied in various classification problems [32]. Training data for SVMs should be represented as labeled vectors in a high-dimensional space where each vector is a set of features that describes one class. For correct classification of tissues, this representation is constructed to preserve as much information as possible about the features. SVMs are trained with the optimized features identified by the cuckoo search algorithm. Trained SVMs are used for classification of the tissues of various types, such as normal, benign, and malignant.

13.4   Results and Discussions

The performance of the proposed optimized system for breast cancer diagnosis is validated and evaluated using the sample MIAS database. The data set consists of 322 images of 161 patients, where 51 were diagnosed to be malignant, 4 were benign, and the rest (207) were said to be normal. The empirical analysis of the proposed system is tested by implementing the same using MATLAB 2013a in an Intel Core i5 processor with 8 GB RAM. The output of the entire process of the proposed optimization system for breast cancer diagnosis is given in Figures 13.2, 13.3, and 13.4.

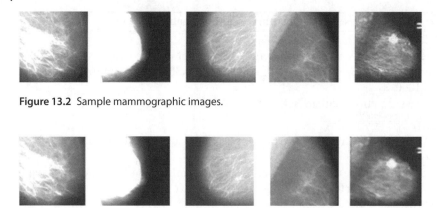

Figure 13.2 Sample mammographic images.

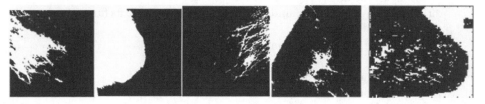

Figure 13.3 Enhanced mammographic images.

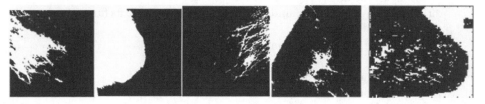

Figure 13.4 Otsu's thresholded and morphological segmented breast tissues.

From the segmented images, the 23 features such as shape, texture, invariant moments, and statistical features required for classification, as stated in Section 13.2, are calculated for the further extraction of nodules. In order to obtain the optimized features for classification, the mentioned 23 features are given as an initial solution to the cuckoo search optimization algorithm. The optimization algorithm is used to select the features that are appropriate for classification using the fitness function. First, random selection of features is done and evaluates the fitness function. The fitness function is based on the Mathew correlation coefficient (MCC), which performs the correlation between the true-positive rate and false-positive rate. The MCC value varies between −1 and +1 [33].

This process of computing the fitness function is repeated until the decided accuracy has been obtained with optimized features. As a result, this further improves the selection of features for classification. These features and the data set are given as input to the SVM for training. Once the SVM is trained, it is fit to use with the actual data set that is to be classified. The SVM classifier will read all the recorded data features and classify them according to the knowledge that it has after training. The SVM is trained with these limited optimized features and produces better classification with limited computational cost and time. In this experiment, 85% of the images of each class were used to form the training set, and the other 15% of the images were used to form the test set. The performance of the proposed breast cancer diagnosis system, compared with the existing methods such as FSVM [18], DWT-Neural Networks [16], and DWT-NeuroFuzzy Logic [22], in terms of the classification accuracy, sensitivity, and specificity are given by:

$$ClassificationAccuracy = \frac{TP + TN}{TP + FP + TN + FN} \qquad (13.16)$$

$$Sensitivity = \frac{TP}{TP + FN} \qquad (13.17)$$

$$Specificity = \frac{TN}{FP + TN} \qquad (13.18)$$

where TP (true positive) is the probability that the person with a cancer is found to be having cancer; FP (false positive) is the probability that the detection value for the healthy person is found to be cancer; TN (true negative) is the probability that the person with cancer is found to be healthy; and FN (false negative) is explained as a person without cancer, or healthy, is found to be having cancer. The same is shown in Table 13.1.

From Table 13.1 values, we can say that classification accuracy, sensitivity, and specificity of the proposed Cuckoo search and SVM-based breast cancer diagnosis system are high when compared to the existing approaches such as FSVM, DWT-neural networks, and DWT-neuro fuzzy logic-based breast cancer diagnosis systems. The sensitivity, specificity, and accuracy values for the two classes show that the classifier is capable of efficiently discriminating breast cancer tissues as either benign or malignant from the other segmented region.

In Table 13.2, we have recorded the average execution time of both the training and the test operations. By comparing the execution time of each approach, we found that the proposed system recorded the lowest time when compared to the existing systems.

Table 13.1 Comparison of classification accuracy of proposed approach with the existing methods

Sample no.	Method	Class	Classification accuracy	Average accuracy	Average sensitivity	Average specificity
1	Fuzzy SVM	Normal/benign/ malignant	100/85/89	91.4	86.43	91.69
2	DWT neural networks	Normal/benign/ malignant	100/88/84	90.5	85.33	90.99
3	DWT neuro fuzzy logic	Normal/benign/ malignant	100/99.9/90	93.9	91.82	93.33
4	Proposed	Normal/benign/ malignant	100/95.79/94.36	96.72	98.88	93.68

Table 13.2 Comparison of computational costs

Sample no.	Method	Class	Avg. computation time (sec)
1	Fuzzy SVM	Normal/benign/ malignant	143.9
2	DWT neural networks	Normal/benign/ malignant	191.5
3	DWT neuro fuzzy logic	Normal/benign/ malignant	123.6
4	Proposed	Normal/benign/ malignant	110.08

13.5 Conclusion

In this chapter, an optimized breast cancer diagnosis system using a cuckoo search algorithm and SVM classifier is proposed. The morphological segmentation algorithm is used to segment the tissue of interest from the mammographic breast images. From the experimental results, we found that the proposed segmentation approach detects tissues of different shapes and sizes efficiently and accurately. The more appropriate features, such as texture, geometrical, statistical, and invariant moments, required for classification from the segmented nodules are computed. The number of features used to classify the breast cancer nodules is further reduced in our proposed system by using a Cuckoo search optimization algorithm without compromising the classification accuracy. The optimized features are trained in the SVM classifier, and the same can be used to classify the test sample for benign and malignant nodules. The classification accuracy of the proposed system is validated against the MIAS database. From the experimental results, we infer that the classification accuracy of the proposed system is 96.72%, which is higher than that of the existing breast cancer diagnosis system.

13.6 Future Work

The performance of the proposed system was tested with mammogram images only from the MIAS database. In a future work, the proposed system will be tested with other well-known breast cancer database images. The classification accuracy of the proposed system needs to improve by identifying suitable techniques in the segmentation and feature extraction stages.

References

1 World Health Organization. Cancer. Fact sheet no. 297. Available from http://www.who.int/mediacentre/factsheets/fs297/en.

2 Eltoukhy, M.M., Faye, I., and Samir, B.B. (2012) A statistical based feature extraction method for breast cancer diagnosis in digital mammogram using multiresolution representation. *Computers in Biology and Medicine*, **42** (1), 123–128. Available from http://dx.doi.org/10.1016/j.compbiomed.2011.10.016.

3 Elter, M. and Horsch, A. (2009) Cadx of mammographic masses and clustered microcalcifications: a review. *Medical Physics*, **36** (6), 2052–2068.

4 Giger, M.L., Chan, H.P., and Boone, J. (2008) Anniversary paper: History and status of CAD and quantitative image analysis: the role of medical physics and aapm. *Medical Physics*, **35** (12), 5799–5820, doi: http://dx.doi.org/10.1118/1.3013555. Available from http://scitation.aip.org/content/aapm/journal/medphys/35/12/10.1118/1.3013555.

5 Hong, B.W. and Sohn, B.S. (2010) Segmentation of regions of interest in mammograms in a topographic approach. *IEEE Transactions on Information Technology in Biomedicine*, **14** (1), 129–139, doi: 10.1109/TITB.2009.2033269.

6 Gupta, S., Chyn, P.F., and Markey, M.K. (2006) Breast cancer cadx based on bi-rads™ descriptors from two mammographic views. *Medical Physics*, **33** (6),

1810–1817, doi: http://dx.doi.org/10.1118/1.2188080. Available from http://scitation .aip.org/content/aapm/journal/medphys/33/6/10.1118/1.2188080.

7 van Engeland, S. and Karssemeijer, N. (2007) Combining two mammographic projections in a computer aided mass detection method. *Medical Physics*, **34** (3), 898–905, doi: http://dx.doi.org/10.1118/1.2436974. Available from http://scitation.aip .org/content/aapm/journal/medphys/34/3/10.1118/1.2436974.

8 Paquerault, S., Petrick, N., Chan, H.P., Sahiner, B., and Helvie, M.A. (2002) Improvement of computerized mass detection on mammograms: fusion of two-view information. *Medical Physics*, **29** (2), 238–247.

9 Hajnal, S., Taylor, P., Dilhuydy, M.H., Barreau, B., and Fox, J. (1993) Classifying mammograms by density: rationale and preliminary results. Presented at SPIE Medical Imaging conference. doi: 10.1117/12.148662. Available from http://dx.doi .org/10.1117/12.148662.

10 Taylor, P., Hajnal, S., Dilhuydy, M.H., and Barreau, B. (1994) Measuring image texture to separate "difficult"from "easy"mammograms. *The British Journal of Radiology*, **67** (797), 456–463, doi: 10.1259/0007-1285-67-797-456. Available from http://dx.doi .org/10.1259/0007-1285-67-797-456, pMID: 8193892.

11 Hadjiiski, L., Sahiner, B., Chan, H.P., Petrick, N., and Helvie, M. (1999) Classification of malignant and benign masses based on hybrid art2lda approach. *IEEE Transactions on Medical Imaging*, **18** (12), 1178–1187, doi: 10.1109/42.819327.

12 Mousa, R., Munib, Q., and Moussa, A. (2005) Breast cancer diagnosis system based on wavelet analysis and fuzzy-neural. *Expert Systems with Applications*, **28** (4), 713–723, doi: http://dx.doi.org/10.1016/j.eswa.2004.12.028. Available from http:// www.sciencedirect.com/science/article/pii/S0957417404001757.

13 Wei, L., Yang, Y., Nishikawa, R.M., and Jiang, Y. (2005) A study on several machine-learning methods for classification of malignant and benign clustered microcalcifications. *IEEE Transactions on Medical Imaging*, **24** (3), 371–380. Available from http://dx.doi.org/10.1109/TMI.2004.842457.

14 Székely, N., Tóth, N., and Pataki, B. (2006) A hybrid system for detecting masses in mammographic images. *IEEE Transactions on Instrumentation and Measurement*, **55** (3), 944–952. Available from http://dx.doi.org/10.1109/TIM.2006.870104.

15 Alolfe, M.A., Mohamed, W.A., Youssef, A.B.M., Mohamed, A.S., and Kadah, Y.M. (2009) Computer aided diagnosis in digital mammography using combined support vector machine and linear discriminant analyasis classification, in *ICIP*, IEEE, pp. 2609–2612. Available from conf/icip/2009.

16 Balakumaran, T., Vennila, D.I., and Shankar, C.G. (2013) Microcalcification detection using multiresolution analysis and neural network. Available from http://citeseerx.ist .psu.edu/viewdoc/summary?doi=10.1.1.381.8882; http://ijrte.academypublisher.com/ vol02/no02/ijrte0202208211.pdf.

17 Zhao, H., Xu, W., Li, L., and Zhang, J. (2011) Classification of breast masses based on multi-view information fusion using multi-agent method, in *Bioinformatics and Biomedical Engineering, (iCBBE) 2011 5th International Conference*, pp. 1–4, 10.1109/icbbe.2011.5780304.

18 Moayedi, F. and Dashti, E. (2010) Subclass fuzzy-SVM classifier as an efficient method to enhance the mass detection in mammograms. *Iranian Journal of Fuzzy Systems*, **7** (1), 15–31. Available from http://ijfs.usb.ac.ir/article_158.html.

19 Caramihai, M., Severin, I., Blidaru, A., Balan, H., and Saptefrati, C. (2010) Evaluation of breast cancer risk by using fuzzy logic, in *Proceedings of the 10th WSEAS International Conference on Applied Informatics and Communications, and 3rd WSEAS International Conference on Biomedical Electronics and Biomedical Informatics*, World Scientific and Engineering Academy and Society (WSEAS), Stevens Point, WI, AIC'10/BEBI'10, pp. 37–42. Available from http://dl.acm.org/citation.cfm?id=2170353 .2170366.

20 Dheeba, J. and Selvi, S.T. (2012) A swarm optimized neural network system for classification of microcalcification in mammograms. *Journal of Medical Systems*, **36** (5), 3051–3061, doi: 10.1007/s10916-011-9781-3. Available from http://dx.doi.org/10 .1007/s10916-011-9781-3.

21 Dheeba, J. and Tamil Selvi, S. (2012) An improved decision support system for detection of lesions in mammograms using differential evolution optimized wavelet neural network. *Journal of Medical Systems*, **36** (5), 3223–3232, doi: 10.1007/s10916-011-9813-z. Available from http://dx.doi.org/10.1007/s10916-011-9813-z.

22 Manoharan, P., Prasenjit, N., and Sangeetha, V. (2013) Feature extraction method for breast cancer diagnosis in digital mammograms using multi-resolution transformations and SVM-fuzzy logic classifier. *IJCVR*, **3** (4), 279–292. Available from http://dx .doi.org/10.1504/IJCVR.2013.059102.

23 Guzmán-Cabrera, R., Guzmán-Sepúlveda, J.R., Torres-Cisneros, M., May-Arrioja, D.A., Ruiz-Pinales, J., Ibarra-Manzano, O.G., Aviña-Cervantes, G., and Parada, A.G. (2013) Digital image processing technique for breast cancer detection. *International Journal of Thermophysics*, **34** (8), 1519–1531, doi: 10.1007/s10765-012-1328-4. Available from http://dx.doi.org/10.1007/s10765-012-1328-4.

24 Dheeba, J., Singh, N.A., and Selvi, S.T. (2014) Computer-aided detection of breast cancer on mammograms: a swarm intelligence optimized wavelet neural network approach. *Journal of Biomedical Informatics*, **49**, 45–52. Available from http://dx.doi .org/10.1016/j.jbi.2014.01.010.

25 Pereira, D.C., Ramos, R.P., and do Nascimento, M.Z. (2014) Segmentation and detection of breast cancer in mammograms combining wavelet analysis and genetic algorithm. *Computer Methods and Programs in Biomedicine*, **114** (1), 88–101. Available from http://dx.doi.org/10.1016/j.cmpb.2014.01.014.

26 He, W., Hogg, P., Juette, A., Denton, E.R.E., and Zwiggelaar, R. (2015) Breast image pre-processing for mammographic tissue segmentation. *Computers in Biology and Medicine*, **67**, 61–73. Available from http://dx.doi.org/10.1016/j.compbiomed.2015.10 .002.

27 Sharma, S. and Khanna, P. (2015) Computer-aided diagnosis of malignant mammograms using Zernike moments and SVM. *Journal of Digital Imaging*, **28** (1), 77–90, doi: 10.1007/s10278-014-9719-7. Available from http://dx.doi.org/10.1007/ s10278-014-9719-7.

28 Kuhnigk, J.M., Dicken, V., Bornemann, L., Bakai, A., Wormanns, D., Krass, S., and Peitgen, H.O. (2006) Morphological segmentation and partial volume analysis for volumetry of solid pulmonary lesions in thoracic CT scans. *IEEE Transactions on Medical Imaging*, **25** (4), 417–434, 10.1109/TMI.2006.871547.

29 Kubota, T., Jerebko, A.K., Dewan, M., Salganicoff, M., and Krishnan, A. (2011) Segmentation of pulmonary nodules of various densities with morphological

approaches and convexity models. *Medical Image Analysis*, **15** (1), 133–154, doi: http://dx.doi.org/10.1016/j.media.2010.08.005. Available from http://www.sciencedirect.com/science/article/pii/S136184151000109X.

30 Yang, X.S. and Deb, S. (2010) Engineering optimisation by cuckoo search. *ArXiv e-prints*.

31 Vapnik, V. (1982) *Estimation of dependences based on empirical data*. Springer Series in Statistics (Springer Series in Statistics). Springer-Verlag, New York.

32 Akay, M.F. (2009) Support vector machines combined with feature selection for breast cancer diagnosis. *Expert Systems with Applications*, **36** (2, Pt. 2), 3240–3247, doi: http://dx.doi.org/10.1016/j.eswa.2008.01.009. Available from http://www.sciencedirect.com/science/article/pii/S0957417408000912.

33 Gorodkin, J. (2004) Comparing two k-category assignments by a k-category correlation coefficient. *Computational Biology and Chemistry*, **28** (5-6), 367–374, doi: http://dx.doi.org/10.1016/j.compbiolchem.2004.09.006. Available from http://www.sciencedirect.com/science/article/pii/S1476927104000799.

14

Analysis of Hand Vein Images Using Hybrid Techniques

R. Sudhakar, S. Bharathi, and V. Gurunathan

Department of Electronics and Communication Engineering, Dr. Mahalingam College of Engineering and Technology, Coimbatore, Tamil Nadu, India

14.1   Introduction

Nowadays in this informational society, there is a need of authentication and identification of individuals. Biometric authentication [1] is a common and reliable way to recognize the identity of a person based on physiological or behavioral characteristics. Biometric technology offers promise of a trouble-free and safe method to make an extremely precise authentication of a person. Alternative unique symbols like passwords and identity cards have the susceptibility of being easily misplaced, pooled, or stolen. Passwords may also be easily cracked by means of social engineering and dictionary attacks [2, 3] and offer insignificant protection. As biometric techniques make it a requirement for the client to be actually present during authentication, it prevents the possibility of clients making false repudiation claims at a later stage [4]. Several of the biometric systems installed in real-world applications are unimodal [5–7], which depends on a single source of information for authentication. Some of the restrictions enacted by unimodal biometric systems can be removed by using several sources of information for creating identity. When various sources of information are used for biometric recognition, the system is expected to be more consistent due to the existence of multiple forms of evidence. One such method is recognition of individuals by means of multimodal biometrics. Multimodal biometric recognition [8, 9] is a method of recognizing a person by using two or more biometric modalities.

The significant objective of multimodal biometrics is to improve the accuracy of recognition over a definite method by combining the results of several traits, sensors, or algorithms. The choice of the relevant modality is a demanding task in multimodal biometric systems for the identification of a person. In recent days, human authentication by means of hand vein modalities has attracted the ever-increasing interest of investigators because of its superior performance over other modalities [10–12]. A personal authentication system using finger vein patterns was proposed by Naoto Miura *et al.* [13]. They have used a repeated line-tracking method to extract features from the finger vein images and template matching method to find the similarity between the vein images. The performance of this method was improved by using a

Hybrid Intelligence for Image Analysis and Understanding, First Edition.
Edited by Siddhartha Bhattacharyya, Indrajit Pan, Anirban Mukherjee, and Paramartha Dutta.
© 2017 John Wiley & Sons Ltd. Published 2017 by John Wiley & Sons Ltd.
Companion Website: www.wiley.com/go/bhattacharyya/hybridintelligence

method of calculating local maximum curvatures in cross-sectional profiles [14] of a vein image by the same author. Also, morphological algorithm-based feature extraction for finger vein images was proposed by Mei *et al.* [15], and multifeatures such as local and global features from the same vein patterns were proposed by Yang and Zhang [16].

Lee [17] has proposed a biometric recognition system based on the palm vein. He has extracted the necessary features from the palm vein images based on its texture. A 2D Gabor filter was used by him to extract local features of the palm, and the vein features are encoded by using a directional coding technique. Finally, the similarity between the vein codes was measured by Hamming distance metric. Wu *et al.* [18] have used a directional filter bank to extract line-based features of palm vein, and a minimum directional code was proposed by them to encode the palm vein features. Determination of the nonvein pixels was done by estimating the magnitude of the directional filter, and those pixels were assigned with a non-orientation code. Xu *et al.* [19] proposed a vein recognition system using an adaptive hidden Morkov model in which the parameters are optimized by using a stepper increasing method. They have used radon transform to find the direction and location information of the dorsal vein image. They concluded that the recognition time of their system was sufficient to meet the real-time requirement. Biometric authentication using fast correlation of near-infrared (NIR) hand vein images was proposed by Shahin *et al.* [20]. The performance of the system was verified on their own database of dorsal hand vein images. They concluded that the hand vein pattern is unique for each and every person and is also distinctive for each hand.

Though the veinbased recognition systems discussed here produced satisfactory performance as unimodal systems, their performance can be improved further by combining two or more modalities. Kumar and Prathyusha [21] proposed an authentication system using a combination of hand vein and knuckle shape features. This approach is based on the structural similarity of hand vein triangulation and knuckle shape features. A weighted combination of four different types of triangulation is developed by them to extract key points from the hand vein structure, and knuckle shape features are also simultaneously extracted from the same image. Finally, these two features are combined at the score level using a weighted score-level fusion strategy to authenticate individuals. A multimodal biometric recognition system using simultaneously captured fingerprint and finger vein information from the same device was suggested by Park *et al.* [22]. The fingerprint analysis was done based on the minutiae points of the ridge area, and finger vein analysis was performed based on the local binary pattern (LBP) with the appearance information of the finger area. Their results show that the combined biometrics at the decision level provides improved performance, with a lower false acceptance rate (FAR) and false rejection rate (FRR) than the individual traits.

A contactless palm print and palm vein recognition system was proposed by Goh *et al.* [23]. They have used a linear ridge enhancement (LRE) technique for preprocessing in order to get high-contrast images. Features from palm print and palm vein images are extracted using directional coding techniques, which aim to encode the line patterns of the images. The fusion of these features was carried out by them using a support vector machine. Wang *et al.* [24] proposed a person recognition method based on Laplacian palm by fusing palm print and palm vein images. Image-level fusion was carried out for palm print and palm vein images by an edge-preserving and contrast-enhancing wavelet fusion method. The modified multiscale edges of the palm print and palm vein images are combined by them. Using the locality preserving projections (LPPs), the Laplacian palm feature is extracted from the fused images. They have observed that the results for the fused palm print and palm vein images are better than the results for the palm print and palm vein separately.

This chapter deals with biometric recognition using multiple vein images present in the hand, such as dorsal, palm, finger, and wrist vein images. In multimodal biometric recognition systems, the fusion [25] of various modalities can be done at different levels, such as the sensor level, feature level [26, 27], score level [28], and decision level [29]. The basic block diagram of a multimodal biometric recognition system using hand vein images is shown in Figure 14.1. Hand vein pattern is one of the biometric traits used for secured authentication systems. The shape of the vascular patterns in the hand is distinct for each hand. The shape of the veins changes with the change in the length of the physique from childhood, but it almost always remains stable in adult life and so is a reliable source of authentication. Veins are the network of blood vessels beneath a person's skin. Since the veins are hidden under the skin, it is almost impossible for a pretender to duplicate the vein pattern. Vein biometrics has the following advantages:

1. Vein patterns are imperceptible by human eyes due to their concealed nature.
2. Not so easy to forge or steal
3. Captured by noninvasive and contact-less methods
4. Guarantees handiness and cleanliness

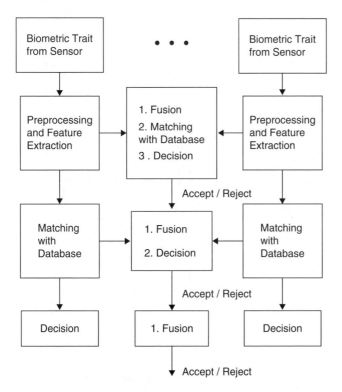

Figure 14.1 The basic block diagram of a multimodal biometric recognition system.

5. Vein pattern can be acquired from a live body only.
6. Natural and substantial evidence that the subject is alive.

This chapter deals with the vein-based biometric recognition system using hybrid intelligence techniques at both the spatial domain and frequency domain.

14.2 Analysis of Vein Images in the Spatial Domain

In this chapter, we first study the analysis of hand vein images in the spatial domain. Here, the processing of vein images is done by direct manipulation of pixels [30] in the image plan itself. The features from the vein images are extracted using a modified 2D Gabor filter [31] in the spatial domain, and they are stored in the database. Then, these extracted features are combined at the feature level using a simple sum rule [32] and at the matching-score level [33] using fuzzy logic. Finally, the authentication rate of the biometric system is evaluated. The flow diagram of the analysis of vein images in the spatial domain is shown in Figure 14.2.

14.2.1 Preprocessing

In order to modify the original input images into the required formats, the preprocessing operation will be carried out on the input images. The hand vein images captured

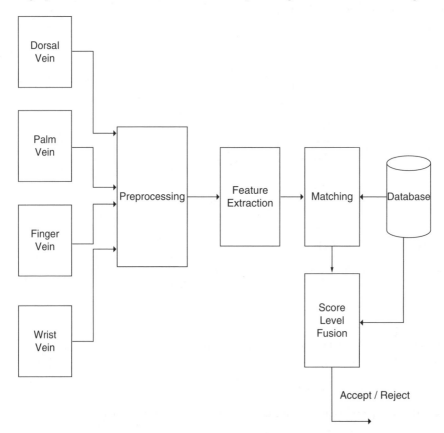

Figure 14.2 Flow diagram of the analysis of vein images in the spatial domain.

by using NIR cameras are of a low-contrast nature, because their blood and bio-cells for NIR lighting have different absorption and scattering coefficients. Hence, these images are to be preprocessed in order to enrich their features for impending analyses such as feature extraction and classification of the individuals. Here, the vein images are enhanced using a hierarchical method called *difference of Gaussian* and contrast-limited adaptive histogram equalization.

Difference of Gaussian [34] comprises subtraction of the distorted version of an original image from another, less distorted version of the original. Subtracting one image from the other conserves spatial information that lies between the range of frequencies that are conserved in the two blurred images. Hence the difference of Gaussian acts as a band pass filter that provides the required spatial information. Contrast-limited adaptive histogram equalization improves the contrast of the image, and it operates on the small regions of the image, called *blocks*, rather than the entire image. Each block's contrast is enhanced by using bilinear interpolation, and the neighboring blocks are then combined to remove artificially created boundaries. The contrast in each block can be limited to avoid the amplification of any kind of noise that exists in the image.

14.2.2   Feature Extraction

Feature extraction is a distinct form of dimensionality reduction in the field of image processing and pattern recognition. When the input data to the system is very large to be processed and it is suspected to be too redundant, then it will be represented as a set of features (also named *feature vectors*). If the features are chosen carefully and extracted, then the desired task will be performed easily on these features rather than on the entire image. From the preprocessed images, the necessary features are extracted here using a modified 2D Gabor filter.

The Gabor filter, named after Dennis Gabor, is a linear filter used for detecting edges in image processing. A 2D Gabor filter is a Gaussian kernel function in the spatial domain that is modulated by a sinusoidal plane wave. Depictions of Gabor filters in terms of frequency and orientation are similar to those of the human visual system, and hence image analysis using these filters is assumed to be similar to the perception of the human visual system. Due to its 2D spectral efficiency for texture as well as its variation with 2D spatial positions, Gabor filters are enormously useful for texture analysis. Daugman developed the 2D Gabor functions, which have good spatial localization, orientation selectivity, and frequency selectivity.

In recent years, Gabor filters have been successfully used in many applications, such as target detection, face recognition, edge detection, texture segmentation, image analysis, and compression. The general form of the Gabor filter is represented as in Equation (14.1).

$$h(x,y,f,\sigma,\theta) = \frac{1}{\sqrt{\pi\sigma_x\sigma_y}} exp\left\{\frac{-1}{2}\left(\frac{x_\theta^2}{\sigma_x^2} + \frac{y_\theta^2}{\sigma_y^2}\right)\right\} exp\{i(f_x x + f_y y)\} \qquad (14.1)$$

where:

$$x_\theta = xcos\theta + ysin\theta; \quad y_\theta = -xsin\theta + ycos\theta$$

$$\sigma_x = \frac{c_1}{f}; \quad \sigma_y = \frac{c_2}{f}; \quad f_x = f\cos\theta; \quad f_y = f\ sin\theta$$

and where the variables x and y are the pixel coordinates of the image, σ_x and σ_y are standard deviations of the 2D Gaussian envelope, c_1 and c_2 are constants, f is the central frequency of the pass band, and θ refers to the spatial orientation. The frequency values f_k are selected in the interval of 0 to π, and orientations (θ_m) are uniformly distributed in the interval of 0 to π.

Numerous Gabor filter designs are obtained by making some modifications to the analytical formula given in Equation (14.1). There are no general methods for selection of Gabor filter parameters. The real (even-symmetry) and imaginary (odd-symmetry) parts of the Gabor filter are used for ridge and edge detections, respectively. The analytical form of an even-symmetric Gabor filter is given in Equation (14.2):

$$h(x, y, f, \sigma, \theta) = \frac{1}{\sqrt{\pi \sigma_x \sigma_y}} exp\left\{\frac{-1}{2}\left(\frac{x_\theta^2}{\sigma_x^2} + \frac{y_\theta^2}{\sigma_y^2}\right)\right\} cos(f_x x + f_y y) \qquad (14.2)$$

The 2D Gabor filter is modified by introducing the smoothness parameter η, which is defined in Equation (14.3). The modified 2D Gabor filter is employed to extract the features from hand vein images because the Gabor filter is invariant to intensity of light, rotation, scaling, and translation. Furthermore, they are immune to photometric disturbances, for instance changes in the illumination of the image and noise in the image.

$$G(x, y) = \frac{f^2}{\pi \gamma \eta} exp\left\{\frac{-x'^2 + \gamma^2 y'^2}{2\sigma^2}\right\} exp\{(j2\pi f x' + \phi)\} \qquad (14.3)$$

where $x' = xcos\theta + ysin\theta$, $y' = -xsin\theta + ycos\theta$, f is the frequency of the sinusoidal factor, θ is the orientation of the Gabor function, ϕ is the phase offset, σ is the standard, deviation, and η is the smoothness parameter of the Gaussian envelope in the direction of the wave and orthogonal to it. The parameters σ and η adjust the smoothness of the Gaussian function, and γ is the spatial aspect ratio that specifies the ellipticity of the Gabor function. The frequency response of the Gabor filter is made to cover the entire frequency plane by changing the values f_{max} and θ simultaneously. The maximum frequency f_{max} can be varied from 0.1 to 2.0, θ can be varied from $\frac{\pi}{8}$ to π, and the aspect ratio from 0 to 1.5. We employed the modified Gabor filter by keeping ω and η as constants and varying the parameters θ and γ, so as to extract the features from the hand vein images; and then these extracted features are stored in the database.

14.2.3 Feature-Level Fusion

In feature-level fusion, the features of multiple biometric traits are amalgamated to form a single feature vector in order to have an appreciable recognition rate. Here, the feature vectors of hand vein images such as dorsal, palm, wrist, and finger vein images are combined using a simple sum rule to generate a single feature vector. The fused images are given in the "Results and Discussion" sections of this chapter (Sections 14.2.5 and 14.3.5).

14.2.4 Score-Level Fusion

In the score-level fusion, there are two approaches for consolidating the scores obtained from different matchers such as classification and combination techniques. In the classification method, a feature vector is created by using matching scores of the individuals, and then it is classified as "Accept" or "Reject." In the combination method, the distinct matching scores are united to generate a single scalar score for making the final decision.

After extracting the necessary features from the hand vein images, the fusion of them is done at the score level using fuzzy logic. In the recognition stage, the matching score of hand vein images is computed by using the Euclidean distance metrics. Let $p_t = (p_{t1}, p_{t2}, \ldots, p_{tn})$ be the feature vector of the test palm vein images, $p_{tr} = (p_{tr1}, p_{tr2}, \ldots, p_{trn})$ be the feature vector of the trained palm vein images, $d_t = (d_{t1}, d_{t2}, \ldots, d_{tn})$ be the feature vector of the test dorsal vein images, $d_{tr} = (d_{tr1}, d_{tr2}, \ldots, d_{trn})$ be the feature vector of the trained dorsal vein images, $w_t = (w_{t1}, w_{t2}, \ldots, w_{tn})$ be the feature vector of the test wrist vein images, $w_{tr} = (w_{tr1}, w_{tr2}, \ldots, w_{trn})$ be the feature vector of the trained wrist vein images, $f_t = (f_{t1}, f_{t2}, \ldots, f_{tn})$ be the feature vector of the test finger vein images, and $f_{tr} = (f_{tr1}, f_{tr2}, \ldots, f_{trn})$ be the feature vector of the trained finger vein images. The corresponding score values of the vein images can be calculated by the Euclidean distance using formulations (14.4) through (14.7):

$$S1 = D(p_t, p_{tr}) = \sqrt{(p_{t1} - p_{tr2})^2 + (p_{t2} - p_{tr2})^2 + \ldots + (p_{tn} - p_{trn})^2} \tag{14.4}$$

$$S1 = D(p_t, p_{tr}) = \sqrt{\sum_{i=1}^{n} (p_{ti} - p_{tri})^2}$$

$$S2 = D(d_t, d_{tr}) = \sqrt{(d_{t1} - d_{tr2})^2 + (d_{t2} - d_{tr2})^2 + \ldots + (d_{tn} - d_{trn})^2} \tag{14.5}$$

$$S2 = D(d_t, d_{tr}) = \sqrt{\sum_{i=1}^{n} (d_{ti} - d_{tri})^2}$$

$$S3 = D(w_t, w_{tr}) = \sqrt{(w_{t1} - w_{tr2})^2 + (w_{t2} - w_{tr2})^2 + \ldots + (w_{tn} - w_{trn})^2} \tag{14.6}$$

$$S3 = D(w_t, w_{tr}) = \sqrt{\sum_{i=1}^{n} (w_{ti} - w_{tri})^2}$$

$$S4 = D(f_t, f_{tr}) = \sqrt{(f_{t1} - f_{tr2})^2 + (f_{t2} - f_{tr2})^2 + \ldots + (f_{tn} - f_{trn})^2} \tag{14.7}$$

$$S4 = D(f_t, f_{tr}) = \sqrt{\sum_{i=1}^{n} (f_{ti} - f_{tri})^2}$$

Decisions based on fuzzy fusion scores: Finally, the person is authorized to compute the final score from the scores of input vein images. Score values of the individual vein images are combined using fuzzy logic to obtain the final score value. Fuzzy logic–based score-level fusion is obtained through the following steps. The first step is fuzzification: the fuzzifier receives the final matching score, and it converts the numerical data into the linguistic values (high, medium, and low) using the membership function. The membership function describes how each score is converted into a membership value (or degree of membership). The Gaussian membership function is used to obtain the degree of membership and is defined as in the following Equation (14.8).

$$Gaussian(x; c, \sigma) = exp \frac{-1}{2} \left(\frac{x - c}{\sigma} \right)^2 \tag{14.8}$$

where x represents the membership function, c represents the center, and σ represents its width.

The next step is rule designing: The main significant module in any fuzzy system is designing fuzzy rules. Four input variables and one output variable are used here for defining fuzzy if-then rules. The rules defined in the rule base are as follows:

If S1, S2, and S3 are HIGH and S4 is LOW, then output is recognized.
If S1, S2, and S4 are HIGH and S3 is LOW, then output is recognized.
If S1, S3, and S4 are HIGH and S2 is LOW, then output is recognized.
If S2, S3, and S4 are HIGH and S1 is LOW, then output is recognized.
If S1, S2, S3, and S4 are HIGH, then output is recognized.

The inference engine: receives the information from the fuzzifier, and it compares the specific fuzzified input with the rule base. Hence, the score is computed, and the inference engine provides the output in the form of linguistic values (high, medium, and low). Finally, with the defuzzifier, the linguistic values from the inference engine are converted to crisp values by the defuzzifier. The crisp value indicates the final fuzzy score and is used for authentication.

14.2.5 Results and Discussion

The experimental results of the proposed method, for analyzing the hand vein images in the spatial domain, are discussed here. The database used for investigation is obtained from the standard databases of the dorsal hand vein (Hand vein database: Prof. Dr. Ahmed M. Badawi, Systems and Biomedical Engineering, Cairo University), finger vein (Finger vein database: http: //mla.sdu.edu.cn/sdumla-hmt.html), and palm vein and wrist images (Palm and Wrist vein database: http://biometrics.put.poznan.pl/vein-dataset). The experimental outcomes of hand vein images at various stages in spatial domain are as follows. The input hand vein images and their respective preprocessed images are illustrated in the Figure 14.3.

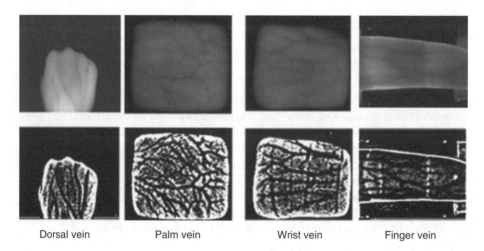

| Dorsal vein | Palm vein | Wrist vein | Finger vein |

Figure 14.3 Input hand vein images (first row). Preprocessed vein images (second row).

14.2.5.1 Evaluation Metrics

For experimental evaluation, FAR, FRR, and recognition accuracy will be considered as evaluation metrics.

False acceptance rate: FAR is the measure of the likelihood that a biometric security system will incorrectly accept an access attempt by an unauthorized user. FAR typically is stated as the ratio of the number of false acceptances divided by the number of identification attempts.

False rejection rate: FRR is the measure of the likelihood that the biometric security system will incorrectly reject an access attempt by an authorized user. FRR typically is stated as the ratio of the number of false rejections divided by the number of identification attempts.

As we have discussed in Section 14.2, the Gabor filter is designed by fixing the value of center frequency $\omega = 0.25$ and varying the value of aspect ratio and orientation simultaneously. The output of the Gabor filter for the above four kinds of hand vein images are illustrated in Figure 14.4–14.6, and 14.7. Fusion of hand vein images at the

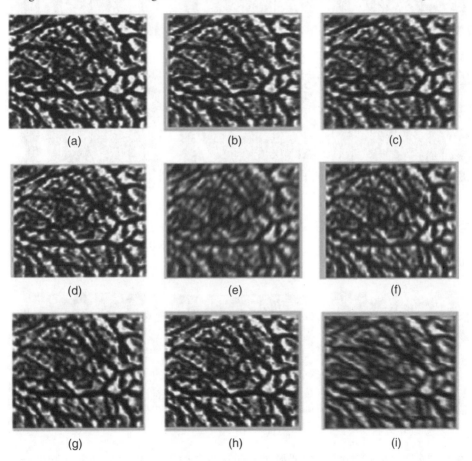

Figure 14.4 Output of Gabor filter for Palm vein images with $\omega = 0.25$: (a) $\theta = \frac{\pi}{8}, \gamma = 0.15$; (b) $\theta = \frac{2\pi}{8}$, $\gamma = 0.15$; (c) $\theta = \frac{3\pi}{8}, \gamma = 0.15$; (d) $\theta = \frac{\pi}{8}, \gamma = 0.55$; (e) $\theta = \frac{2\pi}{8}, \gamma = 0.55$; (f) $\theta = \frac{7\pi}{8}, \gamma = 0.55$; (g) $\theta = \frac{3\pi}{8}, \gamma = 0.55$; (h) $\theta = \frac{5\pi}{8}, \gamma = 0.15$; and (i) $\theta = \frac{7\pi}{8}, \gamma = 0.15$.

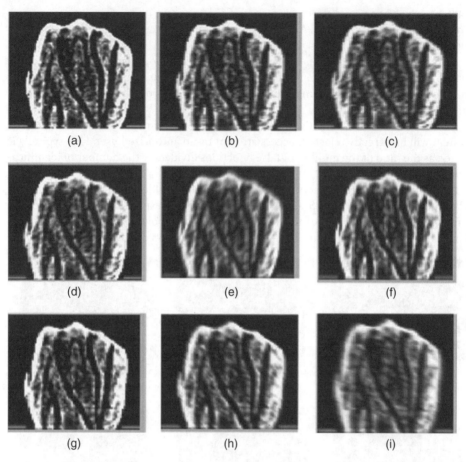

Figure 14.5 Output of Gabor filter for dorsal vein images with $\omega = 0.25$: (a) $\theta = \frac{\pi}{8}, \gamma = 0.15$; (b) $\theta = \frac{2\pi}{8}$, $\gamma = 0.15$; (c) $\theta = \frac{3\pi}{8}, \gamma = 0.15$; (d) $\theta = \frac{\pi}{8}, \gamma = 0.55$; (e) $\theta = \frac{2\pi}{8}, \gamma = 0.55$; (f) $\theta = \frac{7\pi}{8}, \gamma = 0.55$; (g) $\theta = \frac{3\pi}{8}, \gamma = 0.55$; (h) $\theta = \frac{5\pi}{8}, \gamma = 0.15$; and (i) $\theta = \frac{7\pi}{8}, \gamma = 0.15$.

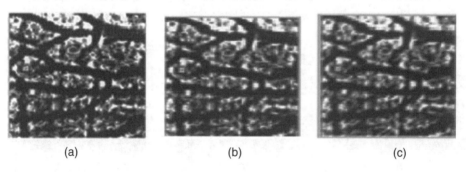

Figure 14.6 Output of Gabor filter for Wrist vein images with $\omega = 0.25$: (a) $\theta = \frac{\pi}{8}, \gamma = 0.15$; (b) $\theta = \frac{2\pi}{8}$, $\gamma = 0.15$; (c) $\theta = \frac{3\pi}{8}, \gamma = 0.15$; (d) $\theta = \frac{\pi}{8}, \gamma = 0.55$; (e) $\theta = \frac{2\pi}{8}, \gamma = 0.55$; (f) $\theta = \frac{7\pi}{8}, \gamma = 0.55$; (g) $\theta = \frac{3\pi}{8}, \gamma = 0.55$; (h) $\theta = \frac{5\pi}{8}, \gamma = 0.15$; and (i) $\theta = \frac{7\pi}{8}, \gamma = 0.15$.

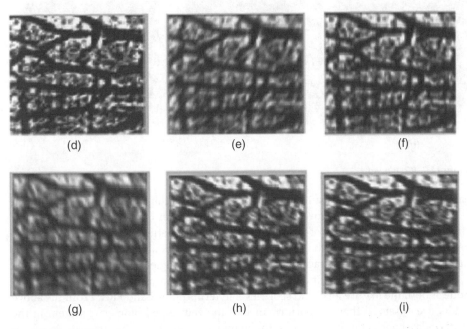

(d)   (e)   (f)

(g)   (h)   (i)

Figure 14.6 (*Continued*)

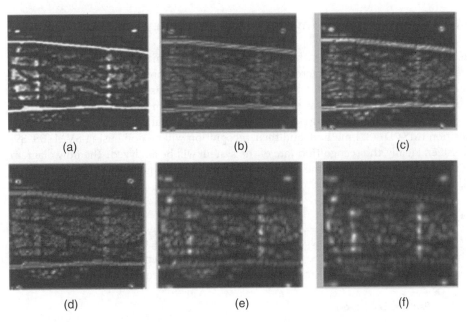

(a)   (b)   (c)

(d)   (e)   (f)

Figure 14.7 Output of Gabor filter for Finger vein images with $\omega = 0.25$: (a) $\theta = \frac{\pi}{8}, \gamma = 0.15$; (b) $\theta = \frac{2\pi}{8}$, $\gamma = 0.15$; (c) $\theta = \frac{3\pi}{8}, \gamma = 0.15$; (d) $\theta = \frac{\pi}{8}, \gamma = 0.55$; (e) $\theta = \frac{2\pi}{8}, \gamma = 0.55$; (f) $\theta = \frac{7\pi}{8}, \gamma = 0.55$; (g) $\theta = \frac{3\pi}{8}, \gamma = 0.55$; (h) $\theta = \frac{5\pi}{8}, \gamma = 0.15$; and (i) $\theta = \frac{7\pi}{8}, \gamma = 0.15$.

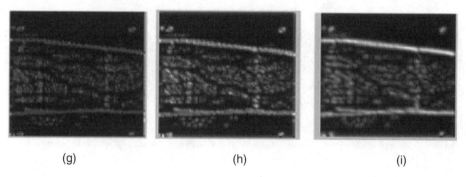

(g)　　　　　　　　　　(h)　　　　　　　　　　(i)

Figure 14.7 *(Continued)*

feature level is carried out using a sum rule, and the resultant fused images are shown in Figures 14.8 to 14.10. Figure 14.8 illustrates the fusion of two modalities of hand vein features, and the fusion of three modalities of vein features is given in Figure 14.9. Finally, Figure 14.10 shows the fusion of all four vein modalities. Table 14.1 indicates the analysis of hand vein images in unimodal mode (single biometric trait) in terms of FAR and FRR. Table 14.2 indicates the analysis of hand vein images in multimodal mode at both the feature level and score level in terms of FAR and FRR. From the above results, we observed that the performance of the vein-based biometric system in the multimodal mode is better compared to the unimodal mode.

14.3　Analysis of Vein Images in the Frequency Domain

In this section, we study how the analysis of hand vein images can be carried out in the frequency domain [35]. Here, after preprocessing the input hand vein images, the necessary features are extracted by applying contourlet transform [36]. These extracted hand vein features in the form of coefficients are stored in the database. Then, these features are fused at the feature level using the multiresolution singular value decomposition (MSVD) [37] method, and their recognition will be done using SVM [38, 39] classifier. Finally, the recognition rate of the system will be analyzed. The flow diagram of the method for analyzing the hand vein images in the frequency domain is shown in Figure 14.11.

14.3.1　Preprocessing

Here, the input hand vein images are enhanced initially using the difference of Gaussian (the same as in the above time domain technique), and then adaptive thresholding is performed in order to make the images more suitable for frequency domain analysis.

14.3.2　Feature Extraction

As seen in the discussion on time domain analysis, the feature extraction is a special form of dimensionality reduction. Here, the necessary features are extracted using the contourlet transform. The contourlet transform is a new 2D transform method for image representations. It has properties of multiresolution, localization, directionality,

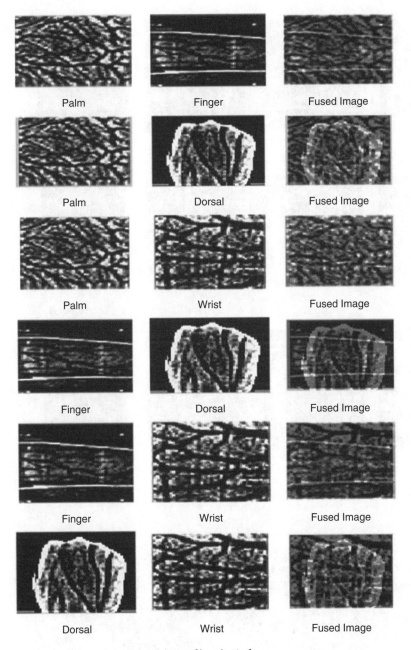

Figure 14.8 Fusion of two modalities of hand vein features.

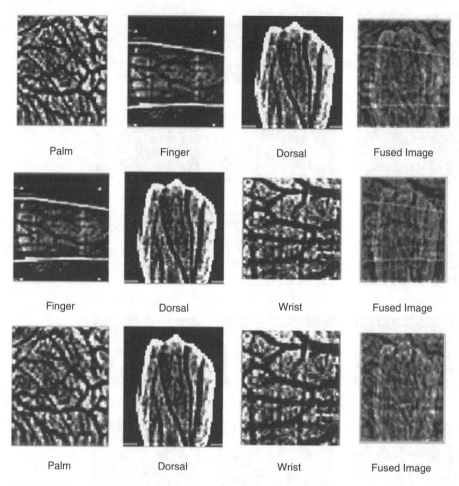

Figure 14.9 Fusion of three modalities of hand vein features.

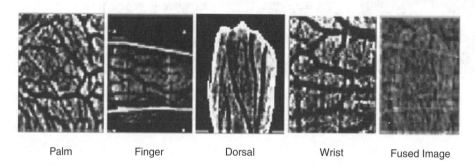

Figure 14.10 Fusion of all four modalities of hand vein features.

Table 14.1 Analysis of hand vein images in unimodal mode

Sample no.	Image	FAR	FRR	Accuracy
1	Dorsal	0.225	0.2	83.21
2	Wrist	0.2	0.15	81.75
3	Palm	0.3	0.175	82.42
4	Finger	0.2	0.125	84.37

Table 14.2 Analysis of hand vein images in multimodal mode

Sample no.	Image	Feature-level fusion		Accuracy	Score-level fusion		Accuracy
		FAR	FRR		FAR	FRR	
1	Palm + finger	0.14	0.12	87.18	0.12	0.11	88.45
2	Palm + dorsal	0.12	0.2	86.75	0.12	0.12	89.36
3	Palm + wrist	0.14	0.12	85.5	0.11	0.1	88.12
4	Finger + dorsal	0.16	0.2	85.72	0.1	0.1	90.37
5	Finger + wrist	0.1	0.12	86.5	0.08	0.06	88.75
6	Dorsal + wrist	0.2	0.2	85.75	0.1	0.1	89.67
7	Palm + finger + dorsal	0.05	0.04	90.12	0	0.02	94.4
8	Finger + dorsal + wrist	0.04	0.02	88.78	0.012	0.01	93.67
9	Palm + dorsal + wrist	0.02	0.04	91.37	0.012	0	94.12
10	Palm + dorsal + finger + wrist	0.011	0.01	93.75	0.01	0	95.38

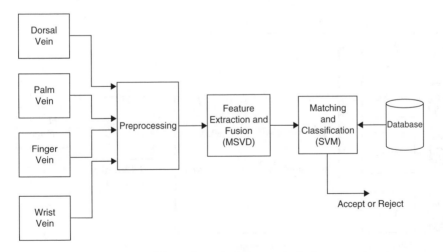

Figure 14.11 Flow diagram of the analysis of vein images in the frequency domain.

critical sampling, and anisotropy. Its basic functions are multiscaling and multidimensioning. A double filter bank structure that comprises a Laplacian pyramid (LP) and directional filter bank (DFB) is used by the contourlet transform in order to get the smooth contours of images. With the combination of the DFP and Laplacian pyramid, the multiscale decomposition is achieved and also the low-frequency components are removed. Hence, the image signals are passed through LP subbands in order to get band pass signals and through DFB to acquire the directional information of the image.

14.3.3 Feature-Level Fusion

In order to combine the features of multiple biometric traits, here the MSVD technique is used. MSVD is very similar to wavelet transform, and here the idea is to replace the finite impulse response filters with singular value decomposition (SVD). The schematic of the fusion for the hand vein images is shown in Figure 14.12.

The vein images to be fused are decomposed into L ($l=1, 2,\ldots, L$) levels using MSVD. At each decomposition level ($l=1, 2,\ldots, L$), the fusion rule will select the sum of the four MSVD detailed coefficients, since the detailed coefficients correspond to sharper intensity changes in the image, such as lines and edges of the vein images and so on. The approximation coefficients are not considered for fusion since the approximation coefficients at the coarser level are the smoothed and subsampled version of the original image and similarly MSVD eigen-matrices. The fused vein image can be represented as Equation (14.9)

$$I_f = \{{}^f\psi_l^v, {}^f\psi_l^H, {}^f\psi_l^L\} \tag{14.9}$$

Here, the contourlet transformed vein images are fused using MSVD so that the improved vein features were achieved. These fused images are classified using SVM classifier in order to know whether the person is authenticated or not.

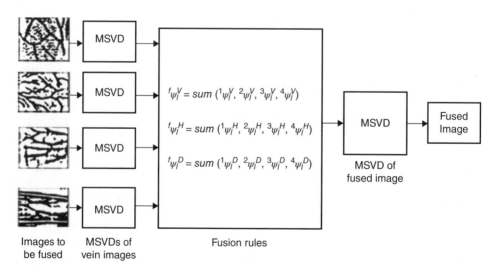

Figure 14.12 Fusion for the hand vein images in the frequency domain.

Figure 14.13 A linear support vector machine.

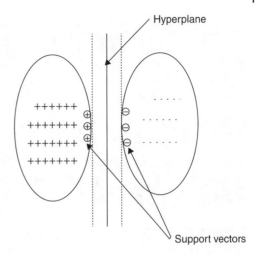

14.3.4 Support Vector Machine Classifier

SVMs are one of the supervised learning methods used for classification, regression, and outlier detection. The concept of a hyperplane is used by the classifier to calculate a border between groups of points. The objective of the SVM is to find not only a separation between the classes but also the optimum hyperplane. Figure 14.13 shows the principle of a linear SVM.

Let the training samples be represented as pairs $(\vec{x}_i - y_i)$, where $\vec{x}_i$ is the weighted feature vector of the training sample and $y_i \in (-1, +1)$ is the label. For linearly separable data, we can determine a hyperplane $f(x) = 0$ that separates the data, as shown in Equation (14.10):

$$f(x) = \sum_{i=1}^{n} wx_i + b = 0 \tag{14.10}$$

where w is an n-dimensional vector and b is a scalar value. The vector w and the scalar b determine the position of the separating hyperplane. For each i, either is as shown in Equation (14.11):

$$\begin{cases} wx_i - b \geq 1 \ \textit{for one class} \\ wx_i - b \leq -1 \ \textit{for another class} \end{cases} \tag{14.11}$$

14.3.5 Results and Discussion

The experimental results of the proposed method, for analyzing the hand vein images in the frequency domain, are discussed here. As mentioned in Section 14.2.5, the database utilized for our experimentation is taken from the standard databases of dorsal hand vein, palm vein, finger vein, and wrist images. The experimental results of hand vein images at various stages are as follows. Figure 14.14 shows the input hand vein images and their respective preprocessed images.

The feature-extracted hand vein images using contourlet transform are shown in Figure 14.15. These extracted vein features are fused by using MSVD, and this fused vector is given to a SVM for classification. Here we have used linear SVM in order

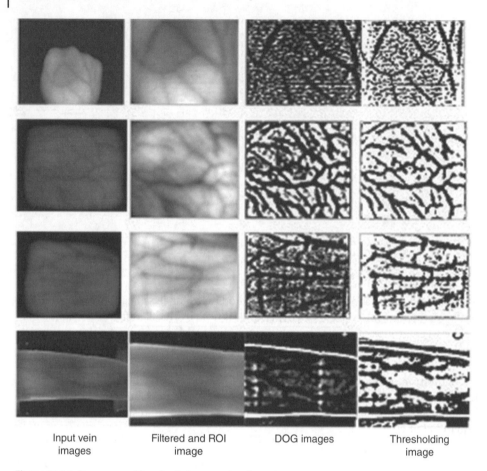

| Input vein images | Filtered and ROI image | DOG images | Thresholding image |

Figure 14.14 Preprocessed hand vein images at various stages.

to classify the features, which in turn enables us to know whether the person is authenticated or not.

Table 14.3 illustrates the frequency domain analysis of hand vein images in terms of sensitivity, specificity, and accuracy. From the table, we observed that the accuracy rate of recognition of the fused image is as high as 96.66% compared to the unimodal (single biometric trait) system. Also, SVM classifier provides a good accuracy rate of above 90% irrespective of the modalities of the biometric system.

14.4 Comparative Analysis of Spatial and Frequency Domain Systems

This section deals with the comparison of spatial and frequency domain systems and the comparative analysis of multimodal biometric systems.

Biometrics is a technology of analyzing human characteristics to authenticate a person in a secured manner. In order to identify a person, the biometric features can be

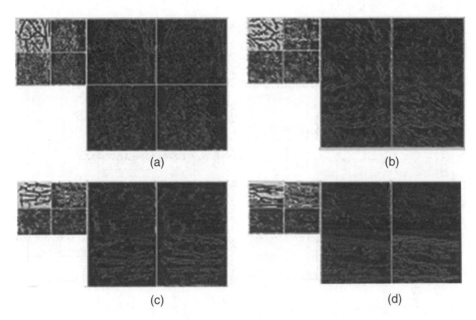

(a)

(b)

(c)

(d)

Figure 14.15 Contourlet-transformed vein images: (a) dorsal hand vein, (b) palm vein, (c) wrist vein, and (d) finger vein.

Table 14.3 Frequency domain analysis of hand vein images

Sample no.	Image	FAR	FRR	Sensitivity	Specificity	Accuracy
1	Dorsal	0.12	0.03	0.909	0.95	92.95
2	Wrist	0.2	0.11	0.88	0.95	91.5
3	Palm	0.15	0.02	0.887	0.9375	91.22
4	Finger	0.11	0.04	0.889	0.9411	91.803
5	Fused vein image	0.001	0	0.9687	0.9642	96.66

extracted and analyzed using spatial and frequency domain methods. There are numerous techniques available in these domains for analyzing the features.

In the spatial domain, the required features are extracted from the biometric traits directly by applying spatial domain techniques on image pixels without converting them into the transform domain. Here, a modified 2D Gabor filter is used to extract the features from the vein modalities in the spatial domain, and the extracted features are combined as a single feature vector at both the feature level and score level for multimodal recognition.

In the frequency domain, initially the biometric traits are converted from the spatial domain to the transform domain, and then the required features are extracted in the form of frequency coefficients. Here, a contourlet transform is used to extract the features from the vein modalities in the frequency domain, and the extracted frequency coefficients are combined at the feature level in order to have multimodal recognition. Features extracted using both the spatial domain and frequency domain techniques can be combined to obtain the hybrid features, and these features can be used for

Table 14.4 Comparative analysis of related work on vein-based biometric recognition

Author	Biometric features	Methodology	No. of users	Performance
Lin et al. [40]	Palm-dorsal vein	Multiresolution analysis and combination	32	$FAR = 1.5$ $FRR = 3.5$ $EER = 3.75$
Kumar [21]	Hand vein and knuckle shape	Matching vein triangulation shape features	100	$FAR = 1.14$ $FRR = 1.14$
Raghavendra et al. [41]	Hand vein and palm print	Log Gabor transform	50	$FAR = 7.4$ $FRR = 4.8$
Raghavendra et al. [41]	Hand vein and palm print	Nonstandard mask	50	$FAR = 2.8$ $FRR = 1.4$
Ferrer et al. [42]	Hand geometry, palm and finger textures, dorsal hand vein	Simple sum rule	50	$FAR = 0.01$ $FRR = 0.01$ $EER = 0.01$
Yang et al. [16]	Finger vein and fingerprint	Supervised local-preserving canonical correlation analysis method	640	$FAR = 1.35$ $FRR = 0$
Li et al. [43]	Finger vein and finger shape	Finger vein network Extraction algorithm	250	$FAR = 2.25$
Rattani et al. [44]	Face and fingerprint	Scale-invariant feature Transform features Minutiae matching	400	$FAR = 4.95$ $FRR = 1.12$
Yuksel [45]	Hand vein	ICA1 ICA2 LEM NMF	100	$EER = 5.4$ $EER = 7.24$ $EER = 7.64$ $EER = 9.17$
Proposed method	Dorsal hand vein, palm vein finger vein, and wrist vein	Gabor filter, sum rule, and fuzzy logic	100	$FAR = 0.01$ Accuracy = 95.38
Proposed method	Dorsal hand vein, palm vein finger vein, and wrist vein	Contourlet transform and MSVD	100	$FAR = 0.001$ $FRR = 0$ Accuracy = 96.67

recognition. The comparative analysis of the related work on vein-based authentication is given in Table 14.4. The results offered in this table are in terms of equal error rate (EER), FAR, and FRR. EER is defined as a point at which FAR is equal to FRR. The lower the values of EER, the better the performance of the system. The performance of the system varies according to the imaging technique, type of biometric trait, number of traits used for fusion, methodologies used for feature extraction and fusion, level of fusion of these features, and number of users in the database.

14.5 Conclusion

Analysis of hand vein images such as dorsal, palm, wrist, and finger vein images is carried out in the spatial domain as well as the frequency domain. In the spatial domain, a modified Gabor filter is employed to extract the features from vein images, and the fusion of these vein features is done at both the feature level and score level. The results show that the score-level fusion of vein features using fuzzy logic provides improved results compared to the feature-level fusion. Similarly, in the frequency domain, a contourlet transform is utilized for feature extraction, and the fusion of these features is done at the feature level using the multiresolution singular value decomposition technique. Finally, these are classified with the help of SVM classifier, and the results show higher accuracy of 96.66%. Hence, by analyzing hand vein images with the help of hybrid intelligence techniques, the improved results in terms of higher recognition accuracy with lower false acceptance and false rejection rates are obtained. Furthermore, the results can be improved with hybrid features, which are obtained by combining both spatial and frequency domain features.

References

1 Jain, A.K., Ross, A. and Prabhakar, S. (2004) An introduction to biometric recognition. *IEEE Transactions on Circuits and Systems for Video Technology*, **14** (1), 4–20.

2 Akhtar, Z., Fumera, G., Marcialis, G. and Roli, F. (2012) Evaluation of multimodal biometric score fusion rules under spoof attacks, in *Biometrics, 2012, Proceedings of the Third IEEE International Conference*, IEEE, pp. 402–407.

3 Jain, A.K., Ross, A. and Prabhakar, S. (2006) Biometrics: a tool for information security. *IEEE Transactions on Information Forensics and Security*, **1** (2), 125–143.

4 Jain, A.K., Nandakumar, K., Lu, X. and Park, U. (2004) Integrating faces, finger-prints and soft biometric traits for user recognition, in *Authentication, 2004. BioAW 2004, Proceedings of ECCV International Workshop*, pp. 259–269.

5 Khan, M.H.-M., Khan, N.M. and Subramanian, R.K. (2010) Feature extraction of dorsal hand vein pattern using a fast modified PCA algorithm based on Cholesky decomposition and Lanczos technique. *World Academy of Science, Engineering and Technology*, **61** (2), 279–283.

6 Wang, J., Yu, M., Qu, H. and Li, B. (2013) Analysis of palm vein image quality and recognition with different distance, in *Digital Manufacturing and Automation, 2013, Proceedings of Fourth International Conference*, pp. 215–218.

7 Liu, Z. and Song, S.L. (2012) An embedded real-time finger-vein recognition system for mobile devices. *IEEE Transactions on Consumer Electronics*, **58** (2), 522–527.

8 Sanjekar, P.S. and Patil, J.B. (2013) An overview of multimodal biometrics. *Signal & Image Processing: An International Journal*, **4** (1), 57–64.

9 Zhu, L. and Zhang, S. (2010) Multimodal biometric identification system based on finger geometry, knuckle print and palm print. *Pattern Recognition Letters*, **31**, 1641–1649.

10 Al-Juboori, A.M., Bu, W., Wu, X. and Zhao, Q. (2013) Palm vein verification using Gabor filter. *International Journal of Computer Science Issues*, **10** (1), 678–684.

11 Zhang, Y.-B., Li, Q., You, J. and Bhattacharya, P. (2007) Palm vein extraction and matching for personal authentication. *Advances in Visual Information Systems*, **4781**, 154–164.

12 Yang, J. and Shi, Y. (2012) Finger-vein ROI localization and vein ridge enhancement. *Pattern Recognition Letters*, **33** (12), 1569–1579.

13 Miura, N., Nagasaka, A. and Miyatake, T. (2004) Feature extraction of finger-vein patterns based on repeated line tracking and its application to personal identification. *Machine Vision and Applications*, **15** (4), 194–203.

14 Miura, N., Nagasaka, A. and Miyatake, T. (2005) Extraction of finger-vein patterns using maximum curvature points in image profiles, in *Machine Vision Applications, 2005, Proceedings of IAPR Conference*, pp. 347–350.

15 Mei, C.-L., Xiao, X., Liu, G.-H., Chen, Y. and Li, Q-A. (2009) Feature extraction of finger-vein image based on morphologic algorithm, in *Fuzzy Systems and Knowledge Discovery, 2009, Proceedings of IEEE International Conference*, IEEE, pp. 407–411.

16 Yang, J. and Zhang, X. (2010) Feature-level fusion of global and local features for finger-vein recognition, in *Signal Processing, 2010, Proceedings of IEEE International Conference*, IEEE, pp. 1702–1705.

17 Lee, J.C. (2012) A novel biometric system based on palm vein image. *Pattern Recognition Letters*, **33** (12), 1520–1528.

18 Wu, K.-S., Lee, J.-C., Lo, T.-M., Chang, K.-C. and Chang, C.-P. (2013) A secure palm vein recognition system. *The Journal of Systems and Software*, **86** (11), 2870–2876.

19 Jia, X., Cui, J., Xue, D. and Pan, F. (2012) A adaptive dorsal hand vein recognition algorithm based on optimized HMM. *Journal of Computational Information Systems*, **8** (1), 313–322.

20 Shahin, M.K., Badawi, A. and Kamel, M. (2007) Biometric authentication using fast correlation of near infrared hand vein patterns. *International Journal of Biological and Medical Sciences*, **2** (3), 141–148.

21 Kumar, A. and Prathyusha, K.V. (2009) Personal authentication using hand vein triangulation and knuckle shape. *IEEE Transactions on Image Processing*, **18** (9), 2127–2136.

22 Park, Y.H., Tien, D.T., Lee, H.C., Park, K.R., Lee, E.C., Kim, S.M. and Kim, H.C. (2011) A multimodal biometric recognition of touched fingerprint and finger-vein, in *Multimedia and Signal Processing, 2011, Proceedings of International Conference*, pp. 247–250.

23 Goh, K.O.M., Tee, C. and Teoh, A.B.J. (2010) Design and implementation of a contactless palm print and palm vein sensor, in *Control, Automation, Robotics and Vision, 2010, Proceedings of International Conference*, pp. 1268–1273.

24 Wang, J.G., Yau, W.Y., Suwandy, A. and Sung, E. (2008) Person recognition by fusing palm print and palm vein images based on laplacian palm representation. *Pattern Recognition*, **41**, 1514–1527.

25 Ross, A. and Jain, A. (2003) Information fusion in biometrics. *Pattern Recognition Letters*, **24** (24), 2115–2125.

26 Chin, Y.J., Ong, T.S., Teoh, A.B.J. and Goh, K.O.M. (2014) Integrated biometrics template protection technique based on finger print and palm print feature-level fusion. *Journal of Information Fusion*, **18**, 161–174.

27 Yang, J. and Zhang, X. (2012) Feature level fusion of fingerprint and finger-vein for personal identification. *Pattern Recognition Letters*, **33** (5), 623–628.

28 Hanmandlu, M. Grover, J., Gureja, A. and Gupta, H.M. (2011) Score level fusion of multimodal biometrics using triangular norms. *Pattern Recognition Letters*, **32** (14), 1843–1850.

29 Saleh, I.A. and Alzoubiady, L.M. (2014) Decision level fusion of iris and signature biometrics for personal identification using ant colony optimization. *International Journal of Engineering and Innovative Technology*, **3** (11), 35–42.

30 Gonzalez, R.C. and Woods, R.E. (2009) *Digital Image Processing*, 3rd ed. Prentice Hall, Upper Saddle River, NJ.

31 Padilla, P., Gorriz, J.M., Ramirez, J., Chaves, R., Segovia, F., Alvarez, I., Salas-Gonzalez, D., Lopez, M. and Puntonet, C.G. (2010) Alzheimer's disease detection in functional images using 2D Gabor wavelet analysis. *IET Electronics Letters*, **46** (8), 556–558.

32 De Marsico, M., Nappi, M., Riccio, D. and Tortora, G. (2011) Novel approaches for biometric systems. *IEEE Transactions Systems, Man, and Cybernetics*, **41** (4), 481–493.

33 Sim, H.M., Asmuni, H., Hassan, R. and Othman, R.M. (2014) Multimodal biometrics: weighted score level fusion based on non-ideal iris and face images. *Expert Systems with Applications*, **41** (11), 5390–5404.

34 Yuan, Y., Liu, Y., Dai, G., Zhang, J. and Chen, Z. (2014) Automatic foreground extraction based on difference of Gaussian. *The Scientific World Journal*, **2014**, 1–9.

35 Conti, V., Militello, C., Sorbello, F. and Vitabile, S. (2010) A frequency-based approach for features fusion in fingerprint and iris multimodal biometric identification systems. *IEEE Transactions Systems, Man, and Cybernetics*, **40** (4), 384–395.

36 Do, M.N. and Vetterli, M. (2005) The contourlet transform: an efficient directional multiresolution image representation. *IEEE Transactions on Image Processing*, **14** (12), 2091–2106.

37 Naidu, V.P.S. (2011) Image fusion technique using multi-resolution singular value decomposition. *Defence Science Journal*, **61** (5), 479–484.

38 Kumar, G.S. and Devi, C.J. (2014) A multimodal SVM approach for fused biometric recognition. *International Journal of Computer Science and Information Technologies*, **5** (3), 3327–3330.

39 Wu, J.D. and Liu, C.T. (2011) Finger vein pattern identification using SVM and neural network technique. *Expert Systems with Applications*, **38**, 14284–14289.

40 Lin, C.L. and Fan, K.C. (2004) Biometric verification using thermal images of palm-dorsa vein patterns. *IEEE Transactions on Circuits and Systems for Video Technology*, **14** (2), 199–213.

41 Raghavendra, R., Imran, M., Rao, A. and Kumar, G.H. (2010) Multi modal biometrics: analysis of hand vein and palm print combination used for personal verification, in *Emerging Trends in Engineering and Technology, 2010, Proceedings of IEEE International Conference*, pp. 526–530.

42 Ferrer, M.A., Morales, A., Travieso, C.M. and Alonso, J.B. (2009) Combining hand biometric traits for personal identification, in *Security Technology, 2009, Proceedings of IEEE International Carnahan Conference*, IEEE, pp. 155–159.

43 Li, Z., Sun, D., Liu, D. and Liu, H. (2010) Two modality-based bi-finger vein verification system, in *Signal Processing, 2010, Proceedings of IEEE 10th International Conference*, IEEE, pp. 1690–1693.

44 Rattani, A., Kisku, D.R., Gupta, P. and Sing, J.K. (2007) Feature level fusion of face and fingerprint biometrics, in *Biometric: Theory, Applications and Systems, 2007, Proceedings of International Conference*, pp. 1–6.

45 Yuksel, A., Akarun, L. and Sankur, B. (2010) Biometric identification through hand vein patterns, in *Emerging Techniques and Challenges for Hand Based Biometrics, 2010, Proceedings of International Workshop*, pp. 1–6.

15

Identification of Abnormal Masses in Digital Mammogram Using Statistical Decision Making

Indra Kanta Maitra[1] and Samir Kumar Bandyopadhyay[2]

[1]*Department of Information Technology, B.P. Poddar Institute of Management and Technology, Kolkata, West Bengal, India*
[2]*Department of Computer Science and Engineering, University of Calcutta, Salt Lake Campus, Kolkata, West Bengal, India*

15.1   Introduction

Great fear and apprehension are related with the word *cancer*; although multidisciplinary scientific studies are undertaking best efforts to combat this disease, the advent of curative medicine for a perfect cure has not yet been discovered [1]. Around 5000 years ago, ayurvedic doctors treated patients with abnormal growths or tumors [23]. Two eminent ayurvedic classics, Charaka [29] and Sushrutasanhitas [3], defined cancer as inflammatory or non-inflammatory swelling, and mentioned them as either Granthi (minor neoplasm) or Arbuda (major neoplasm). According to the modern medical point of view, the definition of cancer is referred to a large number of conditions, characterized by abnormal, progressive, and uncontrolled cell division that destroys surrounding healthy tissue. They have the ability to metastasize and spread throughout the body [23].

A lump or mass, called a *tumor*, is developed by most types of cancer, which are termed after the organ of the body part where the tumor initiates. Breast cancer starts in breast tissue, which is made up of milk producing glands called lobules and ducts that link lobules to the nipple. A tumor with malignancy has the capacity to spread beyond the breast to other organs of the body via the lymphatic system and bloodstream [32].

15.1.1   Breast Cancer

Breast cancer is in the second and fifth positions related to occurrence and death, respectively, among all types of cancer around the world. So, it is an important public health problem in the world for women, specially aged women. Breast cancer is growing throughout the world; about one million women are diagnosed with breast cancer annually, and more than 400,000 die from it [8, 30]. The scenario of developing countries is more serious, where the incident has increased as much as 5% per year [9, 30]. As reported by Tata Memorial Hospital, the second most common cancer in Indian females is breast cancer. The frequency is more in developed parts of urban areas than in the countryside, and it is predominant in the advanced socioeconomic section

Hybrid Intelligence for Image Analysis and Understanding, First Edition.
Edited by Siddhartha Bhattacharyya, Indrajit Pan, Anirban Mukherjee, and Paramartha Dutta.
© 2017 John Wiley & Sons Ltd. Published 2017 by John Wiley & Sons Ltd.
Companion Website: www.wiley.com/go/bhattacharyya/hybridintelligence

due to rapid urbanization and westernization of lifestyle. Only the initial detection and diagnosis are the means of control, but they comprise a major hurdle in India due to absence of consciousness and lethargy among Indian females toward healthcare and systematic checkup, and the unavailability of appropriate healthcare setups, especially for breast cancer treatment.

15.1.2 Computer-Aided Detection/Diagnosis (CAD)

Early and efficient detection, followed by appropriate diagnosis, is the most effective way to have successful treatment and reduce mortality. A specialized breast screening is performed to detect the presence of abnormalities such as tumors and cysts in women's breasts, whereas biopsy is the process to identify malignancies. Several screening techniques are available for examination of the female breast, including commonly used methods such as ultrasound imaging (i.e. a high-frequency band of sound waves that investigates the breast); magnetic resonance imaging (MRI), which examines the breast by the use of powerful magnetic fields; and mammography, which yields X-ray images of the breast; among others. Among all screening associated with clinical breast examination, digital mammography has been shown to be the most efficient and consistent screening method for breast tumor detection at the nascent stage.

Computer technology has had a remarkable impact on medical imaging. CAD is a comparatively young interdisciplinary subject containing components of artificial intelligence and computer vision along commonly used methods with radiological image processing. CAD techniques can reduce the time of medical image analysis and evaluation by radiologists substantially. The evaluation of medical images, however, is still almost entirely dependent on human intervention, but in near future the scenario is expected to change. Computers will be applied more often for image evaluation. CAD is typically applied in areas to detect visible structures and segments more precisely and economically. Today, CAD has been applied in routine clinical exercises and used as a "second opinion" in assisting radiologists in image interpretation, including breast tumor detection.

15.1.3 Segmentation

In CAD, image segmentation is an important step, which segregates the medical image into dissimilar non-overlapping regions such that each section is nearly homogeneous and ideally resembles some anatomical part or region of interest (ROI). The precision of CAD to identify the presence of abnormal masses on mammogram images involves an accurate segmentation algorithm. This is regarded as one of the primary step toward image analysis. The process of segmentation refers to the decomposition of a screen into its constituent parts, so that the problem to be solved depends on each level of subdivision. Segmentation is discontinued once the ROI of a particular application has been segregated, and edge points of an image are obtained. The segmentation technique can be generally categorized into five main classes: threshold based, edge based, region based, clustered based, and water-shaded based. The two main features of a gray-level image are edge and region. The radiological images, specifically DICOM images, are typically grayscale images. Segmentation algorithms for gray images primarily depend on two main features (i.e. values of image intensity; discontinuity and similarity).

An initial approach is to decompose an image based on abrupt variations in intensity, which is the primary feature of image edge detection. By definition, edges comprise a process to identify significant discontinuity (i.e., edges are substantial local variations of intensity), so the discontinuities mean sudden alterations of pixel intensity of the image. Thus, causes for variations of intensity are due to geometric and non-geometric events: geometric events deal with discontinuity in entropy and/or color depth and texture (i.e., object margin) and discontinuity in plain and/or color, and non-geometric events are direct reflections of light called specularity and inner reflection or shadow from other objects or the same objects. It can also be accurately applied to define size, shape, and spatial correlation between anatomical structures from the intensity distributions exhibited by radiological medical images.

The latter approach is primarily based on dividing the image into regions that are similar according to a set of specific conditions. The region-growing method is an example of latter approach. The region-growing method checks the neighboring pixels within one region having the same or similar value. Seeded region growing is a variation of the commonly used methods region growing method depending on the similarity of pixels within the region. In seeded region growing, seed(s) can be collected automatically or by manual means. A set of seeds is applied by this method for partitioning an image into dissimilar regions. Each seeded region is an associated constituent comprising one or more points and is characterized by a set.

The proposed method described in this chapter is to identify the abnormalities (i.e., the presence of abnormal masses). The fundamental idea behind the research is to segment out the abnormal region(s) from the entire breast regions. In the process of segmentation, two different approaches have been used. Intensity values are used to generate edge from mammogram images to differentiate regions by boundaries within breast ROI, whereas similarity and dissimilarity features are used by the commonly used methods seeded region-growing algorithm to differentiate the anatomical regions including abnormal region(s) by statistical decision making, which is described in subsequent sections of this chapter.

15.2 Previous Works

Segmentation or abnormality detection is the primary step in mammographic CAD. A brief review has been performed on alternative approaches proposed by different authors. The discussion is restricted in segmentation of mammographic masses, describing their main features and highlighting the differences among normal and abnormal masses. The key objective of the discussion is to mention methods in a nut shell and, later, compare different approaches with the proposed method. CAD being an interdisciplinary area of research, people from different disciplines like statistics, mathematics, computer science, and medicine have contributed significantly to enrich the topic.

A substantial amount of research has been performed in order to enhance the accuracy of CAD. As discussed earlier, the primary objective of mammogram CAD is to segment out the abnormalities. Quite a few significant works have been conducted using different methods to improve the segmentation accuracy and effective identification of image features. Among these automated and semi-automated methods, a dynamic weighted constant enhancement algorithm that combines adaptive filtering and edge detection [25], an adaptive multilayer topographic region-growing algorithm [35], a gray-level-based iterative and linear segmentation algorithm [6], a dynamic programming approach [33], dynamic contour modeling [4], and so on are some of the significant approaches to segment mass lesions from surrounding breast tissues.

The spiculated masses have been detected by Kegelmeyer *et al.* [12] using orientation of local edge and laws of texture features, but this is not applicable for the finding of nonspiculated masses. Karssemeijer *et al.* [11] applied a statistical analysis approach to determine pixel orientations to isolate the stellate alterations on mammogram. Markov random fields are used by Comer *et al.* [7] and Li *et al.* [14] to differentiate regions based on texture in a mammogram.

A fully automated method developed by Jiang *et al.* [10] comprises three steps. In the first step, the maximum entropy norms are applied in the ROI that is obtained after correcting the background trend to enhance the initial outlines of a mass. After this, an active-contour model is used to enhance the outlines that were formed earlier. Finally, identification of the spiculated lines attached to the mass boundary is performed with the aid of a special line detector.

Kobatake *et al.* [13] proposed a system focused on the solution of two problems: first, how to identify tumors as apprehensive areas with a very low contrast to their background; and, second, how to isolate features that depict malignant tumors. For the initial problem, they proposed a unique adaptive filter called the iris filter to identify abnormalities, whereas malignant tumors are differentiated from other tumors using proposed typical parameters.

Yang *et al.* [34] introduced a series of preprocessing steps that are designed to enhance the intensity of a ROI, remove the noise effects, and locate apprehensive masses using five texture features generated from the spatial gray-level difference matrix (SGLDM) and fractal dimension. Finally, the mass extraction was performed using entropic thresholding techniques coupled with a probabilistic neural network (PNN).

A two-step procedure was proposed by Özekes *et al.* [24] to specify the ROIs and classify ROI depending on a set of rules. Initially, pixel intensity of mammogram images is applied and reads the pixels in eight directions using various thresholds. Finally, connected component labeling (CCL) is used to label all ROIs, and two rules based on Euclidean distance are applied to classify ROIs as true masses or not.

Campanini *et al.* [5] proposed a multiresolution over-complete wavelet representation that has classified the image with redundant information. The vectors of the massive space acquired are then fed to a first SVM classifier. The detection task is categorized into a two-class pattern recognition problem. At the initial stage, the mass is categorized as suspect or not, by using this SVM classifier, then the false candidates are discarded with a second cascaded SVM. In order to reduce the number of false positives further, an ensemble of experts is used to obtain the suspicious regions by using a voting strategy. Rejani *et al.* [28] proposed a four-step scheme: mammogram enhancement using filtering, a top hat operation, DWT, and segmentation of the tumor area by thresholding, extraction of features from the segmented tumor region(s), and SVM classifier applied to identify abnormalities. Martins *et al.* [22] proposed another SVM-based approach for mass detection and classification on digital mammogram. The K-means algorithm is used to segment image, whereas texture of segmented structure is analyzed by the commonly used methods co-occurrence matrix. The SVM is applied to classify masses, and nonmasses, structure using shape and texture feature.

Here, different automatic and semi-automatic methods of segmentation of mammogram masses are presented and reviewed. Special emphasis is given to the approaches and classification methods. The approaches included in the chapter are statistical analysis–based, active-contour–based, adaptive filter–based, neural network–based, threshold-based, and SVM classification–based techniques. The majority of the research papers have not reported the accuracy estimation of their proposed methods. In this chapter, only those research papers are selected that clearly stated the accuracy estimation, so that later these can be compared with the proposed method.

15.3 Proposed Method

Mammography has been proven to be the most viable and effective method to detect early signs of breast abnormalities like benign and malignant masses. The masses are tissues that are absent in normal breast anatomy, and they are called abnormal masses. The density of these abnormal masses is different, along with their variable size and shape. Mammography is a type of radiography. The ionizing radiation produced by radiography technology determines the internal structure of a person depending on the density of body parts. Due to the heterogeneity of density, abnormal masses leave a unique impression in the mammogram image. The proposed method is a set of sequential algorithms to extract and analyze that impression to identify the abnormality (i.e., presence of abnormal masses in a digital mammogram image). The input of the proposed method is raw digital mammogram images, and the output is the decision whether there is any abnormality present or not. These proposed decision-making systems consist of three sequential distinct subsections, namely, preparation, preprocessing, and identification of abnormal region(s). The schematic diagram of the proposed method is depicted in Figure 15.1.

15.3.1 Preparation

Medical images are challenging to interpret; thus, a preparation method is needed in order to increase the image quality and make the segmentation outcome more precise. Here, the efficiency of the decision making depends on the accuracy of segmentation. The preparation process basically standardizes the image for further processing, and standardization of input is the key to any image-processing algorithm. The preparation phase consists of three different steps, namely, image orientation, artifact removal, and denoising.

The two most common mammographic views are mediolateral oblique (MLO) and craniocaudal (CC). MLO is much accepted because its orientation along the horizontal axis shows the entire gland of the breast, whereas CC shows only the central and inner breast tissue. But in the MLO view, one nonbreast region (i.e., pectoral muscle) may be present in the left or right upper corner of the image of the left and right breast, respectively. Due to this heterogeneous orientation, a standardization algorithm is required to transform the image, so as to place the pectoral muscle at the upper-left corner of the mammogram. To perform left-oriented mammogram images, the right breast mammogram needs to be flipped horizontally at 180°, which is an exact mirror reflection of

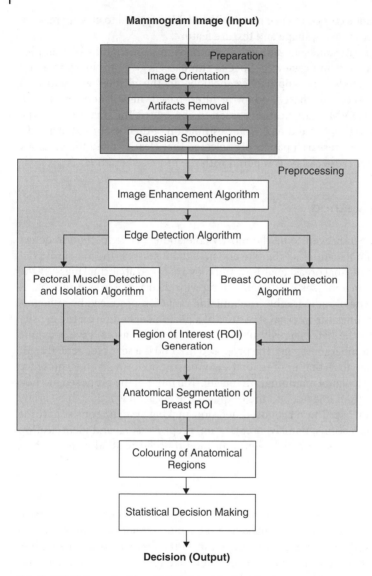

Figure 15.1 Schematic diagram of the proposed method.

the image. This process will allow the breast regions to be compared and analyzed in a similar way, and thus reduce the algorithmic complexity to a great extent. The process of image orientation starts with the identification of mammogram, whether it represents the left or right breast regions. In the case of left, the algorithm will ignore the orientation process, and in the case of right, it will flip the image horizontally [16].

Another additional complexity of mammogram image analysis is the presence of patient-related and hardware-related artifacts. These artifacts provide high-intensity regions on the mammogram and are inconsequential to the investigation of abnormalities within the mammogram. The presence of such artifacts also changes the intensity

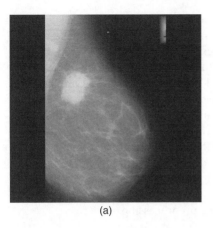
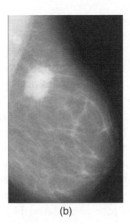

(a)   (b)

Figure 15.2 (a) The original mammogram image and (b) the prepared mammogram image (MIAS mdb184.L). *Source*: Suckling (1994) [31]. Reproduced with permission of Mammographic Image Analysis Society. [31]

levels of the mammogram image significantly, which may affect statistical analysis on the image. An algorithm has been proposed to remove all such artifacts and markings on the nonbreast region of the mammogram, and replace them with the background color. The resulting image is free from any other object except the breast region [16].

The characteristics of high-intensity noise are high values of optical densities, such as shadows that present them as horizontal strips or that are ghost images of previously performed mammography. These noises are embedded in the breast region of the mammogram, thus resulting in loss of information from the breast region. These noises also hinder the detection process to yield false results or negative detection. Such noise must be removed from the image to provide accurate results in the detection process. In this research, the well-known Gaussian filter is used to remove such noise by blurring these salt-and-pepper noises before performing edge detection or other processing on the mammogram images. A Gaussian smoothening operator is applied to blur mammogram images and remove unwanted noise. A Gaussian 2D convolution operator is used as a kernel that is a hump-like structure. Here in the proposed research, a flexible Gaussian smoothening algorithm is used where the size of kernel and value of deviation (Ω) can be adjustable depending on the image quality [17].

The above-mentioned preparation steps enhance the image quality and standardize the mammogram image for those derived from different formats, makes, and versions. In addition to the preparation phase, the mammogram images are trimmed from both left and right sides so that the additional background parts are removed from the image without affecting the breast region (Figure 15.2). It helps to reduce the execution time for further analysis and decision-making algorithms.

15.3.2 Preprocessing

The preprocessing phase consists of mammogram image enhancement and edge detection, isolation and suppression of pectoral muscle, contour determination, and anatomical segmentation. All these processes are mandatory, distinct, and sequential in nature. In order to improve the image quality and to make the resultant segmentation more accurate, the preprocessing steps are mandatory, so that the image is ready for further processing. The two distinctive regions of a mammogram image are the exposed breast region and the unexposed air-background (nonbreast) region, but the principal feature on a mammogram remains as the breast contour, otherwise known as the skin–air interface or breast boundary and the pectoral muscle. On partitioning the mammogram into breast and nonbreast regions and pectoral muscle, we obtain the breast region. On suppression of both nonbreast and pectoral muscle regions, the breast ROI is derived and further processing can be done. By obtaining the breast ROI, anatomical regions need to be differentiated and partitioned, so that abnormal region(s) among normal regions can be identified. The input of preprocessing steps is prepared mammogram images, and the output is anatomical regions within the breast ROI along with abnormal region(s), if present. So, it can be said that non-invasive mapping of breast anatomy can be derived by using preprocessing.

15.3.2.1 Image Enhancement and Edge Detection

The mammogram image enhancement and edge detection are two distinct steps, but here the edge detection algorithm is completely dependent on image enhancement due to use of the homogeneity feature. In the proposed image enhancement method, homogeneity of the image will be substantially increased. In addition to that, the algorithm will quantize color in gray scale up to the applicable level. The homogeneity enhancement is followed by the process of edge detection from a medical image, which is a very crucial step toward comprehending image features. Since edges often develop in those locations demarcated by object boundaries, edge detection is widely used in medical image segmentation where these images are segregated into areas that correspond to separate objects. Here the objects are nothing but different anatomical regions within the breast ROI, and one or more regions are abnormal among these normal regions that need to be identified. The outline of the proposed image enhancement and edge detection method is described in a nutshell. For further knowledge, the research paper on medical edge detection by the authors may be consulted [20].

The proposed method consists of three parts: the first part is to calculate the adaptive threshold, which is the maximum distance threshold (MDT); the subsequent part is to generate an enhanced image; and, finally, the third part is to determine the edge map of the original mammogram.

An initial objective of the proposed process is to determine an adoptive threshold MDT value. The threshold is constant for a particular image, but it varies from image to image depending on their intensity characteristic features like contrast, brightness, and so on. In this research, a modified full and complete binary tree is used to store both the intensity and frequency of the image. The primary objective to use such a tree is to quantize the number of colors, yet preserve the full color palette at every tree level.

In this proposed method, color intensities and their frequencies are extracted from the original image color space to generate a histogram. The extracted data are stored in the leaf nodes of the proposed tree (i.e. each leaf node represents individual disjoint intensity sequentially) (Figure 15.3). So, all the possible color of an image can be represented by the leaf nodes (i.e., if the image contains 2^n number of distinct colors, then the tree will have 2^n leafs at level n). The node structure of the said binary tree will contain pointers for left and right child nodes, along with image data. The data will have two components (i.e., color intensity and its frequency present in the image):

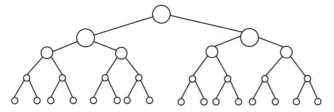

Figure 15.3 Full and complete binary tree.

Node={Node*Left, Node*Right, Int Intensity, Int Frequency}

The frequency of intensity of left child node $f(L)$ and right child node $f(R)$, whichever is greater, will be the intensity (I) of the parent node. The frequency (Fq) for the node will be the summation of $f(L)$ and $f(R)$.

$$Node.I = \begin{cases} f(R).I & \text{if } f(L).Fq < f(R).Fq \\ f(L).I & \text{if } f(L).Fq > f(R).Fq \end{cases} \tag{15.1}$$

Initially, the entire tree is constructed, but only the leaf has image data. In the next phase, the intensity values of image data for all the intermediate nodes, including the root node, are required to be calculated. As per the proposed algorithm, the parent node will hold the intensity value of the child node that has a greater frequency among the two child nodes. To achieve the same, the proposed tree has to be traversed in postorder and the intensity value of the parent node is updated based on the comparison results of the child nodes. The tree structure that is obtained by this procedure contains the histogram of the original image with C gray shades at level n, where $C = 2^n$ and every subsequent upper level contains $n/2$ number of gray shades (i.e., called the level-histogram).

In each level, half of the intermediate color bins are truncated depending on the condition stated above. This intermediate truncation will generate different bin distances in the particular level of the tree (i.e., called the *bin-distance*). The average bin distance (ABD) is the mean of these bin distances. The MDT calculation process first segregates the bins into two categories, namely, *prominent bins* and *truncated bins*. Prominent bins are the points from where sharp changes of intensity values are recorded, whereas truncated bins have an insignificant difference of intensity with their adjacent bins. Prominent bins have a significant role to determine edges of an image. The average bin distance of the prominent bins generates the MDT.

In the second phase, the enhanced image will be generated from the original medical image. The new intensity of a particular pixel has been mapped using the truncated histogram from the tree of a particular level. In the mapping process, a particular intensity of an original image has been selected from the level histogram, and checking has been performed regarding whether the obtained intensity is prominent or not. If the obtained intensity is prominent, then it will be propagated to the image pixel, else the next higher prominent intensity will be selected from the level histogram to propagate.

Finally, the process will scan the enhanced image in row major order. It will be started from the leftmost pixel from the first row and terminated at the rightmost pixel of the last row. Here, two consecutive pixels are compared. If the absolute value of the difference is greater than the MDT, then the corresponding pixel position of the horizontal edge map image will be set to 0 (i.e., black), else N (i.e., white).

$$f(h) = \sum_{i=0}^{r} \sum_{j=0}^{c} P_{i,j} = \begin{cases} 0 & \text{if } |P_{i,j} - P_{i+k,j}| > MDT \\ N & \text{if } |P_{i,j} - P_{i+k,j}| < MDT \end{cases} \tag{15.2}$$

For vertical edge detection, the algorithm is similar to the above horizontal edge detection algorithm except it sets pixels vertically (i.e., in column major order rather than horizontally).

In the previous two steps, horizontal edge map and vertical edge map images have been generated. Now the horizontal and vertical edge maps are superimposed on each other using the logical OR operation.

$$EdgeMap = f(h) \vee f(v) \tag{15.3}$$

The superimposed output image will be the edge of the medical image (i.e., EdgeMapImage). It is the final outcome of the method, which can be applicable to any mammogram image (Figures 15.4 and 15.5).

15.3.2.2 Isolation and Suppression of Pectoral Muscle

The pectoral muscle, a nonbreast region in mammograms, acts like an additional complexity for automated abnormality detection. The pectoral muscle region exhibits a high-intensity region in the majority of MLO projections [27] of mammograms and can influence the outcome of image-processing techniques. Intensity-based methods can produce erroneous results when used to segment high-intensity regions such as suspected masses or fibro-glandular discs because the pectoral muscle has the same opacity as tumors. The authors have published more than one paper on the related topic [18, 19].

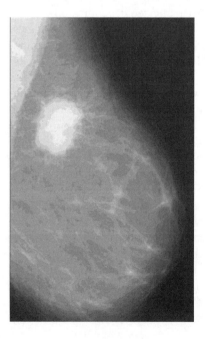

Figure 15.4 Enhanced mammogram image (MIAS mdb184.L). *Source*: Suckling (1994) [31]. Reproduced with permission of Mammographic Image Analysis Society.

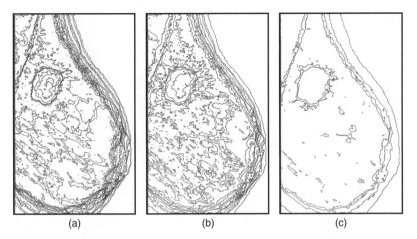

Figure 15.5 (a) Edge map of Level 1, (b) edge map of Level 2, and (c) edge map of Level 3 (MIAS mdb184.L). *Source*: Suckling (1994) [31]. Reproduced with permission of Mammographic Image Analysis Society.

There are two different alternative approaches to isolate pectoral muscle from a mammogram; one is region based and another is edge based. Here the edge-based method is described briefly. The output of the edge detection algorithm is the input of the proposed pectoral muscle isolation method.

As discussed in this chapter, the breast region of an image is differentiated from the background using the edge detection algorithm. The breast region, which is demarcated by breast boundary in between the breast region and the image background, contains outline edges of various breast constituents. For further processing, it is of extreme importance to extract the breast region by eliminating the pectoral muscle region. A new edge-based method is proposed to detect, extract, and isolate the pectoral muscle from the breast region.

As previously discussed, all the mammogram images are homogeneously oriented by the preparation process, so that the pectoral muscle will be placed at the top-left corner for both left and right breast mammogram images and its boundary line traverses from the top margin downward toward the horizontal reference line (i.e., chest wall of the mammogram image), forming an inverted triangle. Since it is a muscular structure, several layers are present. These layers form several impressions in an edge map. Therefore, more than one inverted triangles are in the edge map. The objective is to identify the largest triangle, which is the actual border of pectoral muscle, and divide the pectoral muscle from the breast region, as shown in Figure 15.6.

It is now needed to identify the outermost edge line that constitutes the edge of the pectoral muscle. It is observed that the largest inverted triangle in the edge map starting from the top margin and ending on the vertical baseline (i.e., the horizontal reference line of the breast region on the left side), is the pectoral region. It is important to identify the rightmost pixel of the breast region on the right side and draw a vertical line from the top margin to the bottom margin, parallel to the left vertical baseline. Another line is drawn parallel to the top and bottom margins. The said line originates from the two-thirds position of the horizontal reference line to the right vertical line (Figure 15.7).

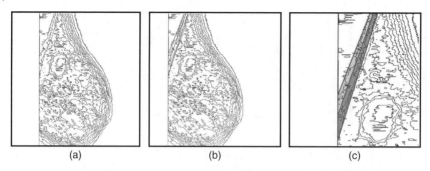

Figure 15.6 (a) The edge map of a mammogram, (b) showing the layers of pectoral muscle, and (c) showing inverted triangles marked by different gray shades (MIAS mdb184.L). *Source*: Suckling (1994) [31]. Reproduced with permission of Mammographic Image Analysis Society.

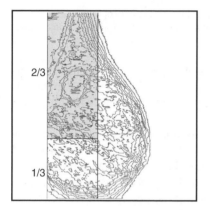

Figure 15.7 Pectoral muscle lies within gray-shaded derived rectangular area (MIAS mdb184.L). *Source*: Suckling (1994) [31]. Reproduced with permission of Mammographic Image Analysis Society.

The pectoral muscle, barring a few mammograms, lies within the rectangle formed by these straight lines (Figure 15.7). The search is limited to the obtained rectangle, thus reducing the processing time instead of considering the entire mammogram image.

The scanning will be right to left, starting from the right vertical line at the first row. If a pixel with black intensity is obtained, that is demarcating the path of the edge. By considering all the surrounding pixels in a clockwise priority, the pixel path is traversed and the highest priority neighboring pixel is considered for further progress. The pixels that surrounded the edge pixel but are of lower priority are stored in a backtrack stack in order to prioritize value, and they are to be used only if the traversal process reaches a dead end. If a dead end is reached, pop from the backtrack stack for a lesser priority pixel and continue with the traversal process. The pixels traversed are stored in a plotting list to be used later for drawing the pectoral boundary. Traversal continues to the next pixel till it reaches the left baseline or the bottom of the rectangle. If the bottom of the rectangle is reached, the path is discarded, the plotting list is erased, and the algorithm resumes by searching the next black pixel at the first row from right to left. If the horizontal reference line is reached that specifies that the path stored in the plotting list is the pectoral boundary, that is drawn in a new image and further processing is terminated.

Figure 15.8 Isolated pectoral boundary (MIAS mdb184.L).
Source: Suckling (1994) [31]. Reproduced with permission of
Mammographic Image Analysis Society.

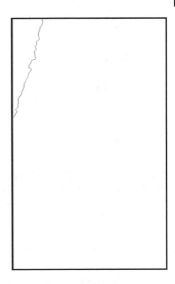

The pectoral boundary obtained (Figure 15.8) is further processed by the pectoral boundary smoothing method. The pectoral boundary image obtained so far is not smooth. A smooth pectoral boundary is obtained by taking pixels at a fixed discrete interval and joining them by drawing a simple curve between the pixel positions to obtain a smooth and enhanced pectoral boundary.

15.3.2.3 Breast Contour Detection

The breast contour is one of the primary characteristics within the breast region. The extracted breast region limited by breast contour is important because it restricts the search for abnormality within a specific region without unwanted influence from the background of the mammogram. It also facilitates enhancements for techniques such as comparative analysis, which includes the automated comparison of corresponding mammograms. The breast boundary contains significant information relating to the symmetry and deformation between two mammograms.

The authors published a research article on a breast segmentation and contour detection method for mammographic images [15], which is an edge-based technique. The brief idea of the contour detection algorithm is cited here.

The input mammogram image of the proposed contour detection algorithm is the edge map after identification and isolation of the pectoral muscle area. Breast contour or boundary is nothing but the outermost edge line of the derived edge of the mammogram. All the mammogram images are homogeneously oriented, and the skin–air interface is placed at the right side of the image. So, the image is scanned from the right side of the image to locate the rightmost point of the edge at the topmost row of the image, which is the starting point of the processing. The algorithm considers all the neighboring pixels except the pixel that is already traversed (Figure 15.9).

The priority of the neighboring pixel is clockwise. It will select the pixel with the highest priority within the edge. It stores the pixel traversed in a plotting list to be used later for drawing the breast boundary. The other pixels that surround that pixel, but are of lower priority, are stored according to their priority order in a backtrack stack to be used only if the traversal process reaches a dead end. If a dead end is encountered, where

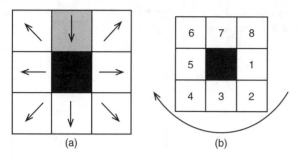

(a)   (b)

Figure 15.9 The technique of traversing process. (a) Black cell represents the current pixel, gray cell is representative of already traversed pixel, and the rest are the path for further traversing. (b) The priority of selection of neighbor is clockwise. *Source*: Suckling (1994) [31]. Reproduced with permission of Mammographic Image Analysis Society.

there are no pixels that have not already been traversed, it pops out from the backtrack stack a lesser priority pixel, creates a new branch, and continues, with the traversal process. The traversal continues to the next pixel till it reaches the baseline (i.e., the bottom of the image or horizontal reference line, i.e., the chest wall, indicating the end of the breast region). The plotting list contains the breast boundary pixels that are plotted on a blank image to obtain the breast boundary for further processing.

The breast boundary image, obtained by implementing the proposed breast boundary detection method, is further processed by the breast boundary smoothing method to obtain the final image. The breast boundary image obtained so far is not smooth. But it is important to get a contour curve that is smooth. The pixel information along with contour is preserved by the previous algorithm in an array. The smoothing algorithm picks up pixels from there with a discrete interval and joins them by drawing a simple curve between two successive pixel positions to obtain a smooth and enhanced breast boundary. The enhanced boundary is single pixel and continuous (Figure 15.10).

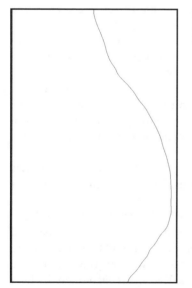

Figure 15.10 Detected breast contour (MIAS mdb184.L). *Source*: Suckling (1994) [31]. Reproduced with permission of Mammographic Image Analysis Society.

15.3.2.4 Anatomical Segmentation

The anatomical segmentation algorithm [21] of the breast ROI is applied on the MLO view of mammogram images devoid of pectoral muscle area and background delineated by breast contour. The principal idea of the proposed algorithm is to differentiate the anatomical breast regions and separate each of the regions with boundary lines.

A clear understanding of breast anatomy and its impression on digital mammograms is the prerequisite to discuss the technical aspects of segmentation of anatomical regions. A mature woman's breast can be divided into four regions: skin, nipple, and subareolar tissues; the subcutaneous region, which contains fat and lymphatics; the parenchyma region, a triangular shape between the subcutaneous and retromammary regions, with the apex toward the nipple; and the retromammary region, which consists of retromammary fat, intercostal muscles, and the pleural reflection [2]. On mammogram images, breast masses, including both noncancerous and cancerous lesions, appear as white regions. Fat appears as black regions on the images. All other components of the breast, like glands, connective tissue, tumors, calcium deposits, and so on, appear as different shades of white on a mammogram.

Any anatomical regions of living beings are of closed structure and bounded by a periphery, and this is also applicable for human breast. Breast comprises bounded anatomical regions. Hence, the edge map indicates various closed structures within the breast region that correspond to the different anatomical regions of the breast, namely, the fatty tissues, ducts, lobules, glands, irregular masses, and calcifications. The objective of the algorithm is to clearly identify these regions on the mammogram image and erase all other unwanted edges, lines, and dots from the edge map for further processing and analysis (Figure 15.11).

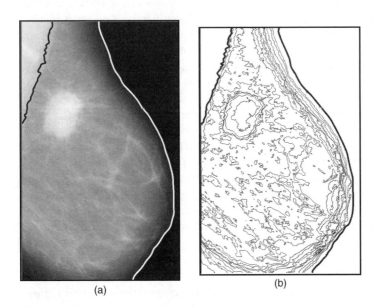

(a) (b)

Figure 15.11 (a) Breast ROI and (b) Boundaries of anatomical regions within breast ROI (MIAS mdb184.L). *Source*: Suckling (1994) [31]. Reproduced with permission of Mammographic Image Analysis Society.

The principal objective of the proposed algorithm is to traverse the edge map to identify and plot all the circular paths forming a closed structure. To achieve the objective, a ladder-like structure is proposed that is delineated by the chest wall of the breast image on the left and right sides by a vertical line passing through the rightmost pixel of the breast ROI. Horizontally, the image is divided by parallel lines subsequently after specific intervals. The detailing of the anatomy depends on the density of the horizontal lines. If the number of horizontal lines is high, all the minute details will be recorded, but it will increase the execution time of the algorithm and complexity of the anatomy, whereas reduction of horizontal lines will miss some tiny structure that may have importance in medical diagnosis. So, the optimal number of horizontal lines have to be applied on a case-to-case basis under the direction of medical expert.

The traversal process of circular edge paths starts from the horizontal lines subsequently and is recorded in the plotting list. This list is plotted on another blank image if the edge path is circular, ends on the last row of the image, or ends on the vertical line representing the left boundary of the breast. The pixel search in horizontal lines is in a left-to-right direction, and it repeats for all the subsequent horizontal lines from top to bottom. So, the algorithm consists of two iterations: the outer one depends on the number of horizontal lines, whereas the inner one searches the starting pixel on the same horizontal line to begin the traveling process. The deliverable of the proposed algorithm is a new image plotted with closed traveling paths that are the boundaries of different internal regions within the breast ROI.

15.3.3   Identification of Abnormal Region(s)

The principal objective of the proposed method is to identify the region(s) with abnormality among normal regions in the breast. Now the questions are: is what is that feature indicating the abnormality, and how can that be identified by a decision-making system? From a medical point of view, masses are tissues that are absent in normal breast anatomy and are called abnormal masses. But from the point of view of medical image processing, density of abnormal masses is different and at the same time their size and shape are variable. Due to heterogeneity, such masses leave a unique high-intensity impression in the mammogram image. The motivation behind the research is to reduce the workload of the radiologist and reduce the percentage of false detection. It is not a replacement for radiologists but can be an assistance tool for second opinions.

The decision-making system is composed of two parts, to initially differentiate the regions depending on the intensity distribution using a modified seeded region-growing algorithm (i.e. coloring of regions), and finally a statistical model is used to take a stand regarding the presence or absence of abnormalities. The inputs of the proposed algorithm are mammogram images after preparation and output of the anatomical segmentation algorithm containing boundary outlines of anatomical regions of the breast ROI (i.e., the concluding result of preprocessing).

15.3.3.1   Coloring of Regions

The segmentation process performed on the edge map differentiates various regions on the breast, depending on their intensity values. Each region has a different intensity value. The fatty tissues, glands, lobules, and ducts exhibit different intensity values, and thus can be segregated into different regions. An abnormality, such as a mass, tumors, or calcifications, that may be present within the breast has noticeably higher intensity values than the normal tissues of the breast. So, it is needed to categorize all the obtained closed structures on the basis of their intensity values. The distribution of pixel intensities also varies within each segmented region, but the majority of the pixels have similar intensity values. So, the respective arithmetic node value is calculated for each region from the original mammogram and replaces those pixels in the region with the computed mode values. To propagate the mode value, a seeded region-growing technique is used. Each region within the mammogram is bounded by a single-pixel boundary as obtained during the edge detection process followed by the process of anatomical segmentation.

During this process, the segmented image is scanned to locate a region that is yet to be colored. The scanning process starts from the first row of the image, proceeds in row major order, and terminates at the right most pixel of the last row. On finding the seed for a region, the coloring process is started for the region by first comparing the pixel intensity of that pixel location on the original mammogram image. For each pixel, the four boundary pixels located north, east, west, and south of the pixel are also checked to find out whether those are colored or the boundary pixel. If the pixels are not colored and not boundary pixels, they again form the seed for further searching. A stack is used to store the seeds to be investigated, while a list is used to store the pixels of the region that have been included in the region and already traversed. All the pixel positions within the list are then searched on the original image to get their intensity values to derive the mode value. The pixel locations of each region are then substituted by the computed mode value intensity (Figure 15.12).

15.3.3.2 Statistical Decision Making

The regions are heterogeneous in color intensity, but there is some degree of homogeneity present. The abnormal regions are present within these regions with some

Figure 15.12 Intensity distribution of regions after coloring (MIAS mdb184.L). *Source*: Suckling (1994) [31]. Reproduced with permission of Mammographic Image Analysis Society.

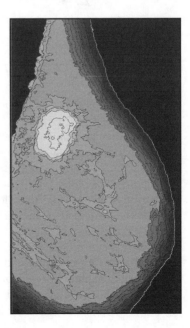

asynchronous characteristics. The objective is to extract these characteristics' features to prove the presence of abnormality. To identify the abnormalities, a statistical decision-making system is applied to analyze the distribution of the colors domain through a step-by-step elimination model.

The proposed SRG algorithm categorized and enumerated each region along with their respective statistical mode value. This data set is used for further statistical analysis. First, the arithmetic mean (μ) for the distribution is calculated to obtain the deviations of each region.

$$\mu = \frac{1}{M} \sum_{i=0}^{h} \sum_{j=0}^{w} Image_{i,j} \qquad \text{where } I_{i,j} > 0 \tag{15.4}$$

where M is the number of pixels; and $Image_{i,j} > 0 \land Image_{i,j} <$Total Number of color. Subsequently, the standard deviation of the data set is calculated.

$$\sigma = \sqrt{\frac{1}{RegCount} \sum_{i=0}^{RegCount} (RegMod_i - \mu)^2} \tag{15.5}$$

where $RegCount$ is the number of regions; and $RegMod$ is the region's mode value.

Then the Z score is calculated to normalize the distribution.

$$Zscore = (Image_{i,j} - \mu)/\sigma \tag{15.6}$$

The regions with negative Z values are truncated because they are insignificant and normal regions. So, only regions with a positive Z score values will be considered for further processing. Now the truncated mean value ($T\mu$) is calculated along with the standard deviation ($T\sigma$) using the truncated data set.

$$T\mu = \frac{1}{n} \sum_{i=0}^{RegCount} Zscore_i \qquad \text{where } Zscore_i > 0 \tag{15.7}$$

where n is the number of regions with positive Z-score values:

$$T\sigma = \sqrt{\frac{1}{n} \sum_{i=0}^{n} (Zscore_i - T\mu)^2} \tag{15.8}$$

Finally, the $2T\sigma$ and $3T\sigma$ of the population are calculated on a truncated data set.

Now, the regions are categorized into four discrete levels according to their color intensity. There are some regions with color values greater than truncated mean ($T\mu$) but less than the truncated mean ($T\mu$) + $T\sigma$ level. Some have color values greater than truncated mean ($T\mu$) + $T\sigma$ but less than the truncated mean ($T\mu$) + $2T\sigma$ level. There are few regions whose color value is beyond the truncated mean ($T\mu$) + $2T\sigma$ level but within the truncated mean ($T\mu$) + $3T\sigma$ level. There are rare regions, if present, with color values over the truncated mean ($T\mu$) + $3T\sigma$ level. According to the algorithm, the fourth category is the most suspicious and uncommon in the frequency of the color region domain. This category is marked by the algorithm as the abnormal region. The third category is also apprehensive and uncommon, and the algorithm marked this category as a suspected region. This category regions are mostly absent in normal mammograms.

Finally, an image is generated depicting the anatomical regions. The colorings of regions are done through a color lookup table. The color lookup table selects the color for a particular region, depending upon the decision generated by the decision-making system as stated above. The algorithm highlighted the abnormal region(s) with a very deep gray shade; the suspected region(s) are marked with gray shade for further analysis by the medical experts. The highly dense regions are colored with a light gray shade, and normal areas are demarcated by white (Figure 15.13). The color lookup table (Table 15.1) describes the color scale of breast regions according to the decision-making system.

The proposed seeded region-growing algorithm scans the image in row major order according to the boundary defined by the ROI. Assume that height is n and width is m. So, the running time of the proposed SRG will be $O(n*m)$; if $n=m$, then the running time complexity will be $O(n^2)$. The proposed image mean value calculation will also run in $O(n^2)$. The Z score calculation followed by the truncated mean calculation will work in

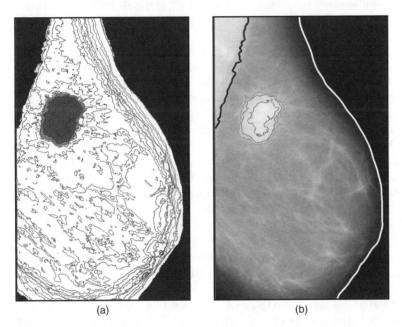

| | (a) | | (b) | |

Figure 15.13 (a) Highlighted regions with abnormal masses and (b) boundary of abnormal regions (MIAS mdb184.L). *Source*: Suckling (1994) [31]. Reproduced with permission of Mammographic Image Analysis Society.

Table 15.1 Color lookup table

Categories	Intensity limit	Type	Color
Category 01	$< \mu$	Normal	White
Category 02	$< T\mu$	Normal	White
Category 03	$< T\mu + T\sigma$	Normal	White
Category 04	$< T\mu + 2T\sigma$	Dense	Light gray
Category 05	$< T\mu + 3T\sigma$	Suspected	Gray
Category 06	$\geq T\mu + 3T\sigma$	Abnormal	Very deep gray

linearly constant time depending on the number of regions present in the image. So, the cumulative complexity of the method is $O(n^2)$.

15.4 Experimental Result

Three different types of mammogram imaging techniques are most commonly used in clinical diagnosis. The first one is the traditional film mammography printed, and the other two are computed radiography (CR) and digital radiography (DR) mammogram systems that store images using electronic signals and return back responses to the requesting machine. Hence, CAD can be deployed on the last two methods, and the DICOM (digital imaging and communication machine) image format is used by both CR and DR mammography. In CR 2^8 and in DR 2^{12} or 2^{16} grayscale color intensity values are used to represent the 8-bit or the 12/16-bit pixel information in the DICOM image format.

Any gray scale image is basically a 2D array of pixel intensities where intensities range from $k=0$ to n. In CR mammogram DICOM images, the gray shade ranges from $k=0$ to 255, whereas in DR it is $k=0$ to 2^{12} or 2^{16}. So the intensity of the image is the primary feature to determine the abnormality. The significant variation is observed in CR and DR mammogram images, at the same time that distinguishable variations of intensity can also be marked in different makes of mammographic machines of the same type as well as different versions of the same make.

The proposed abnormal masses detection algorithm has been extensively tested with two well-known mammogram databases, namely, MIAS (Mammographic Image Analysis Society) digital mammogram database with 322 images in the 8-bit category, whereas in DR (i.e., the 16-bit category), the DEMS (Dokuz Eylul University Mammography Set) database has been considered with 485 images. For experimental purposes, CR images obtained from different medical institutes have been considered, but due to ethical issues and anonymity of the patients, the results are not published here. The MIAS database is 8-bit .png images, whereas DEMS is 16-bit DICOM images.

The proposed algorithm is tested with MIAS, DEMS, and other benchmarked databases of mammograms that are in the public domain. The experimental outcomes are extremely encouraging. Some of the outputs of MIAS belonging to different categories are cited here for demonstration of correctness of the proposed algorithm.

15.4.1 Case Study with Normal Mammogram

The proposed algorithm is applied on MIAS Image 272.L with predominantly fatty tissues, and according to the MIAS observation no abnormality is present. The proposed algorithm agrees with the MIAS observation and detects no abnormalities (Figure 15.14).

15.4.2 Case Study with Abnormalities Embedded in Fatty Tissues

The proposed algorithm is applied on MIAS Image 028.L with predominantly fatty tissues, and according to the MIAS observation, an abnormality is present. The proposed algorithm agrees with the MIAS observation and detects abnormalities (i.e., the presence of a mass) (Figure 15.15).

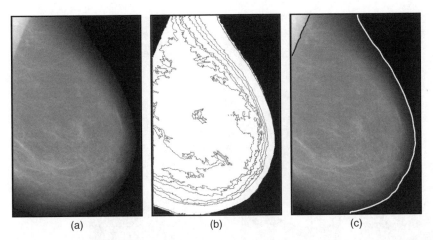

Figure 15.14 MIAS mdb272.L: (a) mammogram image, (b) segmented anatomical regions without highlighted abnormality, and (c) derived image showing absence of boundary of abnormal region(s). *Source*: Suckling (1994) [31]. Reproduced with permission of Mammographic Image Analysis Society.

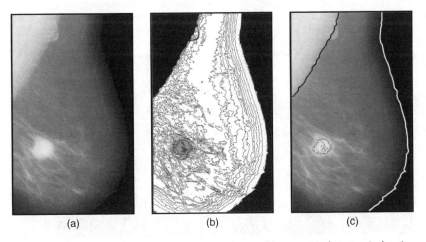

Figure 15.15 MIAS mdb028.L: (a) mammogram image, (b) segmented anatomical regions with highlighted abnormality, and (c) derived image showing boundary of abnormal region(s). *Source*: Suckling (1994) [31]. Reproduced with permission of Mammographic Image Analysis Society.

15.4.3 Case Study with Abnormalities Embedded in Fatty-Fibro-Glandular Tissues

The proposed algorithm is applied on MIAS Image 001.R with predominantly fatty-fibro-glandular tissues, and according to the MIAS observation, an abnormality is present. The proposed algorithm agrees with the MIAS observation and detects abnormalities (i.e., the presence of masses) (Figure 15.16).

15.4.4 Case Study with Abnormalities Embedded in Dense-Fibro-Glandular Tissues

The proposed algorithm is applied on MIAS Image 145.R with predominantly dense-fibro-glandular tissues, and according to the MIAS observation, an abnormality

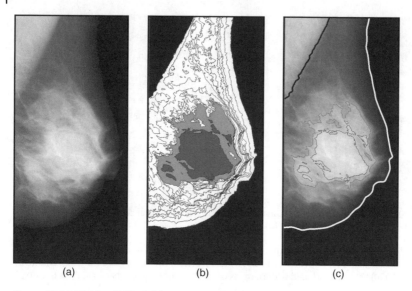

| (a) | (b) | (c) |

Figure 15.16 MIAS mdb001.R: (a) mammogram image, (b) segmented anatomical regions with highlighted abnormality, and (c) derived image showing boundary of abnormal region(s). *Source*: Suckling (1994) [31]. Reproduced with permission of Mammographic Image Analysis Society.

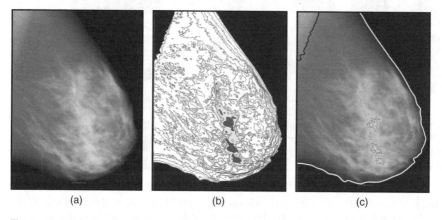

| (a) | (b) | (c) |

Figure 15.17 MIAS mdb145.R: (a) mammogram image, (b) segmented anatomical regions with highlighted abnormality, and (c) derived image showing boundary of abnormal region(s). *Source*: Suckling (1994) [31]. Reproduced with permission of Mammographic Image Analysis Society.

is present. The proposed algorithm agrees with the MIAS observation and detects abnormalities (i.e., the presence of masses) (Figure 15.17).

15.5 Result Evaluation

The benchmarked MIAS and DEMS databases contain different types of case studies that are divided into two distinct classes: one is normal, and another is abnormal. The abnormality can also be subdivided into two categories: the presence of calcification and the presence of a mass or masses. The masses can be well-defined or circumscribed masses, spiculated masses, and other ill-defined masses. The calcification clusters are very tiny regions like dots, whereas masses are well defined and comparatively larger structures. The determination of these abnormalities depends on the segmentation of breast anatomy. The proposed ladder concept is the key deciding factor for determination of calcification and masses. If the number of segments is higher in the ladder structure, the anatomy is more detailed and is able to identify the calcifications. When the number of segments is less, the anatomy of breast regions is clearer; at the same time, unwanted information is completely eliminated from the ROI, but there is inability to isolate the microcalcification clusters in breast anatomy. In both ways, masses can be segregated efficiently. The decision-making system, defined earlier in this chapter, can mark both the high-intensity abnormalities (i.e., microcalcification clusters and masses) proficiently depending upon segregation of the region by the algorithm.

The quantitative analysis of the proposed algorithm is performed in three parts. In the first part, statistical analysis is done to identify and segregate abnormalities, if present, using the Z score analysis. In the subsequent part, ROC analysis is done to measure the accuracy of identification of abnormalities by the proposed method with MIAS and DEMS observations. Finally, the accuracy of the segmented abnormal masses is calculated for the images in the MIAS and DEMS databases where abnormal masses have been identified by the proposed method. In the next section, the accuracy of the abnormal mass detection is compared with results obtained by other comparable methods described in Section 15.2 previous work.

15.5.1 Statistical Analysis

The Z score analysis is used to identify the tumor region(s) and distinguish them from other areas of the mammogram. This analysis is performed on mammogram images in the MIAS and DEMS databases. The graph Z score analyses of the mammograms that are reported in the Experimental Result section (Section 15.4) are shown in Figures 15.18–15.20, and 15.21. An adjustment value of ±0.1 is considered for inference in the 2σ and 3σ levels.

The mammogram images are considered in three broad categories, as already mentioned in this chapter. The fatty mammogram images consist of the least number of regions, whereas the fatty-glandular have higher and dense-glandular have the highest number of regions. So, identification of abnormalities becomes more challenging with the latter two categories. For image mdb272.L, none of the regions exceeds the 3σ level. For image mdb028.L representing fatty, image mdb001.R representing fatty-glandular, and image mdb145.R representing dense-glandular, the breast mammogram shows abnormal regions, that is, equal to or above the 3σ level (±0.1).

15.5.2 ROC Analysis

The ROC analysis has been conducted on both the MIAS and DEMS mammogram databases. Here, the MIAS database is considered as the benchmark due to its clear documentation regarding classification, size, type, and ground truth (GT) of images by their own radiologist. Among the 322 mammogram images, 251 mammogram images of MIAS are classified by their radiologist as normal or containing tumor(s), and among the 251 images, 207 are normal and 44 have abnormal masses. The confusion matrix

Intensity Distribution of Breast Regions

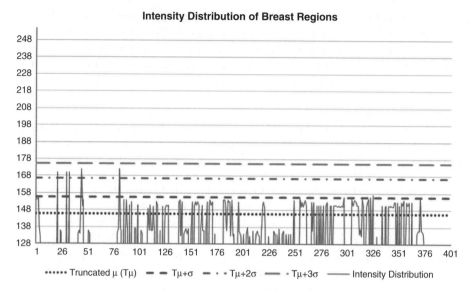

Figure 15.18 The *Z* score analysis graph for mammogram mdb272.L.

Intensity Distribution of Breast Regions

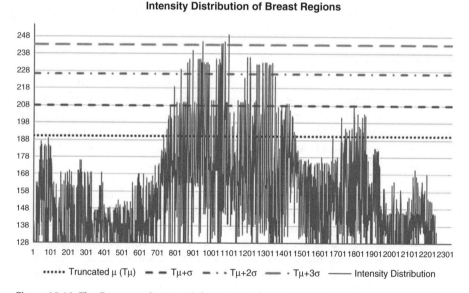

Figure 15.19 The *Z* score analysis graph for mammogram mdb028.L.

(Table 15.2) is obtained after implementing the proposed algorithm on the said 251 images to measure the agreement of the proposed algorithm with the available manual interpretation of the database used (see also Table 15.3).

From the total number of cases of 251, the number of correct detection is 241 with an accuracy value of 96%, sensitivity is 97.6%, and specificity is 88.6%. The total positive cases missed is 5, and negative cases missed is 5. The empiric ROC area obtained is 0.931. The empirical ROC curve is cited in Figure 15.22.

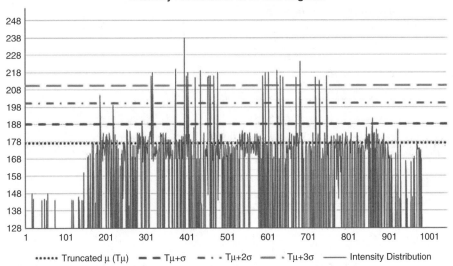

Figure 15.20 The *Z* score analysis graph for mammogram mdb001.R.

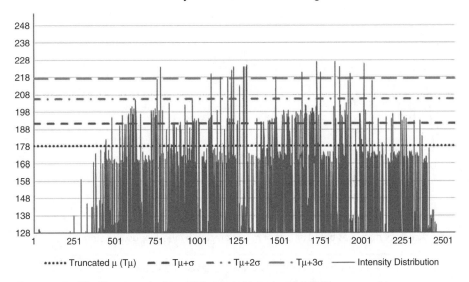

Figure 15.21 The *Z* score analysis graph for mammogram mdb145.R.

Performance evaluation based on the size of tumor obtained by the proposed algorithm and the calculated mass size has been categorized as Table 15.4.

A near-accuracy result is observed in the third and fourth categories due to the large size of well-defined masses. In cases of smaller mass size, the accuracy decreases. For the first category, three cases are missed due to their smaller size. In case of the second category, the intensity for two images shows a very low intensity level for the masses

Table 15.2 Confusion matrix of response data reported from testing

		MIAS truth	
		Tumor	**Normal**
Proposed method	Tumor	39	5
	Normal	5	39.

Table 15.3 Observed operating points

FPF	0.0000	0.1136	1.0000
TPF	0.0000	0.9758	1.0000

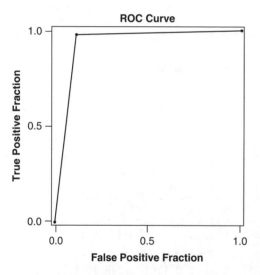

Figure 15.22 Empirical ROC curve for tumor identification.

Table 15.4 Accuracy measures based on size of mass detected by the proposed method

Size	<1.20cm²	1.21cm²−1.80cm²	1.81cm²−3.60cm²	>3.61cm²
MIAS	11	11	12	10
Proposed method	8	9	12	10
Accuracy	86.4%	90.9%	100%	100%
Sensitivity	90.9%	93.9%	100%	100%
Specificity	72.7%	81.8%	100%	100%

and has merged with the adjacent regions, and hence went undetected by the proposed method.

15.5.3 Accuracy Estimation

The performance evaluation of algorithms is done in detail, which endorses the acceptance of results. The quantitative measures are used to prove the accuracy of proposed algorithms by matching the acquired results representing an abnormal mass or masses, as a "mask" with its corresponding GT image. The average results of different quality measures using the MIAS and DEMS database for abnormalities are depicted in Table 15.5. Most of the research articles ignored the boundary accuracy estimation, so it is difficult to compare. It was found that Rabottino *et al.* [26] have a reported boundary accuracy estimation of their research with a sensitivity value of 88.34%, which is less than the sensitivity value of 96.06% achieved by the proposed algorithm.

Table 15.5 Quantitative measures applied to assess the proposed methods

Quantitative measures	Computation	Average result
Accuracy (percentage agreement)	$(|TN|+|TP|)/(|TN|+|TP|+|FP|+|FN|)$	0.9990
Dice similarity coefficient (DSC)	$(2 \times |TP|)/(2 \times |TP|+|FP|+|FN|)$	0.9164
Error rate	$(|FP|+|FN|)/(|FP|+|FN|+|TP|+|TN|)$	0.0009
Sensitivity / completeness (CM)	$|TP|/(|TP|+|FN|)$	0.9606
Correctness (CR)	$|TP|/(|TP|+|FP|)$	0.9521
Specificity (true negative fraction / rate)	$|TN|/(|TN|+|FP|)$	0.9521
False positive fraction / rate	$1 -$ Specificity	0.0478
Underestimation fraction (UEF)	$|FN|/(|TN|+|FN|)$	0.0004
Overestimation fraction (OEF)	$|FP|/(|TN|+|FN|)$	0.0005

Table 15.6 Comparative analysis of proposed method with others

Authors	Accuracy	Sensitivity	Specificity
Kegelmeyer *et al.* [12]	–	100%	82%
Karssemeijer and Te Brake [11]	90%	–	–
Comer *et al.* [7]	100% (abnormal tissues) 58% (stellate lesions)	–	–
Li *et al.* [14]	–	90%	–
Kobatake *et al.* [13]	90.5%	–	–
Yang *et al.* [34]	86%	–	–
Campanini *et al.* [5]	–	80%	–
Jiang *et al.* [10]	66.4%	54.3%	78.3%
Özekes *et al.* [24]	88.37%	–	–
Martins *et al.* [22]	85%	–	–
Rejani *et al.* [28]	–	88.75%	–
Proposed method	96%	97.6%	88.6%

15.6   Comparative Analysis

The comparative analysis is done intensively with the proposed method and other similar algorithms mentioned in the Previous Works section (Section 15.2) for identification of masses. Most of the researchers have not shared the accuracy estimation of their proposed algorithms. Some authors have demonstrated the ways and measurements of accuracy estimations but have used different parameters to describe the accuracy of their algorithms. The most frequently used parameters are accuracy, sensitivity, and specificity. Table 15.6 depicts the comparative analysis of the data gathered from these algorithms with the proposed one.

15.7   Conclusion

The proposed decision-making method comprises three distinct phases, namely, preparation, preprocessing, and isolation of abnormal region(s). Preparation is an essential step toward standardization of medical images that are difficult to interpret. Preprocessing is mandatory and discrete, and it comprises sequential steps like homogeneity enhancement, edge detection, determination of ROI by eliminating nonbreast regions like pectoral muscle, accurate estimation of breast contour, and differention of anatomical regions. Finally, isolation of abnormal regions is a process of segmentation to identify regions with abnormality among other normal regions. Intensity distribution of each region is heterogeneous, yet a degree of homogeneity is present. The proposed method extracts that homogeneity feature of a particular region by using the arithmetic mode value of pixel intensities and colors of that region with that determined mode value. The abnormal regions show unusual intensity distribution among normal regions. The proposed method used a Z score to determine that infrequent intensity distribution and marks it as an abnormality. The proposed method has been tested with standard mammographic databases comprising CR and DR images of different categories and showing comparable results with other related research. The ROC analysis suggests the algorithmic accuracy is 96%, whereas sensitivity and specificity are 97.6% and 88.6%, respectively. The false-positive (FP) and true-negative (TN) cases of mass detection are under an acceptable threshold when the algorithm is tested with a standard data set, whereas the accuracy estimation for boundary of mass shows an accuracy value of 99%, sensitivity 96%, and specificity 95.2%. In conclusion, the proposed method can be incorporated to CAD for a mass screening program toward isolation of abnormalities due to its algorithmic simplicity, efficiency, and accuracy.

Acknowledgments

The authors are obliged to Mammographic Image Analysis Society (MIAS) and Dokuz Eylul University for their public mammographic data sets dedicated for research and development. They are also thankful to Pradip Saha, MD, Radiology, for his continuous participation in the proposed work and to Soma Chakraborty, MD, Radiology (Specialist in Mammography), for her expert opinions and comments. The authors are also appreciative to Sangita Bhattacharjee, Sumit Das, and Sanjay Nag for their contributions related to algorithm development, coding, and documentation of the proposed work. The authors are especially grateful to Sisir Chatterejee for his expert opinions regarding mathematical and statistical modeling related to the proposed method.

References

1 Balachandran *et al.* (2005) Cancer – an ayurvedic perspective. *Pharmacological Research*, **51**, pp. 19–30.

2 Bassett *et al.* (1989) Breast sonography: technique equipment and normal anatomy. *Seminars in Ultrasound CT and MR*, **10** (2), pp. 82–89.

3 Bhishagratha (1991) Sushruta samhita. Choukhamba Orientalia, Varanasi.

4 Brake *et al.* (2001) Segmentation of suspicious densities in digital mammograms. *Medical Physics*, **28**, pp. 259–266.

5 Campanini *et al.* (2004) A novel featureless approach to mass detection in digital mammograms based on support vector machines. *Physics in Medicine and Biology*, **49** (6), pp. 961–975.

6 Catarious *et al.* (2004) Incorporation of an iterative linear segmentation routine into a mammographic mass CAD system. *Medical Physics*, **31**, pp. 1512–1520.

7 Comer *et al.* (1996) *Statistical segmentation of mammograms*. Digital Mammography, International Congress Series. Elsevier, Amsterdam, pp. 471–474.

8 Ferlay *et al.* (2004) *GLOBOCAN 2002: cancer incidence mortality and prevalence worldwide*. IARC Cancer Base, **2** (5).

9 International Agency for Research on Cancer Working Group. (2002) *The evaluation of cancer preventive strategies*. Handbooks of Cancer Prevention, Breast Cancer Screening, vol. 7. IARC Press, Lyon, France.

10 Jiang *et al.* (2008) Automated detection of breast mass spiculation levels and evaluation of scheme performance. *Academic Radiology*, **15** (12), pp. 5341–1544.

11 Karssemeijer *et al.* (1996) Detection of stellate distortions in mammograms. *The Institute of Electrical and Electronics Engineers Transactions on Medical Imaging*, **15**, pp. 611–619.

12 Kegelmeyer *et al.* (1994) Computer-aided mammographic screening for spiculated lesions. *The Journal of Pathology*, **191**, pp. 331–337.

13 Kobatake *et al.* (1999) Computerized detection of malignant tumors on digital mammograms. *IEEE Transactions on Medical Imaging*, **18** (5), pp. 369–378.

14 Li *et al.* (1995) Markov random field for tumor detection in digital mammography. *The Institute of Electrical and Electronics Engineers Transactions on Medical Imaging*, **14**, pp. 565–576.

15 Maitra *et al.* (2011) Accurate breast contour detection algorithms in digital mammogram. *International Journal of Computer Applications*, **25** (5), pp. 1–13.

16 Maitra *et al.* (2011) Artefact suppression and homogenous orientation of digital mammogram using seeded region growing algorithm. *International Journal of Computer Information Systems*, **3** (4), pp. 32–38.

17 Maitra *et al.* (2011) Automated digital mammogram segmentation for detection of abnormal masses using binary homogeneity enhancement algorithm. *Journal of Computer Science and Engineering (IJCSE)*, **2** (3), pp. 416–427.

18 Maitra *et al.* (2011) Detection and isolation of pectoral muscle from digital mammogram: an automated approach. *International Journal of Advance Research in Computer Science*, **2** (3), pp. 375–380.

19 Maitra *et al.* (2012) Technique for pre-processing of digital mammogram. *Computer Methods and Programs in Biomedicine Elsevier (CMPB)*, **107** (2), pp. 175–188.

20 Maitra *et al.* (2015) A tree based approach towards edge detection of medical image using MDT. *International Journal of Computer Graphics*, **6** (1), pp. 37–56.

21 Maitra *et al.* (2011) Anatomical segmentation of digital mammogram to differentiate breast regions. *International Journal of Research and Reviews in Computer Science (IJRRCS)*, **2** (6), pp. 1327–1330.

22 Martins *et al.* (2009) Detection of masses in digital mammograms using K-means and support vector machine. *Electronic Letters on Computer Vision and Image Analysis*, **8** (2), pp. 39–50.

23 Mehta, (2012) Ayurveda & Cancer. Available from pdayurvedatoday.com.

24 Özekes *et al.* (2005) Computer aided detection of mammographic masses on digital mammograms. *Istanbul Ticaret Üniversitesi Fen Bilimleri Dergisi*, **4** (8) pp. 87–97.

25 Petrick *et al.* (1996) Automated detection of breast masses on mammograms using adaptive contrast enhancement and texture classification. *Medical Physics*, **23**, pp. 1685–1696.

26 Rabottino *et al.* (2008) Mass contour extraction in mammographic images for breast cancer identification. Exploring New Frontiers of Instrumentation and Methods for Electrical and Electronic Measurements, pp. 1–6.

27 Rangayyan (2005) *Biomedical image analysis*. CRC Press, New Delhi.

28 Rejani *et al.* (2009) Early detection of breast cancer using SVM classifier technique. *International Journal on Computer Science and Engineering*, **1** (3), pp. 127–130.

29 Sharma (1981) Charaka samhita. Choukhamba Orientalia, Varanasi.

30 Stewart *et al.* (2003) *World cancer report*. IARC Press, Lyon, France.

31 Suckling *et al.* (1994) The Mammographic Image Analysis Society Digital Mammogram Database. Exerpta Medica, International Congress Series 1069, pp. 375–378.

32 Tata Memorial Hospital. (2013) Available from tmc.gov.in

33 Timp *et al.* (2004) A new 2D segmentation method based on dynamic programming applied to computer aided detection in mammography. *Medical Physics*, **31**, pp. 958–971.

34 Yang *et al.* (2005) A computer-aided system for mass detection and classification in digitized mammograms. *Biomedical Engineering applications, Basis & Communications*, **17** (5), pp. 215–229.

35 Zheng *et al.* (1995) Computer detection of masses in digitized mammograms using single-image segmentation and a multilayer topographic feature analysis. *Academic Radiology*, **2**, pp. 959–966.

16

Automatic Detection of Coronary Artery Stenosis Using Bayesian Classification and Gaussian Filters Based on Differential Evolution

Ivan Cruz-Aceves[1], Fernando Cervantes-Sanchez[2], and Arturo Hernandez-Aguirre[2]

[1] *CONACYT – Centro de Investigación en Matemáticas (CIMAT), A.C., Jalisco S/N, Col. Valenciana, Guanajuato, México*
[2] *Centro de Investigación en Matemáticas (CIMAT), A.C., Jalisco S/N, Col. Valenciana, Guanajuato, México*

16.1 Introduction

Automatic detection of coronary artery stenosis represents one of the most important challenges in cardiology. In clinical practice, the specialists perform a visual examination over the X-ray angiograms obtained by a procedure involving cardiac catheterization. Subsequently, a manual detection of potential cases of blood vessel stenosis is carried out. Two cases of coronary stenosis are illustrated in Figure 16.1.

In systems that perform computer-aided diagnosis, the detection of vessel stenosis is commonly addressed in the steps of vessel detection, vessel segmentation, and classification of stenosis. In the first stage of vessel detection, the vessel-like structures are enhanced while removing noise from the input image. The segmentation step is used to discriminate vessel and nonvessel pixels from the enhanced image as white and black regions, respectively. Finally, the classification step to detect coronary stenosis is carried out by selecting a set of independent features in order to apply a classification strategy.

In the literature, the automatic detection and segmentation of blood vessels have been addressed by using different perspectives in the spatial image domain. Some of the state-of-the-art methods are based on mathematical morphology [1–6] and the Hessian matrix [7–15]. Another technique based on the convolution of a template with the input image is the Gaussian matched filter (GMF) proposed by Chaudhuri *et al.* [16]. The GMF represents one of the most commonly applied detection techniques for different types of blood vessels.

The GMF assumes that the shape of blood vessels can be approximated by a Gaussian curve as a matching template at different orientations. The generation of this Gaussian template is governed by the parameters of spread of the Gaussian function, the width and length of the template, and the number of oriented filters to form a directional bank. Due to the efficiency and ease of implementation of the GMF, it has been applied successfully in different clinical studies [17–19]. The main disadvantage of the GMF is the selection of the optimal parameter values for each particular application. To detect blood vessels in retinal fundus images, Cinsdikici and Aydin [20] proposed experimentally

Hybrid Intelligence for Image Analysis and Understanding, First Edition.
Edited by Siddhartha Bhattacharyya, Indrajit Pan, Anirban Mukherjee, and Paramartha Dutta.
© 2017 John Wiley & Sons Ltd. Published 2017 by John Wiley & Sons Ltd.
Companion Website: www.wiley.com/go/bhattacharyya/hybridintelligence

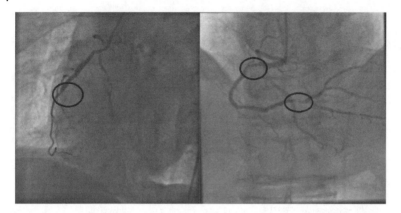

Figure 16.1 Two X-ray angiograms with detection of coronary stenosis performed by a cardiologist.

determined parameter values, and Al-Rawi *et al.* [21, 22] proposed to introduce a search space for each parameter using an exhaustive global search and genetic algorithms as search strategies. On the other hand, to detect coronary arteries, Kang *et al.* [23–25] proposed different empirical values for the spread of the Gaussian profile and the number of directional filters. Cruz *et al.* [26, 27] proposed a new search space for the GMF parameters, and also an empirically determined objective function with an estimation of distribution algorithm to perform the optimization task. In our previous work [28], a comparative analysis of four nature-inspired algorithms was performed over the training stage of the GMF technique. In that analysis, the algorithm of differential evolution (DE) obtained the highest detection performance over a training set of 40 angiograms.

In the present chapter, a novel method for the automatic detection of vessel stenosis in X-ray coronary angiograms is introduced. The proposed method uses GMFs tuned by DE for the detection of coronary arteries, and an iterative thresholding method for the segmentation of the filter response. To evaluate the performance of these two steps, the receiver operating characteristic (ROC) curve and the accuracy measure have been adopted. Finally, in the segmented coronary angiograms, the naive Bayes classifier is applied over a 3D feature vector obtained from the results of a second-order derivative operator in order to identify potential cases of coronary stenosis.

The remainder of this chapter is organized as follows. In Section 16.2, the fundamentals of the GMFs, DE, and the naive Bayes classifier are described in detail. In Section 16.3, the proposed method consisting of the steps of coronary artery detection and segmentation, along with the process to detect potential cases of coronary stenosis, is analyzed. The experimental results are presented in Section 16.4, and conclusions are given in Section 16.5.

16.2 Background

This section introduces the fundamentals of the GMFs for vessel detection, DE as an optimization strategy, and the naive Bayes classifier to identify coronary stenosis, which are of interest in the present work.

16.2.1   Gaussian Matched Filters

The GMF method is a template-based approach proposed by Chaudhuri *et al.* [16] for detecting blood vessels. GMF works on the assumption that the shape of vessel-like structures in the spatial image domain can be computed by a Gaussian curve as a template. This Gaussian matching template can be defined as follows:

$$G(x, y) = -\exp\left(-\frac{x^2 + y^2}{2\sigma^2}\right), \quad |y| \le L/2, \tag{16.1}$$

where L represents the length in pixels of the vessel segment to be enhanced, and σ is a continuous parameter that represents the average width of the vessel-like structures. Since the Gaussian curve has infinitely long double-sided trails, they are usually truncated at $u = \pm 3\sigma$, by introducing a discrete parameter T to determine a neighborhood N, and the position in the Gaussian template where the curve trails will cut as follows:

$$N = \{(u, v), |u| \le T, |v| \le L/2\}. \tag{16.2}$$

To detect vessel-like structures at different orientations, the Gaussian kernel $G(x, y)$ is rotated by applying a geometric transformation in order to form a directional filter bank of $\kappa = 180°/\theta$, evenly spaced filters in the range $[-\frac{\pi}{2}, \frac{\pi}{2}]$, as follows:

$$\kappa = \begin{bmatrix} \cos\theta_i & -\sin\theta_i \\ \sin\theta_i & \cos\theta_i \end{bmatrix}, \tag{16.3}$$

where $\theta_i = (i\pi/\kappa), i = \{1, 2, \ldots, \kappa\}$ is the angular resolution, and κ the number of directional filters. Finally, these directional Gaussian matching templates are convolved with the input image, and for each pixel in the image, the maximum response over all orientations is preserved to generate the filter response.

Since the performance of the GMF depends on three discrete parameters (L, T, κ) and one continuous parameter (σ), the selection of the optimal parameter values for the GMF plays a vital role for each specific application. In order to illustrate the effects of the parameter values for the GMF, in Figure 16.2a and 16.2b, an X-ray angiogram along with its ground-truth image are introduced. Figure 16.2d and 16.2g present different Gaussian templates, along with the obtained filter response in Figure 16.2f and 16.2i, respectively, using the angiogram in Figure 16.2a.

16.2.2   Differential Evolution

DE represents a stochastic real-parameter strategy proposed by Storn and Price [29, 30] for solving numerical global optimization problems. Similar to evolutionary algorithms, DE uses a set of Np potential solutions (also called individuals) $X = \{x_1, x_2, \ldots, x_{Np}\}$, which are iteratively improved using mutation, crossover, and selection operators, and also evaluated according to a fitness function. The mutation step is used to generate a mutant vector $V_{i,g+1}$ for each generation g based on the spatial distribution of the whole population $\{X_{i,g} | i = 1, 2, \ldots, Np\}$ as follows:

$$V_{i,g+1} = X_{r1,g} + F(X_{r2,g} - X_{r3,g}), \quad r1 \ne r2 \ne r3 \ne i, \tag{16.4}$$

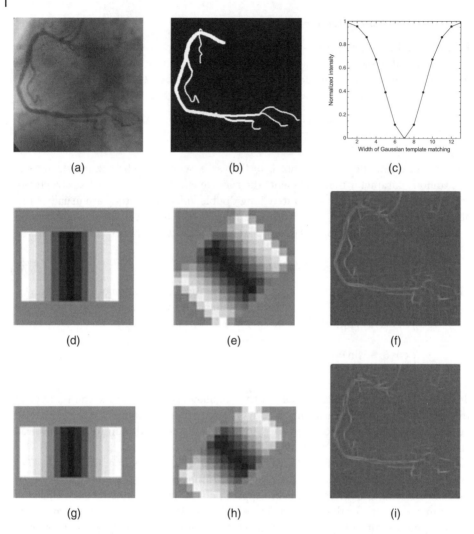

Figure 16.2 (a) X-ray coronary angiogram. (b) Ground-truth image of angiogram in (a). (c) Gaussian profile of the method of Chaudhuri *et al.* [16]. Second row: Template using $\sigma = 2.0, L = 9, T = 13$, with $\theta = 0°$, and $\theta = 45°$, respectively, and resulting filtered image in (f). Last row: Matching template using $\sigma = 2.0, L = 7, T = 15$, with $\theta = 0°$, and $\theta = 45°$, respectively, and resulting filtered image in (i).

where $r1$, $r2$, and $r3$ correspond to the indexes of three different individuals from the population $\{1, \ldots, Np\}$ in the current generation g; and F is the mutation parameter, also called the differentiation factor. In the second step, a crossover operator is applied to create a trial vector $U_{i,g+1}$ as follows:

$$U_{i,g+1} = \begin{cases} V_{i,g+1}, & \text{if } r \leq CR \\ X_{i,g}, & \text{if } r > CR \end{cases} \tag{16.5}$$

where r is a uniform random value on the interval $[0, 1]$, which is generated to be compared with the crossover rate, $CR \in [0, 1]$. The CR parameter is used to control the diversity of solutions in the population, and the rate of information that is copied from the mutant vector. If the random value r is bigger than the parameter CR, the current individual $X_{i,g}$ is conserved; otherwise, the mutant vector information $V_{i,g+1}$ is copied to the trial vector $U_{i,g+1}$ to be used in the selection stage. In the last step of DE, a selection operator is introduced to obtain the best one between the individual $X_{i,g}$ and the trial vector $U_{i,g+1}$ in order to replace the information of the current individual $X_{i,g}$ in the next generation $X_{i,g+1}$.

$$
X_{i,g+1} = \begin{cases} U_{i,g+1}, & \text{if } f(U_{i,g+1}) < f(X_{i,g}) \\ X_{i,g}, & \text{otherwise} \end{cases} \tag{16.6}
$$

According to the above description, the DE technique can be implemented by the following procedure:

1. Initialize the number of generations G and the population size Np.
2. Initialize the mutation factor F and crossover rate CR.
3. Initialize each individual X_i with random values within the search space.
4. For each individual $X_{i,g}$, where $g = \{1, \dots, G\}$:
   A. Compute $V_{i,g+1}$ by the mutation operator (16.4);
   B. Obtain $U_{i,g+1}$ using the crossover operator (16.5);
   C. Update $X_{i,g+1}$, applying the selection operator (16.6).
5. Stop if the convergence criterion is satisfied (e.g., the number of generations).

16.2.2.1 Example: Global Optimization of the Ackley Function

In order to illustrate the implementation details of DE, the optimization of a mathematical function is introduced, which represents the main application of population-based methods. In this section, the Ackley function in two dimensions is presented, which is defined in Equation 16.7 and shown in Figure 16.3. The range for each variable of the

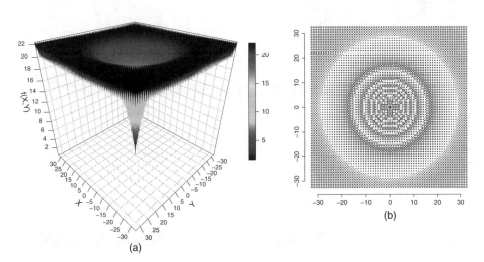

Figure 16.3 Ackley function in two dimensions. (a) Isometric view in X_1 and X_2, and (b) level plot of the function, where the optimal value is located at $X_1 = 0$ and $X_2 = 0$.

Ackley function is $X_1 \in [-32.768, 32.768]$, $X_2 \in [-32.768, 32.768]$, and the optimal value is located on ($X_1 = 0.0, X_2 = 0.0$).

$$f(x) = -20 \cdot exp\left(-0.2 \cdot \sqrt{\frac{1}{n}\sum_{i=1}^{n}(x_i^2)}\right)$$

$$- exp\left(\frac{1}{n}\sum_{i=1}^{n}\cos(2 \cdot \pi \cdot x_i)\right) + 20 + exp. \qquad (16.7)$$

Since DE is a real-coded strategy, the individuals can be randomly initialized in the continuous domain of the particular function. To illustrate the optimization process that DE follows to optimize a mathematical function, in Figure 16.4 a numerical example using DE to minimize the aforementioned Ackley function is presented.

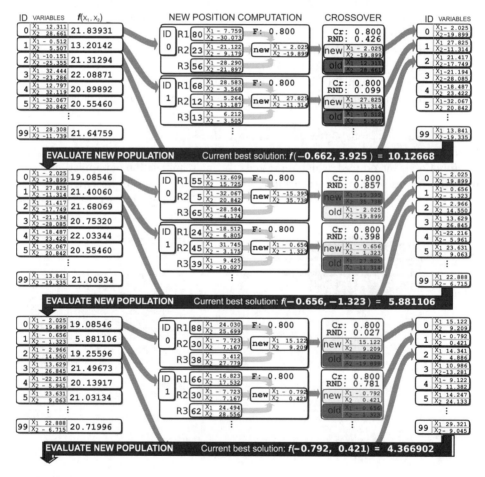

Figure 16.4 Numerical example for solving the 2D Ackley function using DE as an optimization strategy.

16.2.3 Bayesian Classification

To determine the relationship between a feature set and the class variable, the Bayes theorem can be used as a modeling probabilistic strategy. The Bayes theorem has been commonly applied for solving classification problems under a statistical perspective, which can be computed by applying the following formula:

$$P(Y|X) = \frac{P(X|Y)P(Y)}{P(X)} \tag{16.8}$$

where $P(Y|X)$ and $P(Y)$ are the posterior and prior probabilities for Y, respectively; $P(X|Y)$ is the likelihood; and $P(X)$ is the evidence. During the training step, the posterior probabilities $P(Y|X)$ for each combination of X and Y variables must be learned in order to be maximized for correct classification. The process to estimate the posterior probabilities is a challenging problem because it requires a large training set; otherwise, to compute the class-conditional probabilities $P(X|Y)$, the naive Bayes classifier can be applied.

The naive Bayes classifier is used to estimate the class-conditional probability assuming that the features are independent, according to a class label [31]. This independence assumption can be defined as follows:

$$P(X|Y = y) = \prod_{i=1}^{f} P(X_i|Y = y), \tag{16.9}$$

where f denotes the number of features of the set $X = \{X_1, X_2, \ldots, X_f\}$. On the other hand, the process to classify a test record is carried out by computing the posterior probability for each given class Y, as follows:

$$P(Y|X) = \frac{\prod_{i=1}^{f} P(X_i|Y)P(Y)}{P(X)}. \tag{16.10}$$

To work with categorical features, the naive Bayes classifier estimates the rate of training instances in a given class y with a particular feature value x_i. For continuous attributes, a Gaussian distribution has been commonly applied to represent the class-conditional probability. This distribution is defined by two different parameters, its mean (μ) and standard deviation (σ), which can be calculated over the training records that belong to the class y_j as follows:

$$P(X_i = x_i|Y = y_j) = \frac{1}{\sqrt{2\pi}\sigma_{ij}} \exp^{-\frac{(x_i - \mu_{ij})^2}{2\sigma_{ij}^2}}. \tag{16.11}$$

16.2.3.1 Example: Classification Problem

To illustrate the process carried out by the naive Bayes classifier, a training data set for a medical classification problem is introduced in Table 16.1, where the class-conditional probability for each one of the four attributes (one categorical and three continuous) must be computed.

This problem presents a medical test for patients with risk of cardiovascular disease. The test is performed using the following four attributes: antecedents of cardiovascular disease, age of the patient, triglyceride level, and low-density lipoprotein (LDL) cholesterol. The value ranges for the triglyceride level can be categorized as normal for less

Table 16.1 Training records for predicting cardiovascular risk

Patient	Antecedents	Age	Triglyceride level	LDL cholesterol	Risk
1	no	30	150	100	low
2	yes	56	480	190	high
3	yes	60	520	170	high
4	no	32	100	90	low
5	no	24	120	80	low
6	no	48	600	200	high
7	no	52	580	220	high
8	yes	50	540	170	high
9	no	28	135	75	low
10	yes	54	400	210	high

than 150 mg/dL, high between 200 and 499 mg/dL, and very high for above 499 mg/dL. The value ranges for the LDL cholesterol level can be categorized as optimal for less than 100 mg/dL, high between 160 and 189 mg/dL, and very high for above 189 mg/dL.

The class-conditional probabilities for the cardiovascular classification problem are presented here:

For the Antecedents attribute:

$$P(Antecedents=yes|low) = 0$$
$$P(Antecedents=no|low) = 1$$
$$P(Antecedents=yes|high) = \frac{4}{6}$$
$$P(Antecedents=no|high) = \frac{2}{6}.$$

For the Age attribute:

Class low: sample mean = 28.50, sample variance = 11.67
Class high: sample mean = 53.33, sample variance = 18.67.

For the Triglyceride level attribute:

Class low: sample mean = 126.25, sample variance = 456.25
Class high: sample mean = 520, sample variance = 5280.

For the LDL cholesterol attribute:

Class low: sample mean = 86.25, sample variance = 122.92
Class high: sample mean = 193.33, sample variance = 426.67.

To classify the corresponding label of a new record $X = ($ *Antecedents = no, Age = 40, Triglyceride level = 550, LDL cholesterol = 185*$)$, the posterior probabilities need to be computed as follows:

$P(X|low) = P(Antecedents = no|low) \times P(Age = 40|low) \times P(Tri. level = 550|low) \times$
    $P(LDL cholesterol = 185|low) = \mathbf{5.5658e\text{-}110}$

$P(X|high) = P(Antecedents = no|high) \times P(Age = 40|high) \times P(Tri. level = 550|high)$
    $\times P(LDL cholesterol = 185|high) = \underline{\mathbf{2.3491e\text{-}8}}.$

According to the above procedure, since $P(X|high)$ is higher than $P(X|low)$, the prediction of the processed test record reveals that the new patient has a risk of cardiovascular disease.

To perform the naive Bayes classifier in another classification task, the process illustrated in the present example is similar. The posterior probabilities must be computed according to a predefined training set of features.

16.3 Proposed Method

The proposed method consists of three main steps. The first step is used for vessel detection by applying GMFs trained by DE over a predefined search space. The second step is used to segment the GMF response by applying an automatic thresholding technique, and finally, in the segmented vessels, a Bayesian classifier is applied to detect coronary artery stenosis in X-ray angiograms.

16.3.1 Optimal Parameter Selection of GMF Using Differential Evolution

Commonly, the GMF parameter values have been experimentally or empirically determined for each particular application. Chaudhuri *et al.* [16] proposed the set of parameters as $L = 9$, $T = 13$, $\sigma = 2.0$, and $\kappa = 12$ for retinal blood vessels. Kang *et al.* [23–25] proposed an average width of $\sigma = 1.5$ with $\kappa = 6$ directional filters to be applied on coronary angiograms. As part of a system to segment ophthalmoscope images, Cinsdikici and Aydin [20] proposed a new angular resolution using $\theta = 10°$, obtaining $\kappa = 18$ directional templates. Al-Rawi *et al.* [21] introduced an exhaustive search over a new range for the variables as $L = \{7, 7.1, \dots, 11\}$, $T = \{2, 2.25, \dots, 10\}$, and $\sigma = \{1.5, 1.6, \dots, 3\}$ to be applied in retinal fundus images. In this method, the set of parameters with the highest area under the ROC curve in a training step was applied over the test images. Later, Al-Rawi and Karajeh [22] proposed the use of genetic algorithms to replace the exhaustive search, keeping the same search space for retinal images. The GMF trained by genetic algorithms achieves better performance than the methods with fixed values in terms of segmentation accuracy, and also better performance than the exhaustive search in terms of computational time and number of evaluations to obtain the best set of parameters.

Based on the suitable performance of the population-based method over a predefined range of parameters, in the present work, DE has been performed to obtain the optimal set of parameters for blood vessel detection in X-ray coronary angiograms using as a fitness function the area (A_z) under the ROC curve. The search space for the GMF parameters was defined according to the aforementioned methods, and taking into account the features of blood vessels in the angiograms, as can be consulted in our previous work [28]. In the experiments, the search space was set as $L = \{8, 9, \dots, 15\}$, $T = \{8, 9, \dots, 15\}$, and $\sigma = [1, 5]$, keeping constant the number of directional filters as $\kappa = 12$ and obtaining an angular resolution of $\theta = 15°$.

In the optimization process, the set of parameters that maximizes the A_z value using a training set is directly applied over the test set of angiograms. The ROC curve represents a plot between the true-positive fraction (TPF) and false-positive fraction (FPF) of a classification system. The TPF corresponds to the rate of vessel pixels correctly detected by the method, and the FPF to the rate of nonvessel pixels incorrectly classified by the method. The area A_z can be approximated by applying the Riemann-sum method, and it is defined in the range $[0, 1]$, where 1 is perfect classification, and zero otherwise.

16.3.2 Thresholding of the Gaussian Filter Response

To discriminate the vessel-like structures from the background of the Gaussian filter response, a soft classification strategy can be applied. The thresholding method introduced by Ridler and Calvard (RC) [32] assumes that two Gaussian distributions with the same variance are present in the gray-level histogram of an input image. The fundamental idea of the method can be defined by the following:

$$T = \frac{\mu_B + \mu_F}{2} \tag{16.12}$$

where μ_B is the mean of background pixels, and μ_F is the mean of foreground pixels.

This method selects an initial global threshold based on the average gray level of the image to form the two initial classes of pixels. Then, the means of both classes are computed to determine the new global threshold value. This process is iteratively performed until stability is achieved.

According to the above description, the RC method can be implemented as follows:

1. Select an initial threshold value (T).
2. Apply image thresholding using T.
3. Obtain two classes of pixels, B and F.
4. Compute the mean intensity μ_B from B.
5. Compute the mean intensity μ_F from F.
6. Calculate the new T value using Equation 16.12.
7. Repeat until the convergence criterion is satisfied (e.g., $\Delta T < \epsilon$).

After the thresholding is applied over the Gaussian filter response, several isolated regions and pixels can appear in the segmented image, affecting the performance evaluation. To eliminate these isolated regions, a length filter is applied based on the concept of connected components. Since the filter is governed by the parameter of size (number of connected components), this parameter value has to be experimentally determined. In our experiments, the best value for the length filtering was established as *size* = 500, as is illustrated in Figure 16.5.

To evaluate the performance of the segmentation stage, the accuracy measure has been adopted, since it represents the most commonly used metric for binary classification. The accuracy measure reveals the rate of correctly classified pixels (sum of true positives and true negatives) divided by the total number of pixels in the input image, as follows:

$$Accuracy = \frac{TP + TN}{TP + FP + TN + FN} \tag{16.13}$$

16.3.3 Stenosis Detection Using Second-Order Derivatives

To detect potential cases of coronary artery stenosis over the segmented image, a search strategy for local minima points can be implemented. This basic strategy is divided into four different steps. First, a skeletonization operator is applied over the vessel pixels in order to determine the centerline of the coronary arteries. Second, the distance

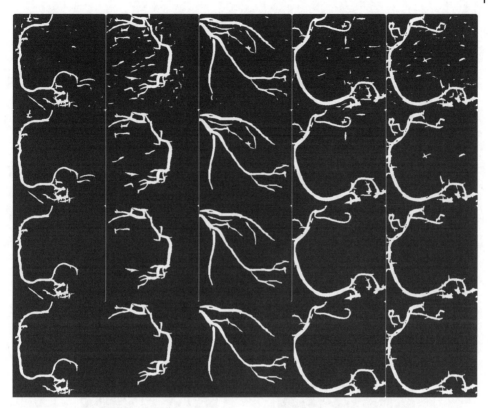

Figure 16.5 First row: Segmentation results using the Ridler and Calvard method. The remaining three rows illustrate the results of length filtering using 100, 200, and 500 pixels as connected components, respectively.

from vessel centerline pixels to the closer boundary pixels of the segmented artery is computed using the Euclidean distance. Subsequently, vessel bifurcations are localized over the morphological skeleton in order to separate the vessel segments in the image. Finally, using the Euclidean distance computed above, a search strategy to detect local minima skeleton pixels (second-order derivatives) is applied over the separate vessel segments. Figure 16.6 illustrates the aforementioned procedure to detect potential cases of stenosis.

It can be observed that the method based on second-order derivatives (local minima) is highly sensitive to false positives. Accordingly, robust methods for classification must be applied; however, the potential cases obtained by this strategy can be used as input for more sophisticated procedures or pattern recognition techniques.

16.3.4   Stenosis Detection Using Bayesian Classification

To apply a Bayesian classification strategy to the problem of vessel stenosis detection, a set of features must be selected. The features represent the independent variables that are evaluated in order to separate the test records according to a corresponding label. In the experiments, the naive Bayes classification technique is applied over three features

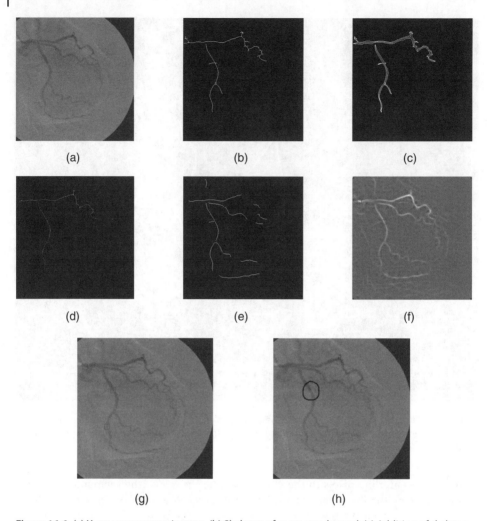

Figure 16.6 (a) X-ray coronary angiogram. (b) Skeleton of segmented vessel. (c) Addition of skeleton and boundary pixels. (d) Skeleton using normalized intensities as Euclidean distance. (e) Separation of vessel segments using bifurcation pixels. (f, g) Detection of local minima points over Gaussian filter response and original angiogram, respectively. (h) Stenosis detection marked in a black circle by cardiologist.

based on the histogram of the vessel width estimation. The three continuous features are:

Sum of vessel widths: The sum of all the amplitudes according to the frequencies
Mean of the vessel width histogram: The mean value of the histogram of the vessel width estimation
Standard deviation of the vessel width histogram: The standard deviation value of the histogram of the vessel width estimation.

To illustrate a case of vessel stenosis and a case of no stenosis, Figure 16.7 presents both cases along with their histogram of vessel width estimation.

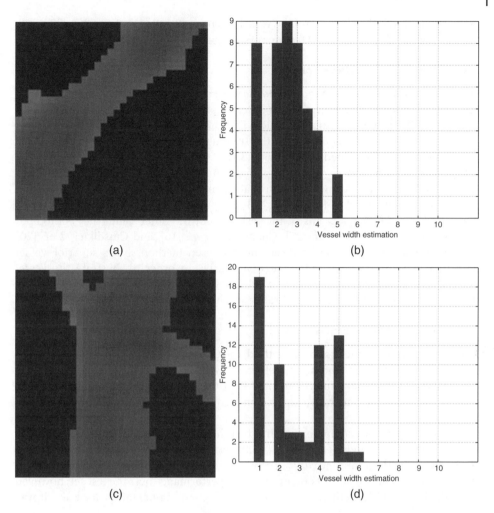

Figure 16.7 (a) Stenosis pattern of 30×30 pixels. (b) Histogram of vessel width estimation of pattern in (a). (c) No stenosis pattern of 30×30 pixels. (d) Histogram of vessel width estimation of pattern in (c).

The feature selection was established taking into account the shape and distribution of the histograms generated with the training set of 20 patterns. Since the stenosis problem represents a binary classification task, the accuracy measure consisting in the fraction of correct predictions divided by the total number of data points has been adopted.

16.4   Computational Experiments

The computational simulations presented in this section were performed with an Intel Core i3, 2.13 GHz processor, and 4 GB of RAM using MATLAB version 2013b. The database of X-ray angiograms was provided by the Mexican Social Security Institute, and it consists of 80 images from different patients. To assess the performance of the proposed method, 20 images are used as a training set and the remaining 60 angiograms as a testing set. For the classification of vessel stenosis, 40 patterns of size 30×30 pixels are used, where the training and testing sets are formed by 20 patterns each.

16.4.1 Results of Vessel Detection

Since the vessel detection step represents the most important task to enhance vessel-like structures, a comparative analysis with different state-of-the-art detection methods is introduced. Because the method of Al-Rawi *et al.* [21] and the proposed method need a training stage, the 20 angiograms of the training set were used for tuning the GMF parameters, as was described in Section 16.3.1.

In Table 16.2, the vessel detection performance in terms of the A_z value of five GMF-based methods is compared with that obtained by the proposed method using the test set. The method of Kang *et al.* [23–25] presents the lowest detection performance using the 60 angiograms of the test set. On the other hand, the methods of Al-Rawi *et al.* [21], Cruz *et al.* [27], Chaudhuri *et al.* [16], and Cinsdikici *et al.* [20] present similar performance; however, the proposed method shows superior vessel detection in terms of area under the ROC curve. In order to illustrate the vessel detection results, Figure 16.8 presents a subset of X-ray angiograms with the Gaussian filter response obtained by the comparative methods.

16.4.2 Results of Vessel Segmentation

In the second stage of the proposed method, the Gaussian filter response is thresholded in order to classify vessel and nonvessel pixels, where different thresholding strategies can be compared. In our experiments, to select the best one thresholding strategy for coronary arteries, a comparative analysis in terms of segmentation accuracy was performed.

Table 16.3 presents a comparison of five thresholding methods over the test set of 60 angiograms. The methods of Kapur *et al.* [33] based on the entropy of the histogram and the method of histogram concavity [34] obtain the lowest performance. The methods of Pal and Pal [35] and RATS [36] obtain a suitable performance over the test set; however, since the iterative RC method [32] achieves the highest segmentation accuracy, it was selected for further analysis.

To perform a qualitative analysis of the segmented blood vessels, Figure 16.9 presents a subset of angiograms with the corresponding ground-truth images. The results

Table 16.2 Comparative analysis of A_z values with the testing set, using the proposed method and five GMF-based methods of the state of the art

Detection method	A_z value
Kang *et al.* [23–25]	0.918
Al-Rawi *et al.* [21]	0.921
GMF entropy [27]	0.923
Chaudhuri *et al.* [16]	0.925
Cinsdikici *et al.* [20]	0.926
Proposed method	**0.941**

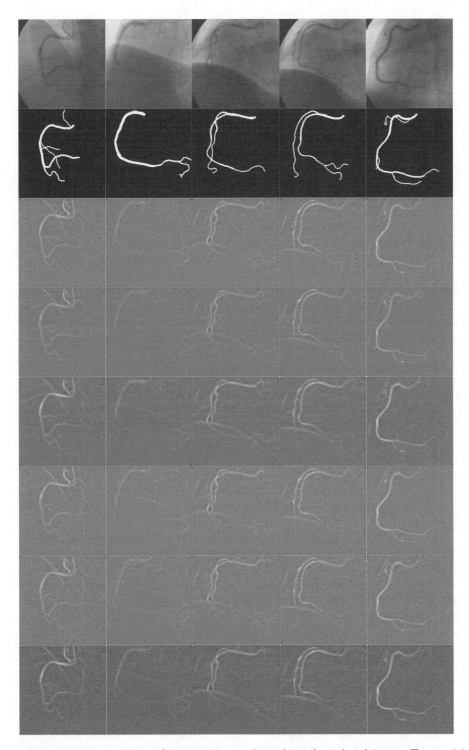

Figure 16.8 First row: Subset of X-ray angiograms. Second row: Ground-truth images. The remaining six rows present the Gaussian filter response of the methods of Kang *et al.* [25], Al-Rawi *et al.* [21], Cruz *et al.* [27], Chaudhuri *et al.* [16], and Cinsdikici *et al.* [20], and the proposed method, respectively.

Table 16.3 Comparative analysis of five automatic thresholding methods over the Gaussian filter response using the test set of X-ray angiograms

Thresholding method	Accuracy value
Kapur *et al.* [33]	0.467
Histogram concavity [34]	0.656
Pal and Pal [35]	0.857
RATS [36]	0.953
Ridler and Calvard [32]	**0.962**

acquired by the methods of Kapur *et al.* [33] and Rosenfeld *et al.* [34] present a low rate of true-positive pixels and the presence of broken vessels, which leads to a low-accuracy performace. The main issue of the method of Pal and Pal [35] is the presence of a high rate of false-positive pixels, which decreases the accuracy measure. The RATS method [36] shows many broken vessels while presenting a high rate of false-positive pixels. Moreover, the RC method [32] shows a suitable detection of true-positive pixels while avoiding broken vessels and obtaining a low rate of false-positive pixels, which is useful to analyze the entire coronary tree for the blood vessel stenosis problem.

16.4.3 Evaluation of Detection of Coronary Artery Stenosis

To perform the last stage of the proposed method, the product between the vessel segmentation result and the input X-ray angiogram is illustrated in Figure 16.10. This product is useful for working with a wide range of features, which is highly desirable for the classification of vessel stenosis.

On the other hand, a subset of the patterns used for the vessel stenosis problem is visualized in Figure 16.11. The previously mentioned three continuous features (*Sum of vessel widths, Mean of the vessel width histogram,* and *Standard deviation of the vessel width histogram*) were computed over the training and testing sets of vessel patterns.

In Table 16.4, the classification values obtained from the naive Bayes classifier using the test set of 20 patterns are shown. From the stenosis test records, the value presented in *Bayes-stenosis* must be bigger than the value shown in the column of *Bayes-no stenosis*, and the opposite for the *No-stenosis* records. It can be observed that the test records of *No-stenosis-2* and *No-stenosis-3* are incorrectly classified (*Status = no*), which decreases the accuracy of the classification system.

Finally, in Table 16.5, the confusion matrix obtained from the classification stage is presented. The overall accuracy using the confusion matrix over the test set of vessel patterns can be computed as $(10 + 8)/(10 + 10) = 0.9$.

In general, a classification task involves a high number of features; however, the performance obtained with the three features based on the histogram of vessel width estimation (90%) can be appropriate for systems that perform computer-aided diagnosis in cardiology.

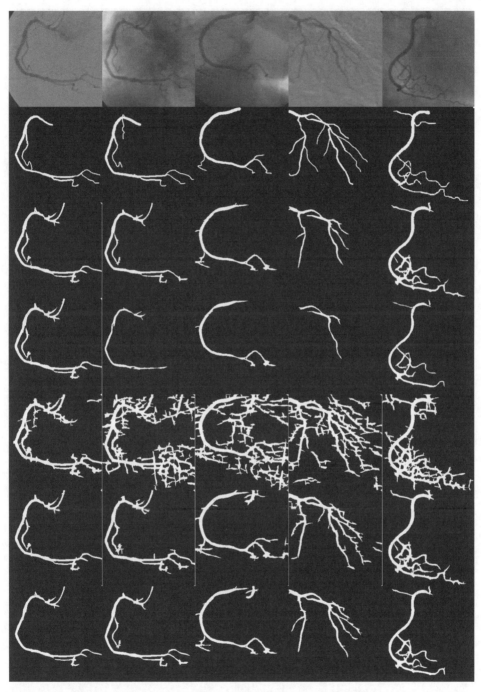

Figure 16.9 First row: Subset of X-ray angiograms. Second row: Ground-truth images. The remaining five rows present the segmentation results of the methods of Kapur *et al.* [33], histogram concavity [34], Pal and Pal [35], RATS [36], and Ridler and Calvard [32], respectively.

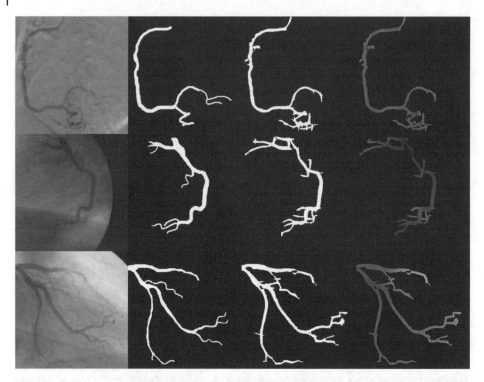

Figure 16.10 First column: Subset of X-ray angiograms. Second column: Ground-truth images. Third column: Segmentation result obtained from the proposed method. Last column: Product between segmentation result and input angiogram.

Figure 16.11 First row: Subset of patterns of no-stenosis cases. Second column: Subset of vessel stenosis patterns.

16.5 Concluding Remarks

In this chapter, a new method for automatic coronary stenosis in X-ray angiograms has been presented. In the detection step, Gaussian-matched filters tuned by differential evolution have shown to be more efficient than five state-of-the-art GMF-based methods, achieving an $A_z = 0.941$ with a test set of 60 angiograms. In the second stage of

Table 16.4 Results of naive Bayes classifier over the test set of 20 records

Test record	Bayes-stenosis	Bayes-no stenosis	Status
Stenosis-1	0.0003521702	2.34985126541991E-05	yes
Stenosis-2	0.0001364722	2.3984168239561E-05	yes
Stenosis-3	0.0011275668	2.0114369609354E-05	yes
Stenosis-4	0.0011395404	1.92517874458132E-05	yes
Stenosis-5	0.0011190882	2.02945979302175E-05	yes
Stenosis-6	0.0001185143	1.20243160639164E-05	yes
Stenosis-7	0.0004880505	1.40674998399305E-05	yes
Stenosis-8	0.0011145618	1.93816551985246E-05	yes
Stenosis-9	0.0003115148	1.30993013377647E-05	yes
Stenosis-10	0.0003339428	1.30571489091502E-05	yes
No-stenosis-1	3.45096065896438E-13	1.56155952678713E-05	yes
No-stenosis-2	0.0001173525	1.15647687625231E-05	no
No-stenosis-3	6.89615692721782E-06	1.55312740490901E-06	no
No-stenosis-4	3.17501486637572E-10	1.92568264381307E-05	yes
No-stenosis-5	5.86365100656046E-06	2.11708957418583E-05	yes
No-stenosis-6	1.83238102570245E-07	2.29001505011661E-05	yes
No-stenosis-7	3.21239037388808E-08	0.000020533	yes
No-stenosis-8	5.0374318162336E-10	1.72396409222254E-05	yes
No-stenosis-9	7.17381447380215E-06	2.23627091900428E-05	yes
No-stenosis-10	2.40013627206018E-12	1.48854193021198E-05	yes

Table 16.5 Confusion matrix for the test set of 20 records

Class	Stenosis	No stenosis	Total
Stenosis	**10**	0	10
No stenosis	2	**8**	10
Total	12	8	**20**

vessel segmentation, an iterative thresholding strategy has shown to be the most efficient compared with four other thresholding methods, obtaining a segmentation accuracy of 0.962 with the test set of 60 angiograms. Finally, in the last step of vessel stenosis classification, the naive Bayes technique applied over a 3D feature vector obtained an accuracy of 0.90 with a test set of 20 patterns. According to the experimental results, the proposed method (consisting of the application of GMFs for enhancement, thresholding for vessel segmentation, and naive Bayes classifier) to detect coronary stenosis has proven to be appropriate for computer-aided diagnosis in cardiology.

Acknowledgment

This research has been supported by the National Council of Science and Technology of México (Cátedras-CONACYT No. 3150-3097).

References

1 Eiho, S. and Qian, Y. (1997) Detection of coronary artery tree using morphological operator. *Computers in Cardiology*, **24**, 525–528.

2 Qian, Y., Eiho, S., Sugimoto, N., and Fujita, M. (1998) Automatic extraction of coronary artery tree on coronary angiograms by morphological operators. *Computers in Cardiology*, **25**, 765–768.

3 Maglaveras, N., Haris, K., Efstratiadis, S., Gourassas, J., and Louridas, G. (2001) Artery skeleton extraction using topographic and connected component labeling. *Computers in Cardiology*, **28**, 17–20.

4 Sun, K. and N. Sang (2008) Morphological enhancement of vascular angiogram with multiscale detected by Gabor filters. *Electronic letters*, **44** (2).

5 Bouraoui, B., Ronse, C., Baruthio, J., Passat, N., and Germain, P. (2008) Fully automatic 3D segmentation of coronary arteries based on mathematical morphology. *5th IEEE International Symposium on Biomedical Imaging (ISBI): From Nano to Macro*, pp. 1059–1062.

6 Lara, D., Faria, A., Araujo, A., and Menotti, D. (2009) A semi-automatic method for segmentation of the coronary artery tree from angiography. *XXII Brazilian Symposium on Computer Graphics and Image Processing (SIBGRAPI)*, pp. 194–201.

7 Lorenz, C., Carlsen, I., Buzug, T., Fassnacht, C., and Weese, J. (1997) A multi-scale line filter with automatic scale selection based on the Hessian matrix for medical image segmentation. *Proceedings of the International Conference on Scale-Space Theories in Computer Vision, Springer LNCS*, **1252**, 152–163.

8 Frangi, A., Niessen, W., Vincken, K., and Viergever, M. (1998) Multiscale vessel enhancement filtering. *Medical Image Computing and Computer-Assisted Intervention (MICCAI'98), Springer LNCS*, **1496**, 130–137.

9 Wink, O., Niessen, W., and Viergever, M. (2004) Multiscale vessel tracking. *IEEE Transactions on Medical Imaging*, **23** (1), 130–133.

10 Salem, N. and Nandi, A. (2008) Unsupervised segmentation of retinal blood vessels using a single parameter vesselness measure. *Sixth Indian Conference on Computer Vision, Graphics and Image Processing, IEEE*, **34**, 528–534.

11 Wang, S., Li, B., and Zhou, S. (2012) A segmentation method of coronary angiograms based on multi-scale filtering and region-growing. *International Conference on Biomedical Engineering and Biotechnology*, pp. 678–681.

12 Li, Y., Zhou, S., Wu, J., Ma, X., and Peng, K. (2012) A novel method of vessel segmentation for X-ray coronary angiography images. *Fourth International Conference on Computational and Information Sciences (ICCIS)*, pp. 468–471.

13 Jin, J., Yang, L., Zhang, X., and Ding, M. (2013) Vascular tree segmentation in medical images using Hessian-based multiscale filtering and level set method. *Computational and Mathematical Methods in Medicine*, **2013** (502013), 9.

14 Tsai, T., Lee, H., and Chen, M. (2013) Adaptive segmentation of vessels from coronary angiograms using multi-scale filtering. *International Conference on Signal-Image Technology and Internet-Based Systems*, pp. 143–147.

15 M'hiri, F., Duong, L., Desrosiers, C., and Cheriet, M. (2013) Vesselwalker: Coronary arteries segmentation using random walks and Hessian-based vesselness filter. *IEEE 10th International Symposium on Biomedical Imaging (ISBI): From Nano to Macro*, pp. 918–921.

16 Chaudhuri, S., Chatterjee, S., Katz, N., Nelson, M., and Goldbaum, M. (1989) Detection of blood vessels in retinal images using two-dimensional matched filters. *IEEE Transactions on Medical Imaging*, **8** (3), 263–269.

17 Hoover, A., Kouznetsova, V., and Goldbaum, M. (2000) Locating blood vessels in retinal images by piecewise threshold probing of a matched filter response. *IEEE Transactions on Medical Imaging*, **19** (3), 203–210.

18 Chanwimaluang, T. and Fan, G. (2003) An efficient blood vessel detection algorithm for retinal images using local entropy thresholding. *Proc. IEEE International Symposium on Circuits and Systems*, **5**, 21–24.

19 Chanwimaluang, T., Fan, G., and Fransen, S. (2006) Hybrid retinal image registration. *IEEE Transactions on Information Technology in Biomedicine*, **10** (1), 129–142.

20 Cinsdikici, M. and Aydin, D. (2009) Detection of blood vessels in ophthalmoscope images using MF/ant (matched filter/ant colony) algorithm. *Computer methods and programs in biomedicine*, **96**, 85–95.

21 Al-Rawi, M., Qutaishat, M., and Arrar, M. (2007) An improved matched filter for blood vessel detection of digital retinal images. *Computers in Biology and Medicine*, **37**, 262–267.

22 Al-Rawi, M. and Karajeh, H. (2007) Genetic algorithm matched filter optimization for automated detection of blood vessels from digital retinal images. *Computer methods and programs in biomedicine*, **87**, 248–253.

23 Kang, W., Wang, K., Chen, W., and Kang, W. (2009) Segmentation method based on fusion algorithm for coronary angiograms. *2nd International Congress on Image and Signal Processing (CISP)*, pp. 1–4.

24 Kang, W., Kang, W., Chen, W., Liu, B., and Wu, W. (2010) Segmentation method of degree-based transition region extraction for coronary angiograms. *2nd International Conference on Advanced Computer Control*, pp. 466–470.

25 Kang, W., Kang, W., Li, Y., and Wang, Q. (2013) The segmentation method of degree-based fusion algorithm for coronary angiograms. *2nd International Conference on Measurement, Information and Control*, pp. 696–699.

26 Cruz-Aceves, I., Hernandez-Aguirre, A., and Valdez-Pena, I. (2015) Automatic coronary artery segmentation based on matched filters and estimation of distribution algorithms. *Proceedings of the 2015 International Conference on Image Processing, Computer Vision, & Pattern Recognition (IPCV'2015)*, pp. 405–410.

27 Cruz-Aceves, I., Cervantes-Sanchez, F., Hernandez-Aguirre, A., Perez-Rodriguez, R., and Ochoa-Zezzatti, A. (2016) A novel Gaussian matched filter based on entropy minimization for automatic segmentation of coronary angiograms. *Computers and Electrical Engineering*.

28 Cruz-Aceves, I., Hernandez-Aguirre, A., and Valdez, S.I. (2016) On the performance of nature inspired algorithms for the automatic segmentation of coronary arteries using Gaussian matched filters. *Applied Soft Computing*, **46**, 665–676.

29 Storn, R. and Price, K. (1995) Differential evolution – a simple and efficient adaptive scheme for global optimization over continuous spaces, *Tech. Rep. TR-95-012*, International Computer Sciences Institute, Berkeley, CA.

30 Storn, R. and Price, K. (1997) Differential evolution – a simple and efficient heuristic for global optimization over continuous spaces. *Journal of Global Optimization*, **11**, 341–359.

31 Tan, P., Steinbach, M., and Kumar, V. (2006) *Introduction to data mining*. Pearson Education, Upper Saddle River, NJ, 227–246.

32 Ridler, T. and Calvard, S. (1978) Picture thresholding using an iterative selection method. *IEEE Transactions on Systems, Man, and Cybernetics*, **8**, 630–632.

33 Kapur, J., Sahoo, P., and Wong, A. (1985) A new method for gray-level picture thresholding using the entropy of the histogram. *Computer Vision, Graphics, and Image Processing*, **29**, 273–285.

34 Rosenfeld, A. and De la Torre, P. (1983) Histogram concavity analysis as an aid in threshold selection. *IEEE Transactions on Systems, Man, and Cybernetics*, **13**, 231–235.

35 Pal, N.R. and Pal, S.K. (1989) Entropic thresholding. *Signal Processing*, **16**, 97–108.

36 Kittler, J., Illingworth, J., and Foglein, J. (1985) Threshold selection based on a simple image statistic. *Computer Vision Graphics and Image Processing*, **30**, 125–147.

17

Evaluating the Efficacy of Multi-resolution Texture Features for Prediction of Breast Density Using Mammographic Images

Kriti[1], Harleen Kaur[1], and Jitendra Virmani[2]

[1] *Electrical and Instrumentation Engineering Department, Thapar University, Patiala, Punjab, India*
[2] *CSIR-CSIO, Sector-30C, Chandigarh, India*

17.1   Introduction

Breast cancer refers to an uncontrolled growth of breast cells. It can start from the cells present in the lobules (glands that produce milk) or ducts (passages to carry milk). Sometimes, the tumors developed in the breast break away and enter the lymph system, from where they can spread to tissues throughout the body. This process is called metastasis [1, 2].

Breast cancer can be broadly categorized as: (1) carcinomas and (2) sarcomas. Carcinoma starts in the cells of the lining of the organ or tissue. This lining is known as the epithelial tissue. The carcinoma developed in the breast tissue is called adenocarcinoma, and it starts in the glandular tissue [3, 4]. Sarcoma is the type of cancer that develops in the non-epithelial tissue or connective tissue like muscles, fat, blood vessels, cartilage, bone, and so on [3].

Other types of breast cancer are:

1. *Ductal carcinoma in situ (DCIS)*: This is the initial stage of breast cancer. It is a noninvasive type of cancer that starts in the inner linings of the milk ducts but cannot spread to other parts of the breast through the duct walls. DCIS can be diagnosed either during a physical examination if the doctor is able to feel any lumps in the breast or during mammography as clusters of calcification inside the ducts [3, 5].
2. *Invasive ductal carcinoma (IDC)*: IDC is the most common type of cancer. This cancer develops in the inner sides of the milk ducts, breaks through the duct walls, and invades the fatty breast tissue from where it can metastasize to other organs and body parts. Some of the early symptoms of IDC include swelling of the breasts, skin irritation, redness or scaliness of the skin, and a lump in the underarm. IDC can be diagnosed by physical examination or different imaging procedures like mammography, ultrasound, breast magnetic resonance imaging (MRI), biopsy, and so on [3, 6].
3. *Invasive lobular cancer (ILC)*: This is the second most common form of breast cancer. It starts developing inside the milk-producing lobules and then metastasizes beyond

Hybrid Intelligence for Image Analysis and Understanding, First Edition.
Edited by Siddhartha Bhattacharyya, Indrajit Pan, Anirban Mukherjee, and Paramartha Dutta.
© 2017 John Wiley & Sons Ltd. Published 2017 by John Wiley & Sons Ltd.
Companion Website: www.wiley.com/go/bhattacharyya/hybridintelligence

the lobules to the breast tissue and other body parts. ILC is much more difficult to locate on a mammogram compared to IDC as the cancer may spread to surrounding stroma. One of the early signs of ILC is hardening of the breast instead of a lump formation. There can also be some change in the texture of the skin. ILC can also be diagnosed during a physical examination or by using some imaging modalities [3, 7].

Although breast cancer has a high mortality rate, the chances of survival are improved if malignancy is detected at an early stage. The best tool used for early detection is mammography [8–16]. A mammogram is an X-ray of the women's breast and is used to check for early signs of breast cancer. There are two types of mammography:

A. Screening mammography
B. Diagnostic mammography.

The screening mammography is used on women who do not experience any symptoms of breast cancer like skin irritation, swelling of a part of a breast, pain in the breast, discharge other than milk, and so on. The diagnostic mammography is used when the patient complains of pain in the breast or some lump formation, or in case an abnormality is detected at the time of screening mammography. The diagnostic mammography enables the radiologist to make an accurate diagnosis and helps in determining the location of the abnormality, its type (e.g., calcification, circumscribed masses, etc.), and the severity of the abnormality.

On the mammograms, the adipose (fatty) tissue and the fibroglandular tissue are displayed along with the present abnormalities. To describe the findings on the mammograms, the American College of Radiology developed a standard system called the Breast Imaging Reporting and Data System (BI-RADS). The categories are described as [17]:

1. *Category 0*: Additional imaging evaluation. If any abnormality is present, it may not be clearly noticeable, and more tests are needed.
2. *Category 1*: Negative. No abnormalities found to report.
3. *Category 2*: Benign finding. The finding in the mammogram is noncancerous, like lymph nodes.
4. *Category 3*: Probably benign finding. The finding is most probably noncancerous but is expected to change over time, so a follow-up is required regularly.
5. *Category 4*: Suspicious abnormality. The findings might or might not be cancerous, so to find the exact nature of the finding, the patient should consider taking a biopsy test.
6. *Category 5*: Highly suggestive malignancy. The finding has more than 95% chance of being cancerous, and biopsy examination is highly recommended for the patient.
7. *Category 6*: Known biopsy-proven malignancy. The findings on the mammogram have been shown to be cancerous by a previous biopsy.

BI-RADS breast density classification is divided into four groups [17]:

1. *BI-RADS I*: Almost entirely fatty breasts. Breasts contain little fibrous and glandular tissue.
2. *BI-RADS II*: Scattered regions of fibroglandular density exists. A few regions of fibrous and glandular tissue are found in the breast.
3. *BI-RADS III*: Heterogeneously dense breasts. The regions of fibrous and glandular tissue are prominent throughout the breast. Small masses are very difficult to observe in mammogram.

4. *BI-RADS IV*: Extremely dense breasts. The whole breast consists of fibrous and glandular tissue. In this situation, it is very hard to find the cancerous region present because the appearance of normal breast tissue and the cancerous tissues tends to overlap.

For a standard mammography examination, the following views are considered: mediolateral oblique view (MLO) and craniocaudal view (CC). The CC view is the top view of the breast in which the entire breast is pictured. The MLO view is an angled or oblique view of the breast. This view allows the largest amount of breast tissue to be visualized. The MLO view is considered to be an important projection as it picturizes most of the breast tissue. Pectoral muscle representation on the MLO view is an important part to evaluate the position of the patient and the film adequacy. The amount of pectoral muscle visible on the image helps in deciding the amount of breast tissue incorporated into the image; this brings out the decisive quality factor that is essential to decrease the number of false negatives and increase the sensitivity of the mammography. Also, as most of the breast pathology happens in the upper external quadrant, that area is very easily visible on MLO views [18]. High breast tissue density is considered to be a prominent indicator of development of breast cancer [19–32]. Radiologists estimate the changes in density patterns by visual analysis of mammographic images. Since this visual analysis is dependent only upon the experience of the radiologist and can be ambiguous, there has been a lot of interest among researchers to implement the CAD systems for classifying the breast tissue as per its density. The CAD system design for breast density classification suggested by various researchers is shown in Figure 17.1. These CAD systems are considered clinically significant, keeping in view the atypical cases of various density

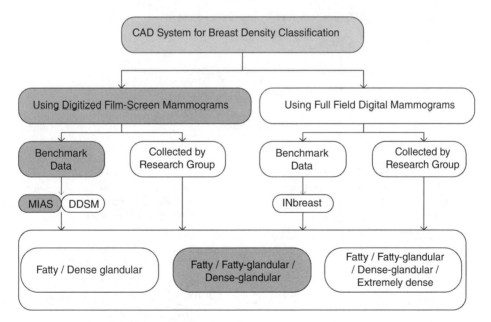

Figure 17.1 CAD system for breast density classification. *Note:* Shaded blocks indicate the steps involved in present work.

patterns exhibited on mammograms. Many researchers have developed different CAD systems that classify breast tissue as per its density into two classes [fatty (F) and dense (D)] [33–50], three classes [fatty (F), fatty-glandular (FG), and dense-glandular (DG)] [50–58], or four classes [fatty (F), fatty-glandular (FG), dense-glandular (DG), and extremely dense] [40, 52, 57]. The breast density classification on the MIAS database, carried out by many researchers, is shown in Figure 17.2. Brief summaries of the studies carried out on the MIAS database for two-class, four-class, and three-class classifications of the breast tissue are depicted in Tables 17.1, 17.2, and 17.3, respectively.

For four-class classification of the MIAS database, a different radiologists suggest a different number of images for different BI-RADS (I to IV) categories. In [52], out of 322 images, 128 images were categorized as BI-RADS I, 80 images as BI-RADS II, 70 images as BI-RADS III, and 44 images as BI-RADS IV. In the study in [40], three experts classified the data: Expert A classified 129 images for BI-RADS I, 78 for BI-RADS II, 70 for BI-RADS III, and 44 for BI-RADS IV. Expert B classified 86 images for BI-RADS I, 112 images for BI-RADS II, 81 images for BI-RADS III, and 43 images for BI-RADS IV. Expert C classified 59 for BI-RADS I, 86 for BI-RADS II, 143 for BI-RADS III, and 34 for BI-RADS IV density categories. In [57], images that fall under the category of fatty (106) were considered to belong to the BI-RADS I category, fatty-glandular (104) as BI-RADS II, while dense (112) was divided into BI-RADS III (95) and BI-RADS IV (37) density categories.

All the related studies [33–60] for classification of breast tissue have been carried out either on segmented breast tissue (SBT) or on some selected region of interest (ROIs).

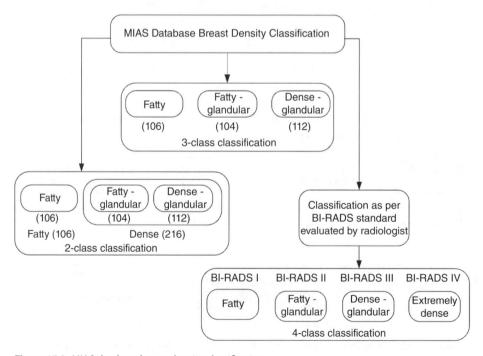

Figure 17.2 MIAS database breast density classification.

Table 17.1 Description of studies carried out for classification of tissue density as fatty or dense on the MIAS database

Investigators	Description			
	Images	SBT/ROI	Classifier	Acc. (%)
Oliver *et al.* (2005) [38]	322	SBT	Bayesian	91.0
Mustra *et al.* (2012) [57]	322	512 × 384	Naïve Bayesian	91.6
Sharma *et al.* (2014) [45]	322	200 × 200	SMO-SVM	96.4
Kriti *et al.* (2015) [49]	322	200 × 200	kNN	95.6
Sharma *et al.* (2015) [46]	212	200 × 200	kNN	97.2
Abdul-Nasser *et al.* (2015) [59]	322	100 × 100	SVM	99.0
Kriti *et al.* (2016) [47]	322	200 × 200	SVM	94.4
Virmani *et al.* (2016) [48]	322	200 × 200	kNN	96.2
Kriti *et al.* (2016) [50]	322	200 × 200	SSVM	94.4

Note: SBT: segmented breast tissue; Acc.: accuracy; SSVM: smooth support vector machine; SVM: support vector machine; SMO: sequential minimal optimization; kNN: *k*-nearest neighbor classifier.

Table 17.2 Description of studies carried out for classification of tissue density as fatty, fatty-glandular, dense-glandular, or extremely dense on the MIAS database

Investigators	Description			
	Images	SBT/ROI	Classifier	Acc. (%)
Bosch *et al.* (2006) [52]	322	SBT	SVM	95.4
Oliver *et al.* (2008) [40]	322	SBT	Bayesian	86.0
Mustra *et al.* (2012) [57]	322	512 × 384	IB1	79.2

Note: SBT: segmented breast tissue; Acc.: accuracy.

The extraction of SBT involves various preprocessing steps, including removal of pectoral muscle and the background. These preprocessing steps can be eliminated by extracting ROIs from the center of mammograms. Exhaustive experimentation carried out in recent studies has indicated that the center region of the breast contains maximum density information [45, 46, 61]. Out of the above studies, some of the studies have been carried out on data sets of mammographic images collected by the individual researchers, while others have been carried out on standard benchmark databases like MIAS and DDSM. It is worth noting that the study in [48] reports two-class breast density classification using wavelet texture descriptors with a maximum accuracy of 96.2 % with a *k*-nearest neighbor kNN) classifier and features derived using a db1 (Haar) wavelet filter. For developing an efficient CAD system, it is necessary that the database is diversified and should contain all possible variations of density patterns belonging to each class. The patterns within each class can be categorized as typical or atypical. The sample images of typical cases present in the MIAS database are shown in Figure 17.3.

Table 17.3 Description of studies carried out for classification of tissue density as fatty, fatty-glandular, or dense-glandular on the MIAS database

Investigators	Description			
	Images	SBT/ROI	Classifier	Acc. (%)
Blot *et al.* (2001) [51]	265	SBT	kNN	63.0
Bosch *et al.* (2006) [52]	322	SBT	SVM	91.3
Muhimmah *et al.* (2006) [53]	321	SBT	DAG-SVM	77.5
Subashni *et al.* (2010) [54]	43	SBT	SVM	95.4
Tzikopolous *et al.* (2011) [55]	322	SBT	SVM	84.1
Li (2012) [56]	42	SBT	KSFD	94.4
Mustra *et al.* (2012) [57]	322	512 × 384	IB1	82.0
Silva *et al.* (2012) [58]	320	300 × 300	SVM	77.1
Abdul-Nasser *et al.* (2012) [59]	322	100 × 100	SVM	85.5
Kriti *et al.* (2016) [50]	322	200 × 200	SVM	86.3
Kriti *et al.* (2016) [60]	322	200 × 200	SVM	87.5

Note: SBT: segmented breast tissue; Acc.: accuracy; DAG: directed acyclic graph; kNN: *k*-nearest neighbor; KSFD:kernel self-optimized fisher discriminant.

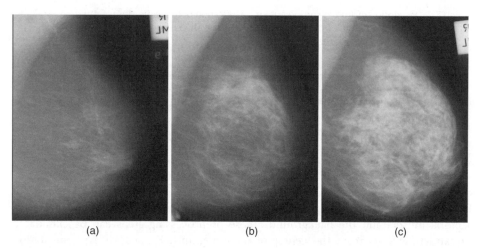

Figure 17.3 Sample mammograms showing typical cases of breast tissue density: (a) typical F tissue *mdb*078, (b) typical FG tissue *mdb*210, and (c) typical DG tissue *mdb*126.

The MIAS database also contains some atypical cases in which identification of density patterns is difficult. The sample images of these atypical cases are shown in Figure 17.4.

It is believed that the texture of the mammographic image gives significant information about the changes in breast tissue density patterns. In the various studies reported above, texture information from the SBT or the ROIs is extracted using statistical and signal processing–based methods carried out on a single scale. However, feature extraction can also be done in the transform domain over various scales by using various multiresolution schemes like wavelets. It is logical to compute texture features in the

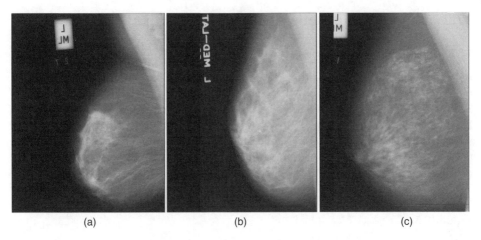

Figure 17.4 Sample mammograms showing atypical cases of breast tissue density: (a) atypical F tissue *mdb*095, (b) atypical FG tissue *mdb*029, and (c) atypical DG tissue *mdb*201.

transform domain as the scale over which feature extraction is carried out is an important characteristic because the visual system of humans is adapted to process the images in a multiscale way [62, 63].

17.1.1 Comparison of Related Methods with the Proposed Method

The work done in the present study can be directly compared with the studies in [50, 57, 58, 60], as these have been carried out on a fixed-size ROI for classification of breast tissue into three classes as per its density. The study proposed by [57] extracted the ROIs of size 512×384 pixels. Texture features were extracted from GLCM along with statistical features and histogram features derived from each ROI. A total of 419 features was extracted. For the three density categories, an accuracy of 82.0% was achieved using kNN as classifier with a value of k equal to 1. In the method proposed by [58], the authors extracted the ROIs of size 300×300 pixels. The method combined the statistical features extracted from the image histogram with those derived from the co-occurrence matrix. The ten-fold cross-validation scheme was used with SVM classifier, and an accuracy of 77.18% was achieved. The method proposed in [59] extracted the ROIs of size 300×300 pixels from each image. The ULDP (uniform local directional pattern) was used as a texture descriptor for breast density classification. The SVM classifier used achieved an accuracy of 85.5% for the three-class classification. In the study in [50], statistical features were extracted from each ROI of size 200×200 pixels. The accuracy of 86.3% was achieved by using only the first four PCs with SVM classifier. In the study proposed by [60], the authors extracted ROIs of size 200×200 pixels from the center of the breast. In the feature extraction module, Laws' mask analysis was used to compute texture features from each ROI image. The principal components analysis (PCA)-SVM classifier was used for the classification task, and an accuracy of 87.5% was achieved for three-class breast density classification.

In the present work, the potential of different multiresolution feature descriptor vectors (FDVs) consisting of normalized energy values computed from different subimages has been evaluated using SVM and smooth SVM (SSVM) classifiers.

17.2 Materials and Methods

17.2.1 Description of Database

In the present work, the algorithms for texture analysis have been tested on the mammograms taken from the mini-MIAS database. The mini-MIAS database consists of 322 film-screen mammograms (digitized database). These images were obtained from 161 women in MLO view only. The original MIAS database is digitized at 50μ pixel edge. The mini-MIAS databse was generated by down-sampling the original database to 200μ pixel and clipping or padding it to a fixed size of 1024×1024 pixels. Out of 322 mammographic images, 106 images belong to the fatty class, 104 images belong to the fatty-glandular class, and 112 images belong to the dense-glandular class. For further details of the database, refer to [64]. The description of the MIAS database is shown in Figure 17.5.

17.2.2 ROI Extraction Protocol

After performing various experiments, it has been asserted that the center of the breast contains maximum density information (i.e., in the region where glandular ducts are prominent) [45, 46, 61]; therefore, fixed-size ROIs (200×200 pixels) are extracted from the central location as represented in Figure 17.6.

A few sample ROIs are shown in Figure 17.7.

17.2.3 Workflow for CAD System Design

CAD systems involve computerized analysis of mammograms, which is then used by the radiologists to validate their diagnosis as these systems tend to detect any lesions that might be missed during subjective analysis [65–80]. The workflow for the prediction of breast density is shown in Figure 17.8.

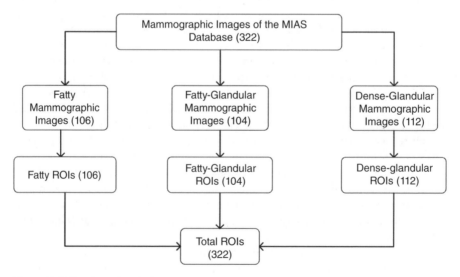

Figure 17.5 Database description.

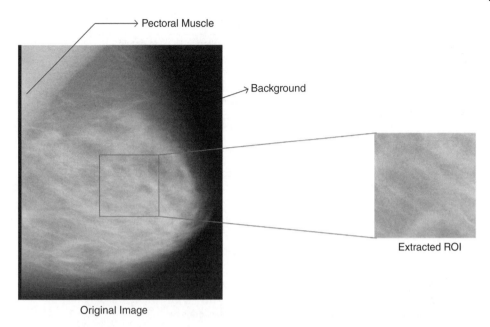

Figure 17.6 ROI extraction protocol *mdb*030.

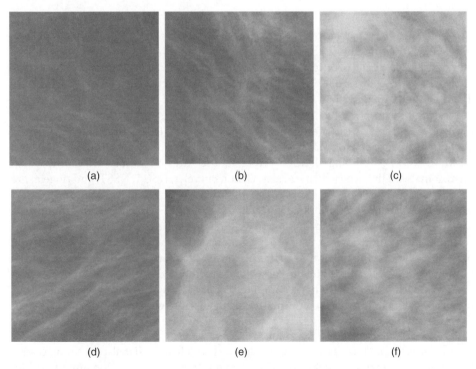

Figure 17.7 Sample ROIs: (a) typical F ROI *mdb*078; (b) typical FG ROI *mdb*210; (c) typical DG ROI *mdb*126; (d) atypical F ROI *mdb*095; (e) atypical FG ROI *mdb*029; and (f) atypical DG ROI *mdb*201.

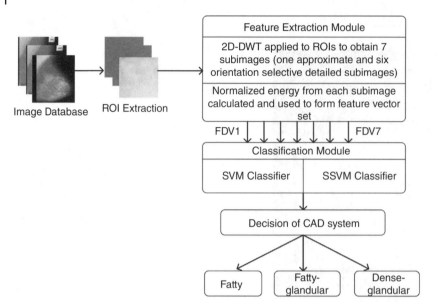

Figure 17.8 Block diagram: workflow for prediction of breast density.

The CAD system design consists of two modules: (a) feature extraction and (b) classification.

17.2.3.1 Feature Extraction

In feature extraction, the texture information present in the image is transformed into numerical values that can be further used in the machine learning algorithms for texture analysis. For the analysis of texture, different feature extraction techniques can be used. These methods are described in Figure 17.9.

The feature extraction can be done on either a single scale or multiple scales. On a single scale, the features are extracted after the spatial interactions between neighborhood pixels are considered for analysis. The examples of such methods are GLCM, GLRLM, NGTDM, SFM, and GLDS. To extract features on various scales, the transform domain methods based on wavelet, Gabor, curvelet, NSCT, NSST, and ridgelet are used. The scale over which feature extraction is carried out is an important characteristic as the visual system of humans is adapted to process the images in a multiscale way [62, 63, 81]. In the transform domain method of feature extraction, signals are converted from the time domain to another domain so that the characteristic information present in it within the time series that is not observable can be easily extracted. For the present work, wavelet-based texture descriptors (computed from ten different compact support wavelet filters) have been computed from each ROI for classifying the breast tissue according to the density information using discrete wavelet transform (DWT).

17.2.3.1.1 Evolution of Wavelet Transform

Fourier transform: Fourier transform is the most widely used signal-processing operation used to convert any signal from time domain to its frequency domain. In Fourier analysis, the signal is represented as a series of sine and cosine waves of varying amplitudes and frequencies. Fourier transform can only provide 2D information about the

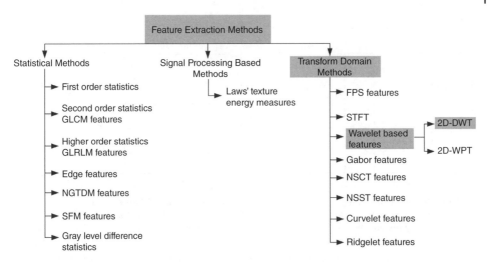

Figure 17.9 Different feature extraction techniques used in texture analysis. Note: GLCM: gray-level co-occurence matrix; GLRLM: gray-level run length matrix; NGTDM: neighborhood gray tone difference matrix; SFM: statistical feature matrix; FPS: Fourier power spectrum; STFT: short-time Fourier transform; 2D-DWT: two-dimensional discrete wavelet transform; WPT: wavelet packet transform; NSCT: non-subsampled countourlet transform; NSST: non-subsampled shearlet transform.

signal (i.e., different frequency components present in it and their respective amplitudes); it gives no information about the time at which different frequency components exist.

Short-Time Fourier Transform (STFT): In STFT, the signal is broken up into small parts of fixed-size lengths, then Fourier transform is taken for each piece of signal. The Fourier transform of each piece of signal provides the spectral information of each time slice. Although STFT provides the time and frequency information simultaneously, there exists the dilemma of resolution. As the window size is fixed, there is poor time resolution for wide windows and poor frequency resolution for narrow windows.

The drawback of Fourier transform can be overcome by using the wavelet transform, which gives 3D information about a signal i.e., different frequency components present in a signal, their respective amplitudes, and the time at which these frequency components exist. In wavelet analysis, the signal is represented as a scaled and translated version of the mother wavelet [82]. According to Heisenberg's uncertainty principle, there can be either high-frequency resolution and poor time resolution or poor-frequency resolution and good temporal resolution. The wavelet analysis solves the resolution problem occurring in STFT. In DWT, the basis function varies both in frequency range and in spatial range. The wavelet transform is designed in such a way that there is good frequency resolution for low-frequency components and good time resolution for high-frequency components. DWT is considered to be better than Fourier transform and STFT because in most of the signals occurring in nature, the low-frequency content is present for longer duration and high-frequency content occurs for short duration. On comparing with time, frequency, and Gabor wavelet-based analysis, it is observed that wavelet analysis uses a time-scale region rather than a time-frequency region to describe the signal, as shown in Figure 17.10.

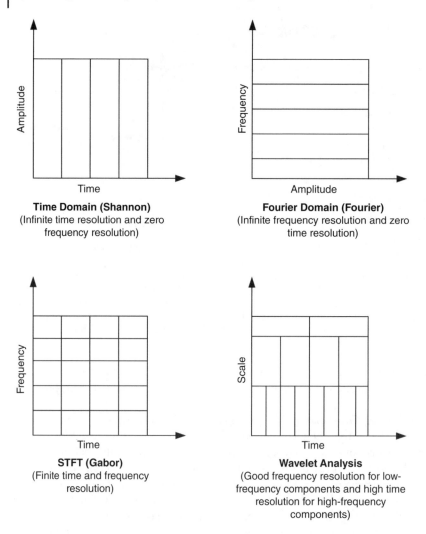

Figure 17.10 Time-domain, frequency-domain, STFT, and wavelet analysis of a signal.

Wavelet transform: The wavelets are waveforms of limited duration that have an average value of zero. These wavelets are basically little waves that are concentrated in time and frequency around a certain point. DWT represents a windowing technique with variable-sized regions. The whole idea of wavelets is to capture the incremental information. Piecewise constant approximation inherently brings the idea of representing the image at a certain resolution. Consider an image consisting of a number of shells in which each shell represents the resolution level. The outermost shell and the innermost shell represent the maximum and minimum resolution levels, respectively. The job of the wavelet is to take out a particular shell. The wavelet translates at maximum resolution and will take out the outermost shell. For next level of resolution, it will take out the next shell, and so on. By reducing the resolution, the wavelet is just peeling off shell by shell using different dilates and translates. Different dilates correspond to different resolution, and different translates track the given resolution level. The process of wavelet analysis of an image is shown in Figure 17.11. A 2D DWT when applied to images can be seen as two 1D transform functions applied to rows and columns of the image separately [83], as shown in Figure 17.12. The 2D wavelet decomposition up to the first level applied to a mammographic image is shown in Figure 17.13. When this operation is applied to an ROI, image it is passed through a set of complementary filters, and at the output one approximate and one detailed sub-image are produced (i.e., the input image

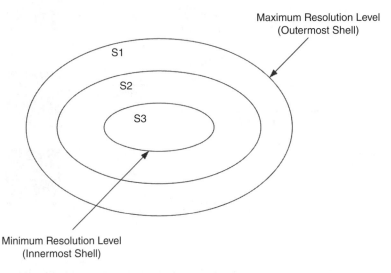

Figure 17.11 Process of wavelet analysis of an image.

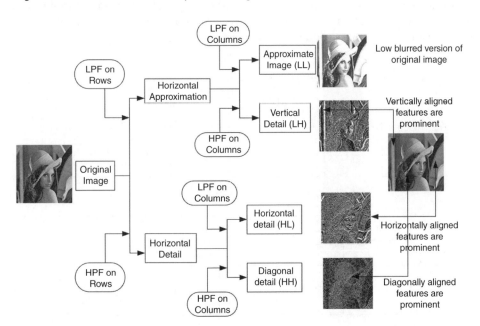

Figure 17.12 Wavelet transform of sample Lena image.

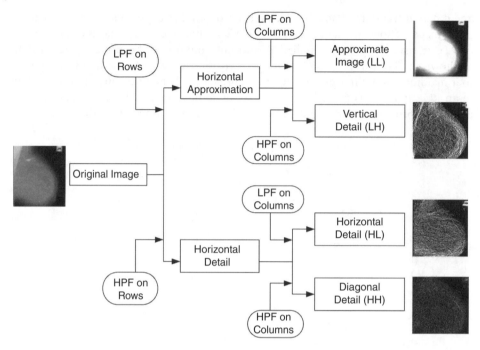

Figure 17.13 Wavelet transform of image F *mdb132*.

is decomposed into two subimages). There can be several levels up to which an image can be decomposed. The wavelet transform at the next level is applied to the approximate sub-image so that it further produces two sub-images [84]. The process of wavelet decomposition up to the second level is shown in Figure 17.14.

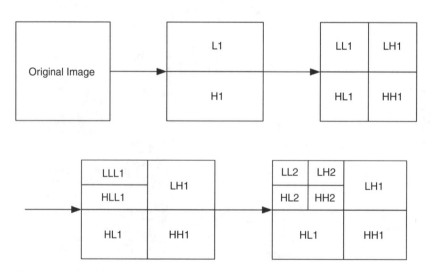

Figure 17.14 Wavelet decomposition of an image up to the second level.

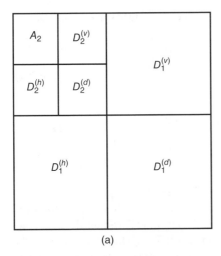

(a)

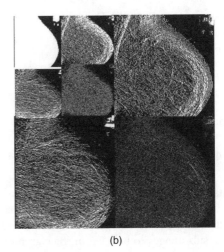

(b)

Figure 17.15 (a) 2D wavelet decomposition of image up to the second level. (b) 2D wavelet decomposition of sample image *mdb*132 using a Haar wavelet filter up to the second level.

In the present work, the decomposition is done up to the second level, and one approximate sub-image A_j and six orientation-selective detailed sub-images $D_j^{(k)}$,k=h,v,d are generated. This wavelet representation of an image is depicted in Figure 17.15.

Selection of wavelet filters: The choice of wavelet filter used for feature extraction is based on some properties that are significant for texture description [81, 85–87]. The properties that are considered for selecting an appropriate wavelet filter include: orthogonality or biorthogonality, support width, shift invariance, and symmetry. The properties of different wavelet filters used are summarized in Table 17.4. From the Table 17.4, it can be observed that the Haar filter is the only filter having the properties of orthogonality, symmetry, and compact support. Wavelet filters that provide compact support are desirable due to their ease of implementation. Compact support filter function has zero value outside the compact specified boundaries. For energy conservation at each level of decomposition, orthogonality is a desirable property. To avoid any dephasing while processing images, symmetry is required. For further details on the properties of wavelet filters, readers are directed to [81, 87]. In the present work, a Haar wavelet filter is used. A Haar wavelet is the first and simplest wavelet, and it was proposed in 1909 by Alfred Haar. The Haar wavelet is discontinuous and resembles a step function. There are two functions that play an important role in wavelet analysis:scaling function ϕ

Table 17.4 Properties of wavelet filters used

Wavelet filter	Biorthogonal	Orthogonal	Symmetry	Asymmetry	Near symmetry	Compact support
Db	✗	✓	✗	✓	✗	✓
Haar	✗	✓	✓	✗	✗	✓
Bior	✓	✗	✓	✗	✗	✓
Coif	✗	✓	✗	✗	✓	✓
Sym	✗	✓	✗	✗	✓	✓

(father wavelet) and wavelet function ψ (mother wavelet). The Haar wavelet function is defined as:

$$\psi(x) = \begin{cases} 1, & 0 \le x < 0.5 \\ -1, & 0.5 \le x < 1 \\ 0, & otherwise \end{cases} \qquad (17.1)$$

The Haar scaling function is defined as $\psi(x) = \phi(2x) - \phi(2x - 1)$:

$$\Phi(x) = \begin{cases} 1, & if \ 0 \le x < 1 \\ 0, & otherwise \end{cases} \qquad (17.2)$$

The Haar wavelet function and scaling function are shown in Figure 17.16.

Selection of texture features: In the transform domain, while using multiresolution methods, features like energy, mean, and standard deviation are frequently extracted from the sub-images to be used in the classification task [85–89]. In the present work, after the generation of approximate and detailed sub-images, normalized energy of each sub-image is calculated. A description of the FDVs used in the present work for the classification task is shown in Table 17.5.

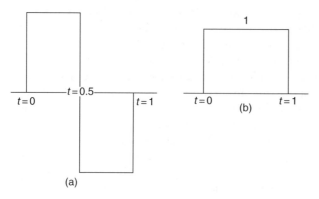

Figure 17.16 (a) Haar wavelet function $\psi(t)$ and (b) Haar scaling function $\phi(x)$.

Table 17.5 Description of FDVs

FDV	Wavelet energy signatures
FDV1	$(NE_2^A, NE_2^h, NE_2^v, NE_2^d, NE_1^h, NE_1^v, NE_1^d)$
FDV2	(NE_1^h, NE_1^v, NE_1^d)
FDV3	$(NE_2^h, NE_2^v, NE_2^d, NE_1^h, NE_1^v, NE_1^d)$
FDV4	$(NE_1^h, NE_1^v, NE_1^d, NE_2^d)$
FDV5	$(NE_2^A, NE_1^h, NE_1^v, NE_1^d)$
FDV6	$(NE_2^A, NE_2^h, NE_2^v, NE_2^d)$
FDV7	(NE_2^h, NE_2^v, NE_2^d)

Note: FDV: feature descriptor vector; NE: normalized energy; A: approximate subimage; h: horizontal subimage; v: vertical subimage; d: diagonal subimage.

The normalized energy is calculated as:

$$NE_y^x = \frac{\|\,subimage\,\|_F^2}{area(subimage)} \tag{17.3}$$

where x: subimage; y: level of decomposition; and $._F$: Frobenius norm.

The Frobenius norm is also known as the Euclidean norm. It is the norm of matrix X of dimension $m \times n$ defined as the square root of the sum of the absolute squares of its elements.

$$\|X\|_F = \sqrt{\sum_{i=1}^{m} \sum_{j=1}^{n} |x_{ij}|^2} \tag{17.4}$$

The Frobenius norm is independent of the manner in which the elements are placed within the matrix. $\left(\text{e.g., if } A1 = \begin{bmatrix} 3 & 4 \\ 0 & 0 \end{bmatrix} \text{ and } A2 = \begin{bmatrix} 3 & 0 \\ 0 & 4 \end{bmatrix}\right)$. The Frobenius norm of both the matrixes comes out as: $\sqrt{(3)^2 + (4)^2 + (0)^2 + (0)^2} = 5$.

NE_y^x represents the normalized energy of a sub-image x at the y level of decomposition (e.g., NE_2^A) represents the normalized energy of an approximate sub-image at the second level of decomposition.

17.2.3.2 Classification

In the classification module, the classifiers like SVM and SSVM are used to classify the mammographic images into F, FG, or DG classes based on the training instances. The SVM classifier is very popular and has been suggested by many researchers to provide good accuracy for mammographic abnormality classifications [90–95] The features were normalized between 0 and 1 before feeding them to the classifier.

SVM classifier: The SVM is a supervised machine learning algorithm used for both classification as well as regression problems. For linearly separable data, the role of SVM is to find the hyperplane that classifies all training vectors in two classes. Multiple solutions exist for linearly separable data. The goal of SVM is to arrive at that optimal hyperplane that segments the data in such a way that there is the widest margin between the hyperplane and the observations. The margin is required to be maximum in order to reduce test error, keep the training error low, and minimize the VC dimension. The samples that touch the decision boundary are called support vectors. In the SVM algorithm, the instances are plotted in an n-dimensional space, where n is length of the feature vector. The working of the SVM algorithm for the two-class problem is shown in Figure 17.17. Consider the two-class problem and let $g(\vec{x}) = \vec{w}^T \vec{x} + w_0$:

$$g(\vec{x}) \geq 1 \ \forall \ \vec{x} \ \epsilon \ class1 \tag{17.5}$$
$$g(\vec{x}) \leq -1 \ \forall \ \vec{x} \ \epsilon \ class2 \tag{17.6}$$

For support vectors, the inequality becomes an equality.
Each training example's distance from the hyperplane is:

$$\vec{z} = \frac{|g(\vec{x})|}{\|\vec{w}\|} \tag{17.7}$$

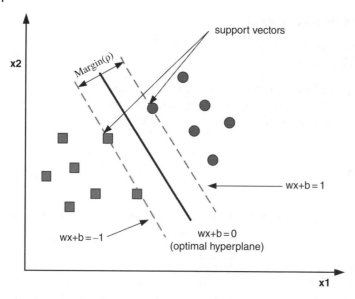

Figure 17.17 SVM classifier for linearly separable data.

The total margin is computed by:

$$\frac{1}{\|\vec{w}\|} + \frac{1}{\|\vec{w}\|} = \frac{2}{\|\vec{w}\|} \tag{17.8}$$

So, the margin is given as:

$$\rho = \frac{2}{\|\vec{w}\|} \tag{17.9}$$

Minimizing the $\vec{w}$ will maximize the separability. Minimizing the $\vec{w}$ is a nonlinear optimization task, solved by Karush-Kuhn Tucker (KKT) conditions, using a Lagrange multiplier λ_i

$$\vec{w} = \sum_{i=0}^{N} = \lambda_i y_i \vec{x}_i \tag{17.10}$$

$$\sum_{i=0}^{N} \lambda_i y_i = 0 \tag{17.11}$$

Figure 17.18 shows an example to illustrate the workings of the SVM algorithm. In this example, x1 and x2 are two features, and only three values are taken to classify the two-class problem. Point (1,1) and point (2,0) belong to class 1 such that g(1,1) and g(2,0) are equal to −1, and the point (2,3) belongs to class 2 such that g(2,3) is equal to 1. From the figure, the weight vector $\vec{w}$=(2,3)−(1,1)=(a,2a), where a is any constant value.

For x=(1,1):

$$a + 2a + w_0 = -1 \tag{17.12}$$

For x=(2,3):

$$2a + 6a + w_0 = 1 \tag{17.13}$$

Figure 17.18 Example to illustrate the SVM algorithm.

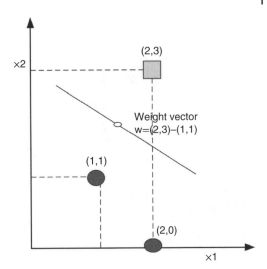

After solving the equations, $a = \frac{2}{5}$ and $w_0 = \frac{-11}{5}$. The value of the weight vector comes out as: $\vec{w} = (\frac{2}{5}, \frac{4}{5})$; these are support vectors as they compose the weight value. The value of g(x) after submission is found to be:

$$g(\vec{x}) = x1 + 2x2 - 5.5 \tag{17.14}$$

This is the equation of the hyperplane that classifies the elements.

In some cases, the clear demarcation cannot be made between the instances of two classes; then the algorithm applies kernel functions to map the nonlinear data points from low-dimensional input space to a higher dimensional feature space. The hyperplane is a linear function of vectors that are drawn from feature space instead of original input space. For nonlinearly separable data, each input vector from input space is mapped into higher dimensional space via some transformation $\phi : x \longrightarrow \varphi(x)$. The value of kernel function provides inner or dot product of vectors in feature space.

$$K(x_i, x_j) = \varphi(x_i)^T \varphi(x_j) \tag{17.15}$$

The basic role of the kernel is to map the nonlinearly separable data space into better representation space where data can easily be separated. Different kernels used in the SVM algorithm are linear, sigmoid, polynomial, and Gaussian radial basis function (GRBF). The linear kernel is represented as:

$$K(x_i, x_j) = x_i^T x_j \tag{17.16}$$

Equation of a polynomial kernel is:

$$K(x_i, x_j) = (1 + x_i^T x_j)^P \tag{17.17}$$

where p is power of polynomial. Equation of Gaussian radial basis function kernel is:

$$K(x_i, x_j) = \exp\left(-\frac{\| x_i - x_j \|^2}{2\sigma^2}\right) \tag{17.18}$$

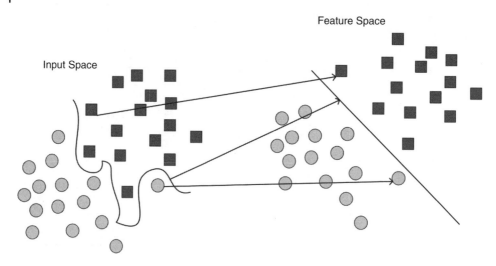

Figure 17.19 SVM for nonlinearly separable data.

The sigmoid kernel is expressed as:

$$K(x_i, x_j) = \tanh(\beta_0 x_i^T x_j + \beta_1) \qquad (17.19)$$

An example of SVM classifier for nonlinearly separable data is shown in Figure 17.19. In the present work, a multiclass SVM classifier is implemented using the LibSVM library [96], and a grid search procedure is used for obtaining the optimum values of the classifier parameters [97–102].

SSVM Classifier: SSVMs can be used to classify highly nonlinear data space, for example checkboard. For large problems, SSVM is found to be faster than SVM. For multiclass implementation of SSVM classifier, the SSVM toolbox has been used [103]. Some important mathematical properties like strong convexity and infinitely often differentiability make it different from SVM. For further information, the readers are directed to [104–106]. In the case of SSVM also, a grid search procedure is carried out to obtain the optimum values for classifier parameters.

17.3 Results

The design of the proposed CAD system for the prediction of breast density is shown in Figure 17.20.

In the present work, rigorous experimentation was performed to obtain the highest accuracy for the proposed CAD system design based on classification performance of the classifiers for predicting the breast density with FDVs derived using the ten compact support wavelet filters and the computational time required for prediction of 161 testing instances. The experiments are described in Table 17.6.

Figure 17.20 Flowchart of proposed CAD system design.

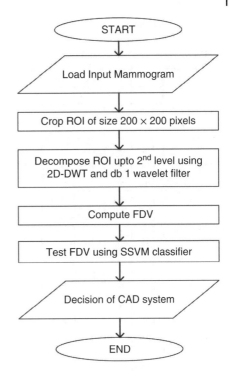

```
        ( START )
            │
            ▼
   / Load Input Mammogram /
            │
            ▼
 [ Crop ROI of size 200 × 200 pixels ]
            │
            ▼
 [ Decompose ROI upto 2nd level using
    2D-DWT and db 1 wavelet filter ]
            │
            ▼
 [ Compute FDV ]
            │
            ▼
 [ Test FDV using SSVM classifier ]
            │
            ▼
   / Decision of CAD system /
            │
            ▼
        ( END )
```

Table 17.6 Experiment descriptions

Experiment	Description
Experiment I	To determine the performance of different FDVs using SVM classifier
Experiment II	To determine the performance of different FDVs using SSVM classifier

Note: FDV: feature descriptor vector.

17.3.1 Results Based on Classification Performance of the Classifiers (Classification Accuracy and Sensitivity) for Each Class

17.3.1.1 Experiment I: To Determine the Performance of Different FDVs Using SVM Classifier

The experiment has been carried out to evaluate the classification performance of all seven FDVs (FDV1–FDV7) derived from each of the ten wavelet filters using the SVM classifier. The results are tabulated in Table 17.7.

It is observed that the maximum CA of 85.0% is achieved using FDV1 and FDV6 derived using the coif1 and db4 wavelet filters, respectively. The texture descriptors used in FDV6 are less than the descriptors in FDV1 and still give the same accuracy, so FDV6 is considered to be the choice for obtaining maximum classification accuracy. The sensitivity values of FDV6 are 90.5%, 71.1%, and 92.8% for F, FG, and DG classes, respectively.

Table 17.7 Classification performance of SVM classifier using different FDVs

FDV (*l*)		F	FG	DG	Max. CA (filter)	*Sen.*$_F$(%)	*Sen.*$_{FG}$(%)	*Sen.*$_{DG}$(%)
		F	FG	DG				
FDV1 (7)	F	47	6	0	85.0 (coif1)	88.6	73.0	92.8
	FG	9	38	5				
	DG	0	4	52				
FDV2 (6)	F	37	13	3	78.2 (db1)	69.8	76.9	87.5
	FG	4	40	8				
	DG	0	7	49				
FDV3 (3)	F	43	7	3	83.2 (db1)	81.1	80.7	87.5
	FG	2	42	8				
	DG	1	6	49				
FDV4 (4)	F	42	8	3	74.7 (db1)	79.2	69.2	78.5
	FG	11	36	5				
	DG	6	6	44				
FDV5 (4)	F	46	6	1	84.4 (db1)	86.7	71.1	94.6
	FG	11	37	4				
	DG	0	3	53				
FDV6 (4)	F	48	5	0	85.0 (db4)	90.5	71.1	92.8
	FG	10	37	5				
	DG	0	4	52				
FDV7 (3)	F	44	8	1	80.7 (db1)	83.0	71.1	87.5
	FG	4	37	11				
	DG	1	6	49				

Note: FDV: feature descriptor vector; *l*: length of FDV; CM: confusion matrix; F: fatty class; FG: fatty-glandular class; DG: dense-glandular class; Max. CA: maximum classification accuracy in % ge; *Sen.*$_F$: sensitivity for fatty class; *Sen.*$_{FG}$: sensitivity for fatty-glandular class; *Sen.*$_{DG}$: sensitivity for dense-glandular class.

17.3.1.2 Experiment II: To Determine the Performance of Different FDVs Using SSVM Classifier

The experiment has been carried out to evaluate the classification performance of all seven FDVs (FDV1–FDV7) derived from each of the ten wavelet filters using the SSVM classifier. The results are tabulated in Table 17.8.

It is observed that the maximum CA of 89.4 % is achieved using FDV6 derived using the db1 (Haar) wavelet filter. A comparable classification accuracy of 88.2 % is achieved from FDV1 derived from the db1 (Haar) wavelet filter. The sensitivity values of FDV6 are 86.7%, 86.5%, and 94.6% for F, FG, and DG classes, respectively.

17.3.2 Results Based on Computational Efficiency of Classifiers for Predicting 161 Instances of Testing Data Set

For both the CAD systems, the time taken for feature extraction is the same. Therefore, the computational efficiency of the CAD systems has been compared with respect to the

Table 17.8 Classification performance of SSVM classifier with different FDVs

FDV (*l*)	CM				Max. CA (filter)	*Sen.$_F$*(%)	*Sen.$_{FG}$*(%)	*Sen.$_{DG}$*(%)
		F	FG	DG				
	F	44	8	1				
FDV1 (7)	FG	3	45	4	88.2 (db1)	83.0	86.5	94.6
	DG	0	3	53				
	F	38	13	2				
FDV2 (6)	FG	0	45	7	80.7 (db1)	71.6	86.5	83.9
	DG	1	8	47				
	F	35	16	2				
FDV3 (3)	FG	0	47	5	81.3 (db1)	66.0	90.3	87.5
	DG	1	6	49				
	F	34	17	2				
FDV4 (4)	FG	1	46	5	76.4 (db1)	64.1	88.4	76.7
	DG	5	8	43				
	F	45	7	1				
FDV5 (4)	FG	8	38	6	84.4 (db1)	84.9	73.0	94.6
	DG	0	3	53				
	F	46	6	1				
FDV6 (4)	FG	5	45	2	89.4 (db1)	86.7	86.5	94.6
	DG	0	3	53				
	F	38	13	2				
FDV7 (3)	FG	1	47	4	81.9 (db1)	71.6	90.3	83.9
	DG	1	8	47				

Note: FDV: feature descriptor vector; *l*: length of FDV; CM: confusion matrix; F: fatty class; FG: fatty-glandular class; DG: dense-glandular class; Max. CA: maximum classification accuracy in %ge; *Sen.$_F$*: sensitivity for fatty class; *Sen.$_{FG}$*: sensitivity for fatty-glandular class; *Sen.$_{DG}$*: sensitivity for dense-glandular class.

time taken for predicting the labels for 161 instances of the testing data set, as shown in Table 17.9.

From the table, it can be observed that the computational time required to predict the labels of testing instances is less in case of SSVM as compared to SVM for both FDV1 and FDV6. SSVM classifier requires 2.0840 seconds to predict the labels of testing instances for FDV6 and 2.0081 seconds to predict the labels of testing instances for FDV1.

17.4 Conclusion and Future Scope

From the experiments conducted, it can be concluded that the feature vector (consisting of normalized energy values computed from all four sub-images obtained at the second level of decomposition using 2D-DWT with a db1 (Haar) wavelet filter along with SSVM classifier) yields an efficient three-class tissue density prediction system.

Table 17.9 Comparison of computational time for prediction of testing instances

Classifier used	FDV	CA (%)	Computational time (s)
SVM	FDV1	85.0	10.3794
	FDV6	85.0	12.1001
SSVM	FDV1	88.2	2.0081
	FDV6	89.4	2.0840

Note: The above computations have been carried out on an Intel Core I3-2310 M, 2.10 GHz with 3 GB RAM. FDV: feature descriptor vector; SSVM: smooth support vector machine; SVM: support vector machine.

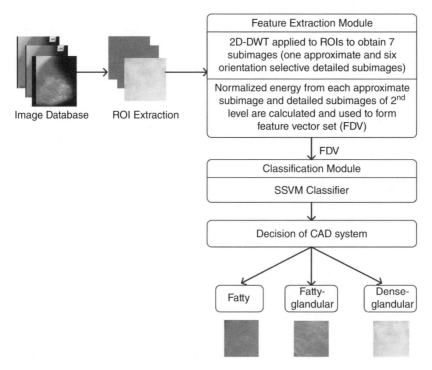

Figure 17.21 Proposed CAD system design.

The block diagram of the proposed CAD system design for predicting breast density is shown in Figure 17.21.

There are certain cases where the density information cannot be clearly determined solely on subjective analysis, so the CAD systems for breast density classification come into play. In the clinical environment, it is essential for a radiologist to correctly identify the density type of the breast tissue, and then double-check the mammogram that has been classified as dense to look for any lesions that might be hidden behind the dense tissue by varying the contrast of the image. Using CAD systems for breast density classification, uncertainties that are present at the time of visual analysis can be removed, and this also improves the diagnostic accuracy by highlighting certain areas of suspicion that may contain any tumors obscured behind the dense tissue.

Following are the recommendations for future work:

1. The proposed system can be tested on a full-field digital mammogram (FFDM) database like the INbreast database [107] consisting of DICOM images.
2. The proposed system can be tested for images acquired from MRI and other imaging modalities.
3. The proposed system can be improved by incorporating an automatic ROI extraction algorithm.

References

1 What is breast cancer. Available from http://www.nationalbreastcancer.org/what-is-breast-cancer.

2 What is breast cancer (2015) Available from http://www.breastcancer.org/symptoms/understand_bc/what_is_bc.

3 Types of breast cancers (2014) Available from http://www.cancer.org/cancer/breastcancer/detailedguide/breast-cancer-breast-cancer-types.

4 Cancer health center. Available from http://www.webmd.com/cancer/what-is-carcinoma.

5 DCIS-ductal carcinoma in situ (2015) Available from http://www.breastcancer.org/symptoms/types/dcis.

6 IDC–invasive ductal carcinoma (2016) Available from http://www.breastcancer.org/symptoms/types/idc.

7 ILC–invasive lobular carcinoma (2016) Available from http://www.breastcancer.org/symptoms/types/ilc.

8 Heine, J.J., Carston, M.J., Scott, C.G., Brandt, K.R., Wu, F.F., Pankratz, V.S., Sellers, T.A., and Vachon, C.M. (2008) An automated approach for estimation of breast density. *Cancer Epidemiology Biomarkers & Prevention*, **17** (11), 3090–3097.

9 Cheddad, A., Czene, K., Eriksson, M., Li, J., Easton, D., Hall, P., and Humphreys, K. (2014) Area and volumetric density estimation in processed full-field digital mammograms for risk assessment of breast cancer. *PLoS One*, **9** (10), e110 690.

10 Colin, C., Prince, V., and Valette, P.J. (2013) Can mammographic assessments lead to consider density as a risk factor for breast cancer? *European Journal of Radiology*, **82** (3), 404–411.

11 Huo, Z., Giger, M.L., and Vyborny, C.J. (2001) Computerized analysis of multiple-mammographic views: potential usefulness of special view mammograms in computer-aided diagnosis. *IEEE Transactions on Medical Imaging*, **20** (12), 1285–1292.

12 Zhou, C., Chan, H.P., Petrick, N., Helvie, M.A., Goodsitt, M.M., Sahiner, B., and Hadjiiski, L.M. (2001) Computerized image analysis: estimation of breast density on mammograms. *Medical physics*, **28** (6), 1056–1069.

13 Oliver, A., Freixenet, J., Martí, R., and Zwiggelaar, R. (2006) A comparison of breast tissue classification techniques, in *Medical Image Computing and Computer-Assisted Intervention–MICCAI 2006*, Springer, pp. 872–879.

14 Jagannath, H., Virmani, J., and Kumar, V. (2012) Morphological enhancement of microcalcifications in digital mammograms. *Journal of the Institution of Engineers (India): Series B*, **93** (3), 163–172.

15 Virmani, J. and Kumar, V. (2010) Quantitative evaluation of image enhancement techniques, in *Proceedings of International Conference on Biomedical Engineering and Assistive Technology (BEATS-2010), NIT Jalandhar, India.*

16 Yaghjyan, L., Pinney, S., Mahoney, M., Morton, A., and Buckholz, J. (2011) Mammographic breast density assessment: a methods study. *Atlas Journal of Medical and Biological Sciiences*, **1**, 8–14.

17 American Cancer Society (2014) Understanding your mammogram report: birads categories. Available from http://www.cancer.org/treatment/understandingy-ourdiagnosis/examsandtestdescriptions/mammogramsand otherbreastimaging procedures/mammograms-and-other-breast-imaging- procedures-mammo-report.

18 Mediolateral oblique view. Available from http://radiopaedia.org/articles/mediolateral-oblique-view.

19 Wolfe, J.N. (1976) Breast patterns as an index of risk for developing breast cancer. *American Journal of Roentgenology*, **126** (6), 1130–1137.

20 Wolfe, J.N. (1976) Risk for breast cancer development determined by mammographic parenchymal pattern. *Cancer*, **37** (5), 2486–2492.

21 Boyd, N., Martin, L., Chavez, S., Gunasekara, A., Salleh, A., Melnichouk, O., Yaffe, M., Friedenreich, C., Minkin, S., and Bronskill, M. (2009) Breast-tissue composition and other risk factors for breast cancer in young women: a cross-sectional study. *Lancet Oncology*, **10** (6), 569–580.

22 Boyd, N.F., Lockwood, G.A., Byng, J.W., Tritchler, D.L., and Yaffe, M.J. (1998) Mammographic densities and breast cancer risk. *Cancer Epidemiology Biomarkers & Prevention*, **7** (12), 1133–1144.

23 Eng, A., Gallant, Z., Shepherd, J., McCormack, V., Li, J., Dowsett, M., Vinnicombe, S., Allen, S., and dos Santos-Silva, I. (2014) Digital mammographic density and breast cancer risk: a case-control study of six alternative density assessment methods. *Breast Cancer Research*, **16** (5), 439–452.

24 Boyd, N.F., Rommens, J.M., Vogt, K., Lee, V., Hopper, J.L., Yaffe, M.J., and Paterson, A.D. (2005) Mammographic breast density as an intermediate phenotype for breast cancer. *Lancet Oncology*, **6** (10), 798–808.

25 Boyd, N.F., Martin, L.J., Yaffe, M.J., and Minkin, S. (2011) Mammographic density and breast cancer risk: current understanding and future prospects. *Breast Cancer Research*, **13** (6), 223–235.

26 Boyd, N.F., Guo, H., Martin, L.J., Sun, L., Stone, J., Fishell, E., Jong, R.A., Hislop, G., Chiarelli, A., Minkin, S. *et al.* (2007) Mammographic density and the risk and detection of breast cancer. *New England Journal of Medicine*, **356** (3), 227–236.

27 Vachon, C.M., Van Gils, C.H., Sellers, T.A., Ghosh, K., Pruthi, S., Brandt, K.R., and Pankratz, V.S. (2007) Mammographic density, breast cancer risk and risk prediction. *Breast Cancer Research*, **9** (6), 1–9.

28 Martin, L.J. and Boyd, N. (2008) Potential mechanisms of breast cancer risk associated with mammographic density: hypotheses based on epidemiological evidence. *Breast Cancer Research*, **10** (1), 1–14.

29 Warren, R. (2004) Hormones and mammographic breast density. *Maturitas*, **49** (1), 67–78.

30 Mousa, D.S.A., Brennan, P.C., Ryan, E.A., Lee, W.B., Tan, J., and Mello-Thoms, C. (2014) How mammographic breast density affects radiologists' visual search patterns. *Academic Radiology*, **21** (11), 1386–1393.

31 Østerås, B.H., Martinsen, A.C.T., Brandal, S.H.B., Chaudhry, K.N., Eben, E., Haakenaasen, U., Falk, R.S., and Skaane, P. (2016) Bi-rads density classification from areometric and volumetric automatic breast density measurements. *Academic Radiology*, **23** (4), 468–478.

32 Ekpo, E.U., Ujong, U.P., Mello-Thoms, C., and McEntee, M.F. (2016) Assessment of interradiologist agreement regarding mammographic breast density classification using the fifth edition of the bi-rads atlas. *American Journal of Roentgenology*, **206** (5), 1119–1123.

33 Miller, P. and Astley, S. (1992) Classification of breast tissue by texture analysis. *Image and Vision Computing*, **10** (5), 277–282.

34 Karssemeijer, N. (1998) Automated classification of parenchymal patterns in mammograms. *Physics in Medicine and Biology*, **43** (2), 365.

35 Bovis, K. and Singh, S. (2002) Classification of mammographic breast density using a combined classifier paradigm, in *Medical Image Understanding and Analysis (MIUA) conference, Portsmouth*, Citeseer.

36 Wang, X.H., Good, W.F., Chapman, B.E., Chang, Y.H., Poller, W.R., Chang, T.S., and Hardesty, L.A. (2003) Automated assessment of the composition of breast tissue revealed on tissue-thickness-corrected mammography. *American Journal of Roentgenology*, **180** (1), 257–262.

37 Petroudi, S., Kadir, T., and Brady, M. (2003) Automatic classification of mammographic parenchymal patterns: a statistical approach, in *Engineering in Medicine and Biology Society, 2003. Proceedings of the 25th Annual International Conference of the IEEE*, vol. 1, IEEE, vol. 1, pp. 798–801.

38 Oliver, A., Freixenet, J., and Zwiggelaar, R. (2005) Automatic classification of breast density, in *Image Processing, 2005. ICIP 2005. IEEE International Conference on*, vol. 2, IEEE, vol. 2, pp. 1258–1261.

39 Castella, C., Kinkel, K., Eckstein, M.P., Sottas, P.E., Verdun, F.R., and Bochud, F.O. (2007) Semiautomatic mammographic parenchymal patterns classification using multiple statistical features. *Academic Radiology*, **14** (12), 1486–1499.

40 Oliver, A., Freixenet, J., Marti, R., Pont, J., Perez, E., Denton, E.R., and Zwiggelaar, R. (2008) A novel breast tissue density classification methodology. *Information Technology in Biomedicine, IEEE Transactions*, **12** (1), 55–65.

41 Vállez, N., Bueno, G., Déniz-Suárez, O., Seone, J.A., Dorado, J., and Pazos, A. (2011) A tree classifier for automatic breast tissue classification based on birads categories, in *Pattern Recognition and Image Analysis*, Springer, pp. 580–587.

42 Chen, Z., Denton, E., and Zwiggelaar, R. (2011) Local feature based mammographic tissue pattern modelling and breast density classification, in *Biomedical Engineering and Informatics (BMEI), 2011 4th International Conference*, vol. 1, IEEE, pp. 351–355.

43 He, W., Denton, E., and Zwiggelaar, R. (2012) Mammographic segmentation and risk classification using a novel binary model based Bayes classifier. *Breast Imaging*, pp. 40–47.

44 Kutluk, S. and Gunsel, B. (2013) Tissue density classification in mammographic images using local features, in *Signal Processing and Communications Applications Conference (SIU), 2013 21st*, IEEE, pp. 1–4.

45 Sharma, V. and Singh, S. (2014) Cfs–smo based classification of breast density using multiple texture models. *Medical & Biological Engineering & Computing*, **52** (6), 521–529.

46 Sharma, V. and Singh, S. (2015) Automated classification of fatty and dense mammograms. *Journal of Medical Imaging and Health Informatics*, **5** (3), 520–526.

47 Kriti, Virmani, J., Dey, N., and Kumar, V. (2016) PCA_PNN and PCA-SVM based CAD systems for breast density classification, in *Applications of Intelligent Optimization in Biology and Medicine*, Springer, pp. 159–180.

48 Virmani, J. and Kriti (2016) Breast tissue density classification using wavelet-based texture descriptors, in *Proceedings of the Second International Conference on Computer and Communication Technologies*, Springer, pp. 539–546.

49 Kriti and Virmani, J. (2015) Breast density classification using Laws' mask texture features. *International Journal of Biomedical Engineering and Technology*, **19** (3), 279–302.

50 Kriti, Virmani, J., and Thakur, S. (2016) Application of statistical texture features for breast tissue density classification, in *Image Feature Detectors and Descriptors*, Springer, pp. 411–435.

51 Blot, L. and Zwiggelaar, R. (2001) Background texture extraction for the classification of mammographic parenchymal patterns, in *Medical Image Understanding and Analysis*, pp. 145–148.

52 Bosch, A., Munoz, X., Oliver, A., and Marti, J. (2006) Modeling and classifying breast tissue density in mammograms, in *Computer Vision and Pattern Recognition, 2006 IEEE Computer Society Conference*, vol. 2, IEEE, pp. 1552–1558.

53 Muhimmah, I. and Zwiggelaar, R. (2006) Mammographic density classification using multiresolution histogram information, in *Proceedings of the International Special Topic Conference on Information Technology in Biomedicine, Ioannina, Greece*, Citeseer.

54 Subashini, T., Ramalingam, V., and Palanivel, S. (2010) Automated assessment of breast tissue density in digital mammograms. *Computer Vision and Image Understanding*, **114** (1), 33–43.

55 Tzikopoulos, S.D., Mavroforakis, M.E., Georgiou, H.V., Dimitropoulos, N., and Theodoridis, S. (2011) A fully automated scheme for mammographic segmentation and classification based on breast density and asymmetry. *Computer Methods and Programs in Biomedicine*, **102** (1), 47–63.

56 Li, J.B. (2012) Mammographic image based breast tissue classification with kernel self-optimized fisher discriminant for breast cancer diagnosis. *Journal of Medical Systems*, **36** (4), 2235–2244.

57 Muštra, M., Grgić, M., and Delač, K. (2012) Breast density classification using multiple feature selection. *AUTOMATIKA*, **53** (4), 362–372.

58 Silva, W. and Menotti, D. (2012) Classification of mammograms by the breast composition, in *Proceedings of the International Conference on Image Processing, Computer Vision, and Pattern Recognition (IPCV)*, The Steering Committee of The World Congress in Computer Science, Computer Engineering and Applied Computing (WorldComp), pp. 1–6.

59 Abdel-Nasser, M., Rashwan, H.A., Puig, D., and Moreno, A. (2015) Analysis of tissue abnormality and breast density in mammographic images using a uniform local directional pattern. *Expert Systems with Applications*, **42** (24), 9499–9511.

60 Kriti and Virmani, J. (2016) Comparison of CAD systems for three class breast tissue density classification using mammographic images, in *Medical Imaging in Clinical Applications: Algorithmic and Computer-Based Approaches* (ed. N. Dey, V. Bhateja, and E.A. Hassanien), pp. 107–130.

61 Li, H., Giger, M.L., Huo, Z., Olopade, O.I., Lan, L., Weber, B.L., and Bonta, I. (2004) Computerized analysis of mammographic parenchymal patterns for assessing breast cancer risk: effect of ROI size and location. *Medical Physics*, **31** (3), 549–555.

62 Daugman, J.G. (1993) An information-theoretic view of analog representation in striate cortex, in *Computational Neuroscience*, MIT Press, Cambridge, MA, pp. 403–423.

63 Virmani, J., Kumar, V., Kalra, N., and Khandelwal, N. (2013) Prediction of liver cirrhosis based on multiresolution texture descriptors from b-mode ultrasound. *International Journal of Convergence Computing*, **1** (1), 19–37.

64 Suckling, J., Parker, J., Dance, D., Astley, S., Hutt, I., Boggis, C., Ricketts, I., Stamatakis, E., Cerneaz, N., Kok, S. *et al.* (1994) The mammographic image analysis society digital mammogram database, in *Exerpta Medica. International Congress Series*, vol. 1069, pp. 375–378.

65 Doi, K. (2007) Computer-aided diagnosis in medical imaging: historical review, current status and future potential. *Computerized Medical Imaging and Graphics*, **31** (4), 198–211.

66 Doi, K., MacMahon, H., Katsuragawa, S., Nishikawa, R.M., and Jiang, Y. (1999) Computer-aided diagnosis in radiology: potential and pitfalls. *European Journal of Radiology*, **31** (2), 97–109.

67 Giger, M.L., Doi, K., MacMahon, H., Nishikawa, R., Hoffmann, K., Vyborny, C., Schmidt, R., Jia, H., Abe, K., and Chen, X. (1993) An "intelligent" workstation for computer-aided diagnosis. *Radiographics*, **13** (3), 647–656.

68 Kumar, I., Bhadauria, H., and Virmani, J. (2015) Wavelet packet texture descriptors based four-class birads breast tissue density classification. *Procedia Computer Science*, **70**, 76–84.

69 Virmani, J., Kumar, V., Kalra, N., and Khadelwal, N. (2011) A rapid approach for prediction of liver cirrhosis based on first order statistics, in *Multimedia, Signal Processing and Communication Technologies (IMPACT), 2011 International Conference*, IEEE, pp. 212–215.

70 Zhang, G., Wang, W., Moon, J., Pack, J.K., and Jeon, S.I. (2011) A review of breast tissue classification in mammograms, in *Proceedings of the 2011 ACM Symposium on Research in Applied Computation*, ACM, pp. 232–237.

71 Tourassi, G.D. (1999) Journey toward computer-aided diagnosis: role of image texture analysis 1. *Radiology*, **213** (2), 317–320.

72 Hela, B., Hela, M., Kamel, H., Sana, B., and Najla, M. (2013) Breast cancer detection: a review on mammograms analysis techniques, in *Systems, Signals & Devices (SSD), 2013 10th International Multi-Conference*, IEEE, pp. 1–6.

73 Liu, Q., Liu, L., Tan, Y., Wang, J., Ma, X., and Ni, H. (2011) Mammogram density estimation using sub-region classification, in *Biomedical Engineering and Informatics (BMEI), 2011 4th International Conference*, vol. 1, IEEE, pp. 356–359.

74 Llobet, R., Pollán, M., Antón, J., Miranda-García, J., Casals, M., Martínez, I., Ruiz-Perales, F., Pérez-Gómez, B., Salas-Trejo, D., and Pérez-Cortés, J.C. (2014) Semi-automated and fully automated mammographic density measurement and

breast cancer risk prediction. *Computer Methods and Programs in Biomedicine*, **116** (2), 105–115.

75 Papaevangelou, A., Chatzistergos, S., Nikita, K., and Zografos, G. (2011) Breast density: Computerized analysis on digitized mammograms. *Hellenic Journal of Surgery*, **83** (3), 133–138.

76 Mustra, M., Grgic, M., and Delac, K. (2010) Feature selection for automatic breast density classification, in *ELMAR, 2010 Proceedings*, IEEE, pp. 9–16.

77 Cheng, H., Shan, J., Ju, W., Guo, Y., and Zhang, L. (2010) Automated breast cancer detection and classification using ultrasound images: a survey. *Pattern Recognition*, **43** (1), 299–317.

78 Virmani, J., Kumar, V., Kalra, N., and Khandelwal, N. (2011) Prediction of cirrhosis from liver ultrasound b-mode images based on laws' masks analysis, in *Image Information Processing (ICIIP), 2011 International Conference*, IEEE, pp. 1–5.

79 Virmani, J., Kumar, V., Kalra, N., and Khandelwal, N. (2011) Prediction of cirrhosis based on singular value decomposition of gray level co-occurence marix and a neural network classifier, in *Developments in E-systems Engineering (DeSE), 2011*, IEEE, pp. 146–151.

80 Virmani, J., Kumar, V., Kalra, N., and Khandelwal, N. (2014) Neural network ensemble based CAD system for focal liver lesions from b-mode ultrasound. *Journal of Digital Imaging*, **27** (4), 520–537.

81 Virmani, J., Kumar, V., Kalra, N., and Khandelwal, N. (2013) SVM-based characterization of liver ultrasound images using wavelet packet texture descriptors. *Journal of Digital Imaging*, **26** (3), 530–543.

82 Sifuzzaman, M., Islam, M., and Ali, M. (2009) Application of wavelet transform and its advantages compared to Fourier transform. *Journal of Physical Sciences*, **13**, 121–134.

83 Li, X. and Tian, Z. (2006) Wavelet energy signature: comparison and analysis, in *Neural information processing*, Springer, pp. 474–480.

84 Mohideen, S.K., Perumal, S.A., and Sathik, M.M. (2008) Image de-noising using discrete wavelet transform. *International Journal of Computer Science and Network Security*, **8** (1), 213–216.

85 Yoshida, H., Casalino, D.D., Keserci, B., Coskun, A., Ozturk, O., and Savranlar, A. (2003) Wavelet-packet-based texture analysis for differentiation between benign and malignant liver tumours in ultrasound images. *Physics in Medicine and Biology*, **48** (22), 3735.

86 Wan, J. and Zhou, S. (2010) Features extraction based on wavelet packet transform for b-mode ultrasound liver images, in *Image and Signal Processing (CISP), 2010 3rd International Congress*, vol. 2, IEEE, pp. 949–955.

87 Tsiaparas, N.N., Golemati, S., Andreadis, I., Stoitsis, J.S., Valavanis, I., and Nikita, K.S. (2011) Comparison of multiresolution features for texture classification of carotid atherosclerosis from b-mode ultrasound. *Information Technology in Biomedicine, IEEE Transactions*, **15** (1), 130–137.

88 Chang, T. and Kuo, C.J. (1993) Texture analysis and classification with tree-structured wavelet transform. *Image Processing, IEEE Transactions*, **2** (4), 429–441.

89 Avci, E. (2008) Comparison of wavelet families for texture classification by using wavelet packet entropy adaptive network based fuzzy inference system. *Applied Soft Computing*, **8** (1), 225–231.

90 Bazzani, A., Bevilacqua, A., Bollini, D., Brancaccio, R., Campanini, R., Lanconelli, N., Riccardi, A., Romani, D., and Zamboni, G. (2000) Automatic detection of clustered microcalcifications in digital mammograms using an SVM classifier, in *ESANN*, pp. 195–200.

91 Kamra, A., Jain, V., Singh, S., and Mittal, S. (2016) Characterization of architectural distortion in mammograms based on texture analysis using support vector machine classifier with clinical evaluation. *Journal of Digital Imaging*, **29** (1), 104–114.

92 de Oliveira, F.S.S., de Carvalho Filho, A.O., Silva, A.C., de Paiva, A.C., and Gattass, M. (2015) Classification of breast regions as mass and non-mass based on digital mammograms using taxonomic indexes and SVM. *Computers in Biology and Medicine*, **57**, 42–53.

93 Rejani, Y. and Selvi, S.T. (2009) Early detection of breast cancer using SVM classifier technique. *arXiv preprint arXiv:0912.2314*.

94 Campanini, R., Dongiovanni, D., Iampieri, E., Lanconelli, N., Masotti, M., Palermo, G., Riccardi, A., and Roffilli, M. (2004) A novel featureless approach to mass detection in digital mammograms based on support vector machines. *Physics in Medicine and Biology*, **49** (6), 961.

95 Guo, Q., Shao, J., and Ruiz, V. (2005) Investigation of support vector machine for the detection of architectural distortion in mammographic images, in *Journal of Physics: Conference Series*, vol. 15, IOP Publishing, p. 88.

96 Chang, C.C. and Lin, C.J. (2011) Libsvm: a library for support vector machines. *ACM Transactions on Intelligent Systems and Technology (TIST)*, **2** (3), 27.

97 Virmani, J., Kumar, V., Kalra, N., and Khandelwal, N. (2013) A comparative study of computer-aided classification systems for focal hepatic lesions from b-mode ultrasound. *Journal of Medical Engineering & Technology*, **37** (4), 292–306.

98 Li, S., Kwok, J.T., Zhu, H., and Wang, Y. (2003) Texture classification using the support vector machines. *Pattern Recognition*, **36** (12), 2883–2893.

99 Virmani, J., Kumar, V., Kalra, N., and Khandelwal, N. (2013) Svm-based characterisation of liver cirrhosis by singular value decomposition of GLCM matrix. *International Journal of Artificial Intelligence and Soft Computing*, **3** (3), 276–296.

100 Azar, A.T. and El-Said, S.A. (2014) Performance analysis of support vector machines classifiers in breast cancer mammography recognition. *Neural Computing and Applications*, **24** (5), 1163–1177.

101 Virmani, J., Kumar, V., Kalra, N., and Khandelwa, N. (2013) Pca-svm based cad system for focal liver lesions using b-mode ultrasound images. *Defence Science Journal*, **63** (5), 478.

102 Virmani, J., Kumar, V., Kalra, N., and Khandelwal, N. (2013) Characterization of primary and secondary malignant liver lesions from b-mode ultrasound. *Journal of Digital Imaging*, **26** (6), 1058–1070.

103 Lee, Y. and Mangasarian, O. (2015) SSVM toolbox. Available from http://research.cs .wisc.edu/dmi/svm/ssvm/.

104 Lee, Y.J. and Mangasarian, O.L. (2001) SSVM: a smooth support vector machine for classification. *Computational Optimization and Applications*, **20** (1), 5–22.

105 Purnami, S.W., Embong, A., Zain, J.M., and Rahayu, S. (2009) A new smooth support vector machine and its applications in diabetes disease diagnosis. *Journal of Computer Science*, **5** (12), 1003.

106 Lee, Y.J., Mangasarian, O.L., and Wolberg, W.H. (2003) Survival-time classification of breast cancer patients. *Computational Optimization and Applications*, **25** (1-3), 151–166.

107 Moreira, I.C., Amaral, I., Domingues, I., Cardoso, A., Cardoso, M.J., and Cardoso, J.S. (2012) Inbreast: toward a full-field digital mammographic database. *Academic Radiology*, **19** (2), 236–248.

Index

1-level decomposition   55
2-level decomposition   55
2D PCA   266, 267, 271, 274, 275, 278, 279
3σ method   356

a

ABD   347
Abnormal mass   340, 341, 343, 354, 358, 361, 365
Accuracy   315, 323, 332, 335, 341, 343, 361–363, 365, 366, 370, 378, 381, 382, 384, 387
Accuracy rate   65
Action bank representation   165, 167–169, 172
Activation function   131
Adaboost   191–194, 202
adaptive thresholding   326
age   103–106, 113, 114, 118–120
Ais   268
AlexNet   158
Algorithm   340–354, 356–363, 365, 366
Anatomical regions   353
Anatomical segmentation   353
Angiograms   369, 370, 377, 381, 382, 386
Anomaly detection   252
Area under the ROC curve (AUC)   68, 71
Artifacts   344
Artifical neuron   130
authentication   315–317, 335

b

Back-propagation   136, 137
Band selection   263, 265, 271, 275, 276, 278

Basis vectors   55
BI-RADS   392, 394
biometric trait   335
Bonferroni–Dunn test   68
Breast cancer   297–301, 303, 304, 307–310, 339, 340, 391
Breast contour   351
Breast tumor   340

c

CA   411, 412
CAD   340, 341, 358, 366, 393–395, 398, 400, 410, 412, 414
Caffe   148, 150
Cancer   339, 353
CC   343
CC view   393
Ccd   235, 236
CCL   342
Chain rule   137
character recognition   25–27, 38
Classification   191, 369, 375, 376, 378, 379, 381, 382, 384, 387
classification   26, 27, 32–34, 400, 407
Clonal selection algorithm   265, 266, 271, 276, 278, 279
clustering   283–285, 287, 288, 290–293
Cmos   235, 236
compactness   26, 29, 35, 37, 38
Confusion Matrix   70, 384
Contour Signature   86, 88, 89, 94, 95, 98
Control classifier   68
Convolution   139
Convolution kernel   139

Hybrid Intelligence for Image Analysis and Understanding, First Edition.
Edited by Siddhartha Bhattacharyya, Indrajit Pan, Anirban Mukherjee, and Paramartha Dutta.
© 2017 John Wiley & Sons Ltd. Published 2017 by John Wiley & Sons Ltd.
Companion Website: www.wiley.com/go/bhattacharyya/hybridintelligence

Convolution layer   142
Convolutional Neural network   127, 129, 141, 166, 167, 169, 170, 182
Coronary Stenosis   369, 370, 377, 378, 384, 386
Correlation coefficient   8
Cost function   136
CR mammogram   358
Critical difference   68
Crossover   7
Cuckoo search algorithm   306, 307, 310
Cuckoo search optimization algorithm   302–304, 306, 308, 310

d

Data sets   265, 276, 279
database   322, 331
Deep learning   127
Degrees of freedom   67
DEMS   358
Descriptors   188, 190, 191, 201
DICOM   340, 358
Differential Evolution   370, 371, 377, 386
Digitization   221, 222
Discrete wavelet transform   53, 54
distinctness   29, 35, 37, 38
dorsal vein   316, 321
DR mammogram   358
Dropout   147
DWT   401–403, 413

e

Edge detection   341, 342, 345, 346, 348, 349, 355, 366
Edge map   348
EFCM   27, 29, 35, 37, 38
Empirical measure $Q(I)$ 8
Euclidean distance   5, 28–30, 34, 35, 38, 248, 249, 254
Evolutionary algorithm   167, 168, 170–172, 182, 183

f

F-ratio   90, 91, 93, 97, 98
False positive (FP)   366

False positive rate (FPR)   68
FAR   316, 323, 326, 335
FCM   283, 284, 287, 288, 290–293
FDVs   397, 406, 410–412
Feature enhancement   89, 90, 93, 96, 98
Feature extraction   26, 27, 31, 32, 35, 239, 247, 248, 265, 266, 271, 275, 278, 319, 326, 335, 396, 400
feature level fusion   335
Feature mining   247
Feature selection   247, 248, 251, 265, 266, 271
Feature Vector   319–321, 333, 370, 384, 387
Features   187, 190–193
Feed-forward network   135
Filter Response   370, 371, 378, 382
finger vein   315, 316, 320–322, 331, 335
Forward propagation   137
Fourier transform   400, 401
frequency domain   318, 326, 331–333, 335
Friedman test   66
FRR   316, 323, 326, 335
Fully connected layer   146
fusion   316, 317, 320, 321, 326, 330, 335
Fuzzy *c*-means (FCM)   205, 208, 217, 224, 226, 227
Fuzzy *c*-means (FCM) algorithm   2, 4, 5, 9
Fuzzy KNN   266, 268, 269, 271, 277, 279
Fuzzy logic   318, 321, 335

g

Gabor filter   316, 318–320, 323, 333, 335
Gaussian filter   345
Gaussian Matched filters   369–371, 377, 386
Gaussian radial basis function kernel   409
gender   103–106, 111, 113–120
Generations   373
Generic point   188
Genetic algorithms (GAs)   3, 6
Geometric moment   80, 86, 87, 94, 95, 97, 98
GIS   206
GoogleNet   159
Gradient descent   136
GT   361

h

Haar   395, 405, 406, 412, 413
Haar wavelet transform   55, 56
hand vein images   316, 318, 320–323, 326, 330, 331, 335
Handedness   103–106, 113, 114, 117–120
handwriting   103–106, 108, 113, 118, 120
Hh component   56
Hiddden Morkov Model   316
Hierarchical and temporal machine   158
hierarchical method   319
High-pass filter   55
Hl component   55
Homogeneity enhancement   346, 366
Horizontal projection   58
Hotspot   205–209, 215, 217, 218, 224, 226–229
HSI   83
Human action recognition   165–169, 182
Hybrid techniques   234
Hyper-parameters   143
Hyperspectral image   263–265, 271, 276, 279
Hyperspectral imaging   233, 235, 240, 256

i

Iinear mixing model   241–243
illumination   80–85, 92–94, 97, 99
Image analysis   263, 265, 283, 284, 287, 293, 319
Image classification   234, 247, 248, 250, 251
Image Segmentation   1–3
Iman and Davenport test   67
Indiana dataset   271, 272, 274, 277, 279
ISODATA   207, 226, 227

k

K-means clustering   84, 93, 94, 98
Kappa statistics   68
kernel   409
Kernel function   66
Krawtchouk moment   80, 86–88, 94–99

l

L1 regularization   147
L2 regularization   147

LCS   86, 88, 89, 94, 95, 98
Lda   264, 278
LeNet   158
Leptourtic curve   62
Lh component   55
linear kernel   409
Ll component   56
longest run   27, 32–34, 38
Low-pass filter   55

m

Mahalanobis distance   248, 252
Mahalanobish distance   84, 93, 94, 98
Malignancy   339
Mammographic Image Analysis Society (MIAS)   300
Mammography   340, 343, 345, 358, 392, 393
Markov random fields   342
MatConvNet   148, 150
MATLAB   148
Matthews correlation coefficient (MCC)   68, 70
Max pooling   145
MDT   346
Mean absolute error (MAE)   69
Medical images   340, 341, 343, 366
Membership degree   5
Mesokurtic curve   62
Metastasize   339
MfGa based FCM algorithm   4, 9, 12, 14, 18
MIAS   358, 394, 395, 398
MIAS database   300, 307, 310
Mixed pixel   240, 241, 244
MLO   343
MLO view   393
Model building time   65, 66
Modified Genetic (MfGa) Algorithm   6, 9
Moment coefficient of kurtosis   62
Moment coefficient of skewness   61
morphological   80, 81, 85, 94
Mother wavelet   54
MRI   340
Multi criteria evaluation (MCE)   207, 228, 229

Multilayer self-organizing neural network (MLSONN) architecture (*contd.*)

Multilayer self-organizing neural network (MLSONN) architecture   4

Multilevel sigmoidal (MUSIG) activation function   4

Multilingual environment   47, 48

Multimodal biometric   315

Multimodal biometric recognition   315–317

Multiple classifiers   66, 69, 75

Mutation   7

n

Naive Bayes   370, 375, 377, 379, 384, 387

Negatively skewed distribution   60, 61

Nemenyi test   68

Neural network   128

Neuron   130

non-uniform zoning   90, 96, 99

Nonlinear mixing model   241–243

normalized energy   397, 406, 407, 413

Null hypothesis   67

o

Object classification   157

Object memorability   157

Optimization   370, 371, 373, 374, 377

Optimized MUSIG (OptiMUSIG) activation function   4

Overfitting   147

p

palm vein   316, 321, 322, 331

Parallel feature fusion   80, 91–93, 97, 98

Parameter sharing   144

Pca   264–267

Pearson coefficient of skewness   60

Pectoral muscle   346, 350

Percentile coefficient of kurtosis   62

Percentile coefficient of skewness   61

Perceptron   128, 134, 135

Platykurtic curve   62

PNN   342

Polynomial   208, 210

polynomial kernel   409

Pooling layer   145

Population initialization   6

Positively skewed distribution   61

Post-hoc tests   66, 68

Precision   70

preprocessing   25–27, 31, 32, 316, 318

q

Quartile coefficient of skewness   61

r

Radiometric calibration   238

RADIUS   207, 228

Radon transform (RT)   57

Recall   70

Receiver Operating Characteristic   370, 377

Receiver operating characteristic (ROC) curve   71

Recognition   187, 193, 194, 197, 199–201

Rectified linear units   131

Recurrent neural networks   135

Region of interest   340–342, 346, 353, 354, 357, 361, 366

Regression analysis   207, 208, 210, 212

ResNet   159

RGB   82, 83, 85, 86

ROC analysis   361

Root mean square error (RMSE)   69

rotation   80, 81, 85, 92, 93, 97, 99

s

Sammon's nonlinear mapping   35

Scaling function   54

score level fusion   320, 335

Script identification   47, 48, 52

Segmentation   340–343, 345, 346, 351, 353–355, 361, 366

segmentation   80–85, 92, 93, 97

Selection   7

Semisupervised method   265, 271, 278

Sensitivity   362, 365, 366

Serial feature fusion   80, 91–93, 98

SGLDM   342

short time fourier transform   401

Sigmoid kernel   410

Skewness   59

Soft biometrics   103–106, 109, 111, 113, 114, 118, 120, 121

Softmax function   132
Spatial   233, 236–238, 240, 245, 247, 255
spatial domain   318, 319, 333, 335
Specificity   362, 366
Spectral   263, 265, 266, 272, 274
Spectral bands   237, 242, 245, 247–249, 253, 254
Spectral calibration   238
Spectral dimension   233
Spectral information   236, 237, 240, 245
Spectral Library   238, 245, 246, 251
Spectral resolution   233
Spectral response   236
Spectral signature   233, 239, 247–249, 253, 254
Spectral unmixing   240, 246
Spectroscopy   233, 234
SRG   341, 354, 355, 357
SSVM   397, 407, 410, 412, 413
SSVM toolbox   410
STAC   207, 227, 228
static hand gesture   79, 80, 82, 83, 85–89, 92, 94, 95, 97, 99
Statistical decision making   341, 355, 356
Statistical features   71
Statistical performance analysis   65
Statistical significance tests   50
Stochastic gradient descent   138
Structure-based methods   50
Subpixel   255, 256
sum rule   318, 320, 326, 334
Supervised classification   234, 247, 248, 250
Supervised method   265, 278
Support Vector Machine (SVM)   65, 131, 298, 303
SVM   277, 279, 298–300, 303, 307, 309, 397, 407–411, 413
SVM classifier   301, 303, 310, 342, 343

t
Target identification   233, 234, 236, 238, 240, 254–256
Tchebichef moment   80, 87, 88, 94–98
Texture-based features   50
Thresholding   377, 378, 382, 387
TNT-mips   207, 226, 227
Torch   148
Tracking   187, 188, 190, 193, 194, 196, 197, 199–202
True negative (TN)   366
True positive rate (TPR)   68
Tumor   339, 340, 342, 348, 353, 354, 361
Two-stage approach   75

u
uncertainty models   284
uniform zoning   90
unimodal   315, 316, 329, 332
Unsupervised method   264–266, 268

v
Vertical projection   58
Vessel Detection   369, 370, 377, 382
Vessel Segmentation   369, 384, 387
VGGNet   150, 151, 159
Visual appearance-based methods   50, 51

w
Wavelet coefficient   54
Wavelet decomposition tree   55
wavelet filters   400, 405
Wavelet transform   401

y
YCbCr   82, 83, 93, 94
YIQ   82, 83

z
Z-score   356